Topics in Current Genetics

20

Series Editor: *Stefan Hohmann*

Francesc Posas · Angel R. Nebreda (Eds.)

Stress-Activated Protein Kinases

With 41 Figures, 15 in Color; and 12 Tables

Springer

Francesc Posas
Cell Signaling Unit
Departament de Ciències
Experimentals i de la Salut
Universitat Pompeu Fabra (UPF)
E-08003 Barcelona
Spain
e-mail: francesc.posas@upf.edu

Angel R. Nebreda
CNIO (Spanish National Cancer Center)
Melchor Fernández Almagro 3
E-28029 Madrid
Spain
e-mail: anebreda@cnio.es

The cover illustration depicts pseudohyphal filaments of the ascomycete *Saccharomyces cerevisiae* that enable this organism to forage for nutrients. Pseudohyphal filaments were induced here in a wild-type haploid MATa Σ1278b strain by an unknown readily diffusible factor provided by growth in confrontation with an isogenic petite yeast strain in a sealed petri dish for two weeks and photographed at 100X magnification (provided by Xuewen Pan and Joseph Heitman).

ISBN 978-3-642-09480-4 e-ISBN 978-3-540-75569-2

DOI 10.1007/978-3-540-75569-2

Topics in Current Genetics ISSN 1610-2096

Cover: WMXDesign GmbH, Heidelberg, Germany

Printed on acid-free paper

9 8 7 6 5 4 3 2 1

springer.com

Topics in Current Genetics publishes review articles of wide interest in volumes that center around specific topics in genetics, genomics as well as cell, molecular and developmental biology. Particular emphasis is placed on the comparison of several model organisms. Volume editors are invited by the series editor for specific topics, but further suggestions for volume topics are highly welcomed.

Each volume is edited by one or several acknowledged leaders in the field, who invite authors and ensure the highest standard of content and presentation. Only solicited manuscript will be considered. All contributions are peer-reviewed. All papers are published online prior to the print version. Individual DOIs (digital object identifiers) make each article fully citable from the moment of online publication.

All volumes of Topics in Current Genetics are part of the Springer eBook Collection. The collection includes online access to more than 3,000 newly released books, book series volumes and reference works each year. In addition to the traditional print version, this new, state-of-the-art format of book publications gives every book a global readership and a better visibility.

Editorial office:
Topics in Current Genetics
Series Editor: Stefan Hohmann
Cell and Molecular Biology
Göteborg University
Box 462
40530 Göteborg, Sweden
FAX: +46 31 7862599
E-mail: editor@topics-current-genetics.se

Table of contents

List of contributors

Aligue, Rosa
Department of Cell Biology, University of Barcelona, 08036 Barcelona, Spain.
aliguerosa@ub.edu

Bohmann, Dirk
Department of Biomedical Genetics, University of Rochester Medical Center, Rochester, New York 14642, USA
Dirk_Bohmann@urmc.rochester.edu

Crump, Nicholas T.
Nuclear Signalling Laboratory, Department of Biochemistry, Oxford University, South Parks Road, Oxford, OX1 3QU, UK

Cuenda, Ana
Departamento de Inmunología y Oncología, Centro Nacional de Biotecnología-CSIC, Campus de Cantoblanco-UAM, 28049-Madrid, Spain and MRC Protein Phosphorylation Unit, School of life Sciences, University of Dundee, Dundee DD1 5EH, Scotland, UK
a.i.cuenda@dundee.ac.uk

Davis, Roger J.
Howard Hughes Medical Institute and Program in Molecular Medicine, University of Massachusetts Medical School, Worcester, MA 01605.

de Nadal, Eulàlia
Cell Signaling Unit, Departament de Ciències Experimentals i de la Salut, Universitat Pompeu Fabra (UPF), E-08003 Barcelona, Spain.
eulalia.nadal@upf.edu

Dolado, Ignacio
CNIO (Spanish National Cancer Center), Melchor Fernández Almagro 3, 28029 Madrid, Spain

Engelberg, David
Department of Biological Chemistry, The Alexander Silvberman Institute of Life Sciences, The Hebrew University of Jerusalem, Jerusalem, 91904, Israel.
engelber@cc.huji.ac.il

Gaestel, Matthias
Department of Biochemistry, Medical School Hannover, Carl-Neuberg-Str. 1, 30625 Hannover, Germany
gaestel.matthias@mh-hannover.de

Han, Ya Ting
Nuclear Signalling Laboratory, Department of Biochemistry, Oxford University, South Parks Road, Oxford, OX1 3QU, UK

Hohmann, Stefan
Department for Cell and Molecular Biology, Göteborg University, Box 462, 40530 Göteborg, Sweden

Hult, Malin
Department of Cell and Molecular Biology, Göteborg University, Box 462, 405 30 Göteborg, Sweden

Iñesta-Vaquera, Francisco
Departamento de Inmunología y Oncología, Centro Nacional de Biotecnología-CSIC, Campus de Cantoblanco-UAM, 28049-Madrid, Spain

Johnson, Gary L.
Department of Pharmacology, CB# 7365, 1108 Mary Ellen Jones Building, University of North Carolina School of Medicine, Chapel Hill, NC 27599-7365, USA
gary_johnson@med.unc.edu

Keyse, Stephen M.
Cancer Research UK Stress Response Laboratory, Ninewells Hospital and Medical School, Dundee DD1 9SY, United Kingdom
Stephen.Keyse@cancer.org.uk

Klipp, Edda
Max Planck Institute for Molecular Genetics, Ihnestr. 63-73, 14195 Berlin, Germany
klipp@molgen.mpg.de

Kotlyarov, Alexey
Department of Biochemistry, Medical School Hannover, Carl-Neuberg-Str. 1, 30625 Hannover, Germany

Krantz, Marcus
The Systems Biology Institute, 6-31-15 Jingumae, M-31 Suite 6A, Shibuya-ku, Tokyo 150-0001, Japan

Kuma, Yvonne
MRC Protein Phosphorylation Unit, School of life Sciences, University of Dundee, Dundee DD1 5EH, Scotland, UK

Lopez-Aviles, Sandra
Department of Cell Biology, University of Barcelona, 08036 Barcelona, Spain.

Maayan, Inbal
Department of Biological Chemistry, The Alexander Silvberman Institute of Life Sciences, The Hebrew University of Jerusalem, Jerusalem, 91904, Israel.

Mahadevan, Louis C.
Nuclear Signalling Laboratory, Department of Biochemistry, Oxford University, South Parks Road, Oxford, OX1 3QU, UK
louiscm@bioch.ox.ac.uk

Muñoz-Cánoves, Pura
Program on Differentiation and Cancer, Center for Genomic Regulation (CRG), UPF/PRBB, Barcelona and CIBER de Enfermedades Neurodegenerativas (CIBERNED), Spain
pura.munoz@crg.es

Nebreda, Angel R
CNIO (Spanish National Cancer Center), Melchor Fernández Almagro 3, 28029 Madrid, Spain
anebreda@cnio.es

Nordlander, Bodil
Department for Cell and Molecular Biology, Göteborg University, Box 462, 40530 Göteborg, Sweden
bodil.nordlander@cmb.gu.se

Pelet, Serge
Institute of Biochemistry, ETH Zürich, CH-8093 Zürich, Switzerland

Perdiguero, Eusebio
Program on Differentiation and Cancer, Center for Genomic Regulation (CRG), UPF/PRBB, Barcelona and CIBER de Enfermedades Neurodegenerativas (CIBERNED), Spain

Peter, Matthias
Institute of Biochemistry, ETH Zürich, CH-8093 Zürich, Switzerland
matthias.peter@bc.biol.ethz.ch

Posas, Francesc
Cell Signaling Unit, Departament de Ciències Experimentals i de la Salut, Universitat Pompeu Fabra (UPF), E-08003 Barcelona, Spain.
francesc.posas@upf.edu

Rodríguez-Gabriel, Miguel A.
Depto. Microbiologia II. Facultad de Farmacia. Universidad Complutense de Madrid. 28040 Madrid, Spain

Rudolf, Fabian
Competence Center for Systems Physiology and Metabolic Disease (CC-SPMD)
Institute of Biochemistry, ETH Zürich, CH-8093 Zürich, Switzerland

Russell, Paul
Dept. Molecular Biology. The Scripps Research Institute. La Jolla, CA 92037. U.S.A.
prussell@scripps.edu

Rincón, Mercedes
Immunology Program, Department of Medicine, University of Vermont, Burlington, VT 05405.
mrincon@uvm.edu

Sabio, Guadalupe
MRC Protein Phosphorylation Unit, School of life Sciences, University of Dundee, Dundee DD1 5EH, Scotland, UK

Schaber, Jörg
Max Planck Institute for Molecular Genetics, Ihnestr. 63-73, 14195 Berlin, Germany

Stalheim, Lisa
Department of Pharmacology, CB# 7365, 1108 Mary Ellen Jones Building, University of North Carolina School of Medicine, Chapel Hill, NC 27599-7365, USA

Sunnerhagen, Per
Department of Cell and Molecular Biology, Göteborg University, Box 462, 405 30 Göteborg, Sweden
Per.Sunnerhagen@cmb.gu.se

Sykiotis, Gerasimos P.
Department of Biomedical Genetics, University of Rochestèr Medical Center, Rochester, New York 14642, USA

MAPK kinase kinase regulation of SAPK/JNK pathways

Lisa Stalheim and Gary L. Johnson

Abstract

SAPK/JNK members of the MAPK family are regulated by at least fourteen known MAPK kinase kinases (MKKKs). In addition to the kinase domain, each MKKK encodes different protein interaction domains and motifs to control their interaction with upstream GTPases such as Rho, Rac and Cdc42, downstream MAPK kinases, and scaffold proteins that assemble the MKKKs into signaling complexes for the control of physiological responses to a plethora of different stimuli. Several MKKKs coordinately regulate the SAPK/JNK pathway with other MAPKs including p38, ERK1/2 and ERK5. It is the diversity of MKKKs within the MAPK signaling network that provides the signaling specificity for activation of MAPKs including SAPK/JNKs and the integration with other signaling pathways within cells.

1 Introduction

SAPKs are MAPKs shown to be activated by many different stress stimuli, hence their name stress-activated protein kinases (SAPKs) (Kyriakis et al. 1994). The same kinases were shown to phosphorylate c-Jun at Ser 64 and 73 (Pulverer et al. 1991; Smeal et al. 1992; Derijard et al. 1994), hence the name Jun N-terminal kinases (JNKs). There are three SAPK/JNK genes (JNK1, JNK2, JNK3). Herein, for simplicity they are referred to as JNKs. JNK1 and JNK2 are expressed ubiquitously while JNK3 has a more limited expression primarily in brain, heart, and testis (Pulverer et al. 1991; Derijard et al. 1994; Kyriakis et al. 1994; Yang et al. 1997). Including c-Jun, several members of the AP-1 transcription factors are substrates for JNKs including JunD, ATF2, and ATF3 (Behrens et al. 1999; Shaulian and Karin 2001). Phosphorylation of AP-1 members by JNKs enhances AP-1 transcriptional control of specific gene expression. The importance of AP-1 in the transcriptional control of many different genes involved in homeostasis and the role of JNKs in regulating AP-1 activity led to an intense study of JNK regulation and it is now clear that JNKs have many substrates in addition to AP-1. The targeted gene disruption of JNK1, JNK2, and JNK3 has defined tissue-specific functions for each isoform including the control of metabolism, apoptosis, motility, proliferation, DNA repair, and the regulation of genes involved in homeostasis

Topics in Current Genetics, Vol. 20
F. Posas, A. R. Nebreda (Eds.): Stress-Activated Protein Kinases
DOI 10.1007/4735_2007_0238 / Published online: 1 August 2007

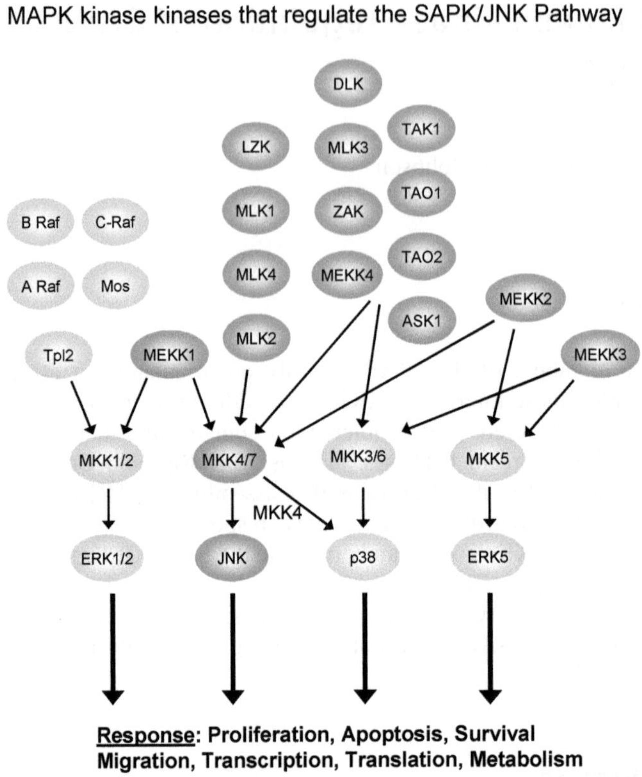

Fig. 1. MKKKs that control the MAPK pathways. There are twenty defined MKKKs known to regulate MAPK pathways. MKKKs phosphorylate and activate specific MKKs. Activated MKKs phosphorylate and activate specific MAPKs. The MKKKs and MKKs that regulate JNKs are highlighted in dark grey.

such as proteases and cytokines (Yang et al. 1997; Kuan et al. 1999; Sabapathy et al. 1999; Chang et al. 2003).

JNKs, like all MAPKs, are part of a three kinase signaling module (Fig. 1). JNKs are phosphorylated and activated by the MAPK kinases, MKK4 and MKK7. MKK4 and MKK7 are phosphorylated and activated by MAPK kinase kinases (MKKKs). Interestingly, whereas there are eleven MAPKs (JNK1/2/3, ERK1/2, p38, α/β/γ/δ, ERK5 and ERK7), there are only seven MKKs and at least twenty MKKKs. It is noteworthy that fourteen of the twenty defined MKKKs activate the MKK4/7→JNK1/2/3 pathway, demonstrating the importance of the JNK signaling pathway in the cellular response to stimuli that frequently involve potentially harmful or lethal consequences for the cell. Such stress stimuli include irradiation, toxins, drugs, osmolarity, temperature, changes in cell shape, adherence, cytoskeletal dynamics, and responses to antigens, growth factors and cytokines. Table 1 shows a partial list of substrates for JNKs that control adaptive responses of the cell to these different stimuli.

Table 1. Phosphorylation substrates for JNKs

Category	Substrate
Transcription factors	c-Jun
	JunD
	ATF2
	ATF3
	Elk-1
	Elk-3
	P53
	NFAT4
	HSF-1
	c-Myc
	Androgen receptor
	RXRα
	RARα
Signaling proteins	IRS-1
	Paxillin
	14-3-3
Microtubule-associated proteins	MAP1
	MAP2A
	Tau
	Doublecortin (DCX)
	Amyloid β precursor protein
Bcl family proteins	Bcl-2
	Bcl-xl
	Mcl-1
	Smac
	Bim
	Bmf
Nuclear core complex	Nup214

2 Organization of the MKKK-MKK4/7-JNK1/2/3 signaling module

Specificity in the organization of JNK signaling modules is controlled in part by recognition motifs for MKKK-MKK4/7 and MKK4/7-JNK1/2/3 interactions. A docking site referred to as DVD (domain for versatile docking) encoded near the C-terminus of MKK4 and 7 interacts with the N-lobe of the kinase domain of the specific MKKK (Takekawa et al. 2005), providing a docking mechanism for selective interaction of MKKKs and MKKs. Docking sites between the JNKs and MKK4 and 7 provide specificity in the interaction of the MKK and MAPK (Jacobs et al. 1999; Sharrocks et al. 2000; Fantz et al. 2001; Ho et al. 2003; Mooney and Whitmarsh 2004; Ho et al. 2006). Similar recognition motifs are present in JNK substrates such as c-Jun and ATF-2 (Sharrocks et al. 2000).

There are also several scaffolding proteins that organize specific JNK signaling modules. These scaffold proteins generally have no catalytic function, but rather

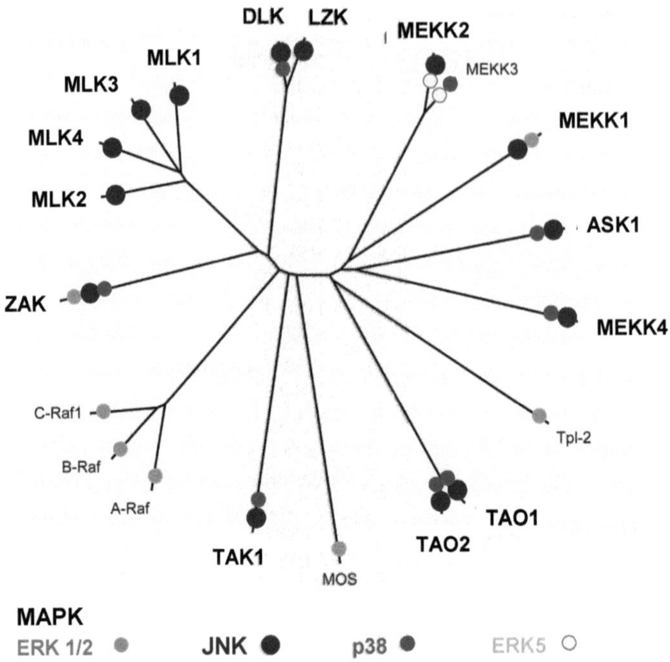

Fig. 2. Dendrogram showing the twenty MKKKs based on homology of their kinase domain primary amino acid sequences. The MAPKs activated by each MKKK are shown in the highlighted circles on the dendrogram.

have docking sites for binding specific MKKKs, MKKs, and MAPKs. Scaffold proteins that regulate the JNK signaling module include the JIP (JNK Interaction Proteins) 1-4 proteins, POSH (Plenty of SH3s), JKAP1 (SKRP), filamin, CrkII and IKAP (Morrison and Davis 2003). Scaffold proteins play an important regulatory role in controlling JNK signaling because they frequently bind specific MKKKs and localize the signaling module within the cell. Thus, scaffold proteins can regulate the spatio-temporal dynamics of JNK signaling.

3 MKKKs as signaling hubs controlling JNK activation

Figure 1 shows the MKKKs that have been defined to regulate MKK-MAPK modules. Of the twenty MKKKs, fourteen have been shown to regulate JNK activity. Six MKKKs regulate the ERK1/2 pathway while only two MKKKs are defined to regulate the ERK5 pathway. Nine MKKKs are known to regulate the p38 pathway. The restricted number of MKKKs regulating the ERK1/2 and ERK5 pathways implies a more restricted response and function for these MAPKs. For example, ERK1/2 is important in regulating cell proliferation in response to tyrosine kinases. The fact that ERK5 has a single MKK and only two defined MKKKs shown to physiologically regulate ERK5 activity suggests a rather restricted function for this MAPK. Physiologically, ERK5 appears important in regulating vascular development and maintenance of the vasculature in adults. In contrast, the large number of MKKKs that regulate JNK and p38 indicates a role for these MAPKs in the response to diverse stress stimuli.

Figure 2 shows a dendrogram for the relationship of the different MKKKs based on sequence homology of their kinase domains. Based on the kinase homologies, the MKKKs can be divided into six groups: MEKK, MLK, Raf, Tao/Tpl2, Mos, and TAK1. Among these twenty known MKKKs, members of the MEKK, MLK, TAO, and TAK1 groups regulate JNK activation. The properties of each group of MKKKs controlling JNK activation is discussed below.

3.1 MLKs (mixed lineage kinases)

The MLK group has seven members that can be further divided into the MLKs (MLK1, 2, 3, 4), DLKs (DLK, LZK), and ZAK (Gallo and Johnson 2002). The members of the MLK subgroup each have an N-terminal Src-homology-3 (SH3) domain, kinase domain, leucine zipper region and a Cdc42/Rac interactive binding (CRIB) domain. DLKs and ZAK have kinase domains and leucine zipper regions but lack the CRIB and SH3 domains. ZAK is structurally similar to the DLKs but also encodes a sterile-alpha motif that mediates homo- or hetero-dimerization.

3.2 MEKKs (MAPK-ERK kinase kinases)

MEKK1, MEKK2, and MEKK4 have each been shown to regulate the JNK pathway in response to different stimuli. MEKK1 is a large 196 kDa protein with complex regulation. MEKK1 appears to be regulated by both Rac/Cdc42 and RhoA GTPases (Fanger et al. 1997; Gallagher et al. 2004). Furthermore, MEKK1 is the only member of the MKKK-MKK-MAPK signaling network that has a caspase 3 cleavage site. MEKK1 is also the only member of the MAPK signaling network to encode a RING domain containing E3 ubiquitin ligase function. The MEKK1 RING domain has been shown to regulate auto-ubiquitination of MEKK1 that inhibits its kinase activity as well as ubiquitinate and stimulate the degradation of ERK1 and c-Jun (Lu et al. 2002; Witowsky and Johnson 2003; Xia et al. 2007). MEKK1 is also one of only a few proteins in defined proteomes to encode a SWIM domain whose function in MEKK1 remains undefined (Makarova et al. 2002).

Whereas MEKK1 regulates both the ERK1/2 and JNK pathways, MEKK2 regulates the ERK5 and JNK pathways. MEKK2 is only one of two MKKKs, the other being MEKK3, which regulate the MEK5-ERK5 pathway (Nakamura and Johnson 2003; Uhlik et al. 2004; Nakamura et al. 2006). MEKK2 and MEKK3 encode PB1 (Phox-Bem1p) domains that selectively heterodimerize with the MEK5 PB1 domain to form a functional MEKK2 (or MEKK3)–MEK5-ERK5 ternary complex (Nakamura and Johnson 2003; Nakamura et al. 2006). The C-terminal moiety of the MEKK2 domain is also capable of binding MKK7 for JNK activation. In contrast, the MEKK3 PB1 domain does not bind MKK7. Thus, MEKK2, but not MEKK3, regulates JNK activation.

MEKK4 selectively phosphorylates MKK3, MKK4, and MKK6 leading to the activation of both JNK and p38. MEKK4 binds Cdc42 and Rac via a CRIB domain and kinase-inactive MEKK4 inhibits Cdc42 and Rac activation of JNK (Fanger et al. 1997; Gerwins et al. 1997). MEKK4 also binds GADD45 α, β and γ proteins (Mita et al. 2002; Chi et al. 2004; Miyake et al. 2007), resulting in activation of MEKK4 kinase activity. MEKK4 and MEKK1 also bind the scaffold protein Axin, leading to JNK activation (Zhang et al. 1999; Zhang et al. 2001; Luo et al. 2003; Wong et al. 2004), suggesting MEKK4 and MEKK1 control JNK activation in the non-canonical Wnt signaling pathway.

3.3 ASK1 (apoptosis signal-regulating kinase 1)

ASK1 binds thioredoxin near its N-terminus and is activated in response to reactive oxygen species that cause the release of thioredoxin (Hayakawa et al. 2006). In addition, ASK1 binds JIP scaffold proteins and phosphorylates MKK4, MKK7, and MKK3, thereby activating both JNK and p38. ASK1 has been found to be activated by LPS and various pro-inflammatory cytokines.

3.4 TAK1 (TGFβ-activated kinase 1)

TAK1 is activated in response to IL-1, TNFα, and LPS stimulation of cells. In response to pro-inflammatory stimuli, TAK1 forms a complex with TAB1, which forms a complex with TAB2, TAB3, and the E3 ubiquitin ligase TRAF6 (Wang et al. 2001; Kanayama et al. 2004). This activated TAK1 complex coordinates the activation of MKK4, MKK6, and IKK leading to the activation of JNK, p38, and NF-κB.

3.5 TAO1 (thousand and one-amino acid kinase 1)

TAO1 and the related kinase TAO2 were cloned using degenerate oligonucleotide-based PCR cloning strategies for kinases related to the yeast Ste20 kinase (Chen et al. 1999; Zhou et al. 2004). Other than overexpression studies showing TAO kinases activate JNKs and p38, little is known about their function.

The different MKKKs described above that regulate the JNK pathway should be thought of in the greater context of the MAPK signaling network. The differing regulatory domains and motifs encoded in each MKKK selectively control their activation, inactivation, and association with regulatory proteins and scaffold proteins to mediate localization within the cell. As we see with most MKKKs that regulate the JNK pathway, MKKKs are often able to phosphorylate more than one MKK and thus regulate more than one MAPK pathway (see Fig. 1). Many

MKKKs that regulate JNK also regulate p38 consistent with a coordinated activation of JNK and p38 in response to stress stimuli. In contrast, MEKK1 activates JNK and ERK1/2, and MEKK2 activates JNK and ERK5. It is the differential control of MKKK activation and the organization of MKKKs with other proteins in signaling complexes that provide an amazing combinatorial diversity for the integration of MAPK networks for control of cellular responses to many different stimuli.

4 Insight into the function of MKKKs regulating the JNK pathway from targeted gene knockouts

The function of MKKKs controlling the JNK pathways have been defined using biochemical and cell biological approaches, but elucidation of physiological functions has mostly come from targeted gene knockouts. Of the MKKKs that regulate the JNK pathway the phenotypes of MEKK1, MEKK2, MEKK4, ASK1, TAK1, and MLK3 knockouts in mice have been reported. A brief description of the phenotypes of these knockout mice demonstrates their selective function in cellular and animal physiology (see Table 2 for summaries of phenotypes).

4.1 MEKK1

Mice having the targeted deletion of MEKK1 generally appear normal at birth and are fertile. In a SVEV129 mouse genetic background, MEKK1-null neonates have open eyes at birth, indicative of an epithelial morphogenesis defect (Yujiri et al. 1998, 2000; Zhang et al. 2003; Xia and Kao 2004; Xia and Karin 2004). In addition, MEKK1 deficient mice have wound healing and homeostasis defects associated with defective tissue remodeling. MEKK1-deficient cells have defective migration due to a loss of calpain activation required for release of focal adhesions at the rear of migrating cells (Cuevas et al. 2003). MEKK1 has also been shown to be required for activin-dependent epithelial cell migration (Zhang et al. 2005). In human breast cancer cell lines and mouse fibroblasts, MEKK1 has also been shown to be a primary regulator of urokinase-type plasminogen activator (uPA), which is required for cell invasion of the extracellular matrix (Witowsky et al. 2003). MEKK1 has been shown to regulate c-Jun, JunB, and Fra-2 expression and degradation and hence is a key regulator of AP-1 function (Cuevas et al. 2005). The role MEKK1 plays in the control of AP-1 function, protease expression, cell migration, and invasion is consistent with its involvement in wound healing, tissue remodeling, and tumor metastasis. In a transgenic model of metastatic breast cancer, MEKK1 deficiency markedly delayed tumor metastasis to the lungs. This was found to be due to a delay in dissemination of tumor cells because of an inability of the tumor to breakdown the basement membrane surrounding the tumor cell-filled ducts of the mammary gland (Cuevas et al. 2006). Recently, MEKK1 deficiency in a C57/Bl6 background has also been shown to be involved in fetal liver

Table 2. Phenotypes of MKKKs Regulating JNK

MKK	Described knockout phenotypes
MEKK1	Epithelial morphogenesis defects (open eyes at birth, defective tissue remodeling)
	Defective cell migration, loss of calpain activation
	Inhibition of urokinase-type plasminogen activator expression
	Suppressed cell invasion and dissemination of tumor cells
	Defective fetal liver hematopoiesis
	CD40 control of germinal center formation and B cell antibody production
	TH2 T cell tolerance
MEKK2	Mice normal and fertile at birth
	Suppressed TNFα, IL1α & β and IL-6 production in response to FGF-2
	Defective IgE-Fc□R1 signaling in mast cells
	Role in osteoclast function
MEKK4	Neural tube closure defects (exencephaly with enhanced apoptosis)
	Neuronal migration defects
	Skeletal defects
ASK1	Viable and fertile
	Fibroblasts have altered response to reactive oxygen species
	Decreased TNFα production in response to LPS
	Resistant to LPS-induced toxic shock
TAK1	Embryonic lethal due to vascularization defects
	Critical for IL-1 and TNFα signaling
MLK3	Viable and fertile
	Mild defect in epidermal tissue of the dorsal midline

hematopoiesis (Bonnesen et al. 2005), CD40 regulation of germinal center formation and B cell antibody production (Gallagher et al. 2007), and a JNK-Itch E3 ubiquitin ligase-mediated TH2 process in TH2 tolerance and lung inflammation (Venuprasad et al. 2006).

4.2 MEKK2

Mice having the targeted deletion of MEKK2 appear normal at birth and are fertile (Kesavan et al. 2004). MEKK2 deficiency has been shown to inhibit activation of ERK5 in response to growth factors such as FGF-2 in fibroblasts and IgE stimulation of the FcεR1 in mast cells (Garrington et al. 2000; Kesavan et al. 2004). Like MEKK1, MEKK2 has been shown to regulate the repertoire of proteins in the AP-1 complex. MEKK2-/- fibroblasts are inhibited in the induction of c-Jun, Fra-1, and Fra-2 mRNA in response to FGF-2 (Kesavan et al. 2004). FGF-2-induced expression of TNFα, IL1α and β, and IL-6 was inhibited in MEKK2-/- fibroblasts as is the expression of specific mast cell cytokines in response to IgE (Garrington et al. 2000). Interestingly, the knockout of MEKK1 and MEKK2 both alter the control of AP-1 components but do so differently, rendering distinct phenotypes. It was also shown that in osteoblasts, TGFβ or bone morphogenesis protein (BMP) activates MEKK2 and JNK. Smurf-1, a HECT domain ubiquitin ligase, binds MEKK2 to promote its degradation and negatively regulate osteoclast function to suppress osteogenic activity (Yamashita et al. 2005).

4.3 MEKK4

The MEKK4 protein is highly expressed in the developing central nervous system. MEKK4 knockout mice have been generated by homologous recombination and MEKK4 knockin mice have been generated by genetic mutation to generate a kinase-inactive MEKK4 (Abell et al. 2005; Chi et al. 2005). The phenotypes of MEKK4 knockout and knockin mice are overlapping but not identical. Prenatal lethality is often seen due to severe defects in both the neural tube and skeleton. Both the knockout and kinase-inactive MEKK4 have exencepahly associated with enhanced apoptosis in the neural tube and loss of both JNK and p38 activity. The phenotype of the MEKK4 knockout and kinase-inactive knockin are extremely similar to the knockout phenotype for the adaptor protein TRAF4 (Abell et al. 2005; Abell 2005). Abell et al. (Abell and Johnson 2005) showed that MEKK4 binds Traf4, promoting MEKK4 oligomerization and activation. This indicates that MEKK4 and TRAF4 are in a common biochemical and genetic pathway. Recently, MEKK4 was shown to be involved in regulating filamin-A expression and controlling neuronal migration in developing forebrain (Sarkisian et al. 2006). Filamin-A is an actin-binding protein essential for cytoskeletal rearrangement and cell locomotion. Loss of MEKK4 expression disrupts filamin-A expression and phosphorylation. The loss of MEKK4 and the resulting phenotype seen in the developing forebrain is similar to that seen with filamin-A mutations that contribute to periventricular heterotopia (PVH), a congenital malformation of the human cerebral cortex.

4.4 ASK1

Mice deficient in ASK1 are viable and fertile. ASK1-/- mouse embryonic fibroblasts respond to reactive oxygen species (H_2O_2) with a normal transient activation of JNK and p38. However, the prolonged activation of these two MAPKs is lost, demonstrating that ASK1 is required for the prolonged phase of JNK and p38 activity in response to H_2O_2 (Tobiume et al. 2001). A similar loss of prolonged JNK and p38 activation is seen in ASK1-/- fibroblasts in response to TNFα. ASK1-/- mice also have diminished TNFα production in response to LPS and are resistant to LPS-induced toxic shock (Matsuzawa et al. 2005). The regulation of ASK1 in response to LPS is through the MyD88/TRAF6 signaling pathway and plays an important role in the control of cytokines for the innate immune response.

4.5 TAK1

The TAK1 knockout mouse is embryonic lethal (Shim et al. 2005; Omori et al. 2006). There is defective vascularization of TAK1-/- embryos and yolk sacs, indicating that TAK1 plays an essential role in vascular development (Jadrich et al. 2006). The role of TAK1 in IL-1 and TNFα signaling suggests it plays an important role in the inflammatory response (Sato et al. 2005, 2006). TAK1-/- fibroblasts show decreased IL-1β-induced IL-6 production (Sato et al. 2005) and TAK1-/- keratinocytes have a reduced survival when challenged with TNFα (Omori et al. 2006). Conditional keratinocyte deletion of TAK1 results in a severe post-natal inflammation with elevated levels of inflammatory cytokines and keratinocyte apoptosis (Omori et al. 2006).

4.6 MLK3

MLK3-/- mice have no obvious phenotype except for a mild defect in the epidermal tissue of the dorsal midline (Brancho et al. 2005). MLK3-/- mice are viable and fertile. TNFα -stimulated JNK activation was partially inhibited in MLK3-/- fibroblasts, but there was no measurable inhibition of ERK1/2 or p38 signaling. The response to UV irradiation, sorbitol, anisomycin, and ceramide was normal in regards to JNK and p38 activation.

Therapeutically, MLKs have been intensely studied. Inhibition of MLK kinase activity by small molecules such as CEP-1347, an indolocarbazole that is a derivative of K252a, protects neurons from apoptosis. The compound was well-tolerated in humans but performed poorly in trials for neurodegenerative diseases including Parkinson's and Alzheimer's disease (Wang et al. 2004).

5 Conclusions

MKKKs are upstream regulators of the MAPKs that integrate MAPK signaling networks with the complex cellular response to many different stimuli. MKKKs that regulate the JNK pathway are clearly involved in neural and vascular development, immune response, and inflammation. MKKKs as therapeutic drug targets are now being explored for small molecule inhibition. Inhibition of a specific MKKK has the potential to selectively inhibit stimulus-specific activation of MAPKs (Johnson et al. 2005). The continued characterization of MKKKs and their function in cells and animal physiology is needed to define their utility as therapeutic targets in human disease.

References

Abell AN, Johnson GL (2005) MEKK4 is an effector of the embryonic TRAF4 for JNK activation. J Biol Chem 280:35793-35796

Abell AN, Rivera-Perez JA, Cuevas BD, Uhlik MT, Sather S, Johnson NL, Minton SK, Lauder JM, Winter-Vann AM, Nakamura K, Magnuson T, Vaillancourt RR, Heasley LE, Johnson GL (2005) Ablation of MEKK4 kinase activity causes neurulation and skeletal patterning defects in the mouse embryo. Mol Cell Biol 25:8948-8959

Behrens A, Sibilia M, Wagner EF (1999) Amino-terminal phosphorylation of c-Jun regulates stress-induced apoptosis and cellular proliferation. Nat Genet 21:326-329

Bonnesen B, Orskov C, Rasmussen S, Holst PJ, Christensen JP, Eriksen KW, Qvortrup K, Odum N, Labuda T (2005) MEK kinase 1 activity is required for definitive erythropoiesis in the mouse fetal liver. Blood 106:3396-3404

Brancho D, Ventura JJ, Jaeschke A, Doran B, Flavell RA, Davis RJ (2005) Role of MLK3 in the regulation of mitogen-activated protein kinase signaling cascades. Mol Cell Biol 25:3670-3681

Chang L, Jones Y, Ellisman MH, Goldstein LS, Karin M (2003) JNK1 is required for maintenance of neuronal microtubules and controls phosphorylation of microtubule-associated proteins. Dev Cell 4:521-533

Chen Z, Hutchison M, Cobb MH (1999) Isolation of the protein kinase TAO2 and identification of its mitogen-activated protein kinase/extracellular signal-regulated kinase kinase binding domain. J Biol Chem 274:28803-28807

Chi H, Lu B, Takekawa M, Davis RJ, Flavell RA (2004) GADD45beta/GADD45gamma and MEKK4 comprise a genetic pathway mediating STAT4-independent IFNgamma production in T cells. EMBO J 23:1576-1586

Chi H, Sarkisian MR, Rakic P, Flavell RA (2005) Loss of mitogen-activated protein kinase kinase kinase 4 (MEKK4) results in enhanced apoptosis and defective neural tube development. Proc Natl Acad Sci USA 102:3846-3851

Cuevas BD, Abell AN, Witowsky JA, Yujiri T, Johnson NL, Kesavan K, Ware M, Jones PL, Weed SA, DeBiasi RL, Oka Y, Tyler KL, Johnson GL (2003) MEKK1 regulates calpain-dependent proteolysis of focal adhesion proteins for rear-end detachment of migrating fibroblasts. EMBO J 22:3346-3355

Cuevas BD, Uhlik MT, Garrington TP, Johnson GL (2005) MEKK1 regulates the AP-1 dimer repertoire via control of JunB transcription and Fra-2 protein stability. Oncogene 24:801-809

Cuevas BD, Winter-Vann AM, Johnson NL, Johnson GL (2006) MEKK1 controls matrix degradation and tumor cell dissemination during metastasis of polyoma middle-T driven mammary cancer. Oncogene 25:4998-5010

Derijard B, Hibi M, Wu IH, Barrett T, Su B, Deng T, Karin M, Davis RJ (1994) JNK1: a protein kinase stimulated by UV light and Ha-Ras that binds and phosphorylates the c-Jun activation domain. Cell 76:1025-1037

Fanger GR, Johnson NL, Johnson GL (1997) MEK kinases are regulated by EGF and selectively interact with Rac/Cdc42. EMBO J 16:4961-4972

Fantz DA, Jacobs D, Glossip D, Kornfeld K (2001) Docking sites on substrate proteins direct extracellular signal-regulated kinase to phosphorylate specific residues. J Biol Chem 276:27256-27265

Gallagher E, Enzler T, Matsuzawa A, Anzelon-Mills A, Otero D, Holzer R, Janssen E, Gao M, Karin M (2007) Kinase MEKK1 is required for CD40-dependent activation of the kinases Jnk and p38, germinal center formation, B cell proliferation and antibody production. Nat Immunol 8:57-63

Gallagher ED, Gutowski S, Sternweis PC, Cobb MH (2004) RhoA binds to the amino terminus of MEKK1 and regulates its kinase activity. J Biol Chem 279:1872-1877

Gallo KA, Johnson GL (2002) Mixed-lineage kinase control of JNK and p38 MAPK pathways. Nat Rev Mol Cell Biol 3:663-672

Garrington TP, Ishizuka T, Papst PJ, Chayama K, Webb S, Yujiri T, Sun W, Sather S, Russell DM, Gibson SB, Keller G, Gelfand EW, Johnson GL (2000) MEKK2 gene disruption causes loss of cytokine production in response to IgE and c-Kit ligand stimulation of ES cell-derived mast cells. EMBO J 19:5387-5395

Gerwins P, Blank JL, Johnson GL (1997) Cloning of a novel mitogen-activated protein kinase kinase kinase, MEKK4, that selectively regulates the c-Jun amino terminal kinase pathway. J Biol Chem 272:8288-8295

Hayakawa T, Matsuzawa A, Noguchi T, Takeda K, Ichijo H (2006) The ASK1-MAP kinase pathways in immune and stress responses. Microbes Infect 8:1098-1107

Ho DT, Bardwell AJ, Abdollahi M, Bardwell L (2003) A docking site in MKK4 mediates high affinity binding to JNK MAPKs and competes with similar docking sites in JNK substrates. J Biol Chem 278:32662-32672

Ho DT, Bardwell AJ, Grewal S, Iverson C, Bardwell L (2006) Interacting JNK-docking sites in MKK7 promote binding and activation of JNK mitogen-activated protein kinases. J Biol Chem 281:13169-13179

Jacobs D, Glossip D, Xing H, Muslin AJ, Kornfeld K (1999) Multiple docking sites on substrate proteins form a modular system that mediates recognition by ERK MAP kinase. Genes Dev 13:163-175

Jadrich JL, O'Connor MB, Coucouvanis E (2006) The TGF beta activated kinase TAK1 regulates vascular development *in vivo*. Development 133:1529-1541

Johnson GL, Dohlman HG, Graves LM (2005) MAPK kinase kinases (MKKKs) as a target class for small-molecule inhibition to modulate signaling networks and gene expression. Curr Opin Chem Biol 9:325-331

Kanayama A, Seth RB, Sun L, Ea CK, Hong M, Shaito A, Chiu YH, Deng L, Chen ZJ (2004) TAB2 and TAB3 activate the NF-kappaB pathway through binding to polyubiquitin chains. Mol Cell 15:535-548

Kesavan K, Lobel-Rice K, Sun W, Lapadat R, Webb S, Johnson GL, Garrington TP (2004) MEKK2 regulates the coordinate activation of ERK5 and JNK in response to FGF-2 in fibroblasts. J Cell Physiol 199:140-148

Kuan CY, Yang DD, Samanta Roy DR, Davis RJ, Rakic P, Flavell RA (1999) The Jnk1 and Jnk2 protein kinases are required for regional specific apoptosis during early brain development. Neuron 22:667-676

Kyriakis JM, Banerjee P, Nikolakaki E, Dai T, Rubie EA, Ahmad MF, Avruch J, Woodgett JR (1994) The stress-activated protein kinase subfamily of c-Jun kinases. Nature 369:156-160

Lu Z, Xu S, Joazeiro C, Cobb MH, Hunter T (2002) The PHD domain of MEKK1 acts as an E3 ubiquitin ligase and mediates ubiquitination and degradation of ERK1/2. Mol Cell 9:945-956

Luo W, Ng WW, Jin LH, Ye Z, Han J, Lin SC (2003) Axin utilizes distinct regions for competitive MEKK1 and MEKK4 binding and JNK activation. J Biol Chem 278:37451-37458

Makarova KS, Aravind L, Koonin EV (2002) SWIM, a novel Zn-chelating domain present in bacteria, archaea, and eukaryotes. Trends Biochem Sci 27:384-386

Matsuzawa A, Saegusa K, Noguchi T, Sadamitsu C, Nishitoh H, Nagai S, Koyasu S, Matsumoto K, Takeda K, Ichijo H (2005) ROS-dependent activation of the TRAF6-ASK1-p38 pathway is selectively required for TLR4-mediated innate immunity. Nat Immunol 6:587-592

Mita H, Tsutsui J, Takekawa M, Witten EA, Saito H (2002) Regulation of MTK1/MEKK4 kinase activity by its N-terminal autoinhibitory domain and GADD45 binding. Mol Cell Biol 22:4544-4555

Miyake Z, Takekawa M, Ge Q, Saito H (2007) Activation of MTK1/MEKK4 by GADD45 through induced N-C dissociation and dimerization-mediated trans-autophosphorylation of the MTK1 kinase domain. Mol Cell Biol 27:2765-2776

Mooney LM, Whitmarsh AJ (2004) Docking interactions in the c-Jun N-terminal kinase pathway. J Biol Chem 279:11843-11852

Morrison DK, Davis RJ (2003) Regulation of MAP kinase signaling modules by scaffold proteins in mammals. Annu Rev Cell Dev Biol 19:91-118

Nakamura K, Johnson GL (2003) PB1 domains of MEKK2 and MEKK3 interact with the MEK5 PB1 domain for activation of the ERK5 pathway. J Biol Chem 278:36989-36992

Nakamura K, Uhlik MT, Johnson NL, Hahn KM, Johnson GL (2006) PB1 domain-dependent signaling complex is required for extracellular signal-regulated kinase 5 activation. Mol Cell Biol 26:2065-2079

Omori E, Matsumoto K, Sanjo H, Sato S, Akira S, Smart RC, Ninomiya-Tsuji J (2006) TAK1 is a master regulator of epidermal homeostasis involving skin inflammation and apoptosis. J Biol Chem 281:19610-19617

Pulverer BJ, Kyriakis JM, Avruch J, Nikolakaki E, Woodgett JR (1991) Phosphorylation of c-jun mediated by MAP kinases. Nature 353:670-674

Sabapathy K, Jochum W, Hochedlinger K, Chang L, Karin M, Wagner EF (1999) Defective neural tube morphogenesis and altered apoptosis in the absence of both JNK1 and JNK2. Mech Dev 89:115-124

Sarkisian MR, Bartley CM, Chi H, Nakamura F, Hashimoto-Torii K, Torii M, Flavell RA, Rakic P (2006) MEKK4 signaling regulates filamin expression and neuronal migration. Neuron 52:789-801

Sato S, Sanjo H, Takeda K, Ninomiya-Tsuji J, Yamamoto M, Kawai T, Matsumoto K, Takeuchi O, Akira S (2005) Essential function for the kinase TAK1 in innate and adaptive immune responses. Nat Immunol 6:1087-1095

Sato S, Sanjo H, Tsujimura T, Ninomiya-Tsuji J, Yamamoto M, Kawai T, Takeuchi O, Akira S (2006) TAK1 is indispensable for development of T cells and prevention of colitis by the generation of regulatory T cells. Int Immunol 18:1405-1411

Sharrocks AD, Yang SH, Galanis A (2000) Docking domains and substrate-specificity determination for MAP kinases. Trends Biochem Sci 25:448-453

Shaulian E, Karin M (2001) AP-1 in cell proliferation and survival. Oncogene 20:2390-2400

Shim JH, Xiao C, Paschal AE, Bailey ST, Rao P, Hayden MS, Lee KY, Bussey C, Steckel M, Tanaka N, Yamada G, Akira S, Matsumoto K, Ghosh S (2005) TAK1, but not

TAB1 or TAB2, plays an essential role in multiple signaling pathways *in vivo*. Genes Dev 19:2668-2681
Smeal T, Binetruy B, Mercola D, Grover-Bardwick A, Heidecker G, Rapp UR, Karin M (1992) Oncoprotein-mediated signalling cascade stimulates c-Jun activity by phosphorylation of serines 63 and 73. Mol Cell Biol 12:3507-3513
Takekawa M, Tatebayashi K, Saito H (2005) Conserved docking site is essential for activation of mammalian MAP kinase kinases by specific MAP kinase kinase kinases. Mol Cell 18:295-306
Tobiume K, Matsuzawa A, Takahashi T, Nishitoh H, Morita K, Takeda K, Minowa O, Miyazono K, Noda T, Ichijo H (2001) ASK1 is required for sustained activations of JNK/p38 MAP kinases and apoptosis. EMBO Rep 2:222-228
Uhlik MT, Abell AN, Cuevas BD, Nakamura K, Johnson GL (2004) Wiring diagrams of MAPK regulation by MEKK1, 2, and 3. Biochem Cell Biol 82:658-663
Venuprasad K, Elly C, Gao M, Salek-Ardakani S, Harada Y, Luo JL, Yang C, Croft M, Inoue K, Karin M, Liu YC (2006) Convergence of Itch-induced ubiquitination with MEKK1-JNK signaling in Th2 tolerance and airway inflammation. J Clin Invest 116:1117-1126
Wang C, Deng L, Hong M, Akkaraju GR, Inoue J, Chen ZJ (2001) TAK1 is a ubiquitin-dependent kinase of MKK and IKK. Nature 412:346-351
Wang LH, Besirli CG, Johnson EM Jr (2004) Mixed-lineage kinases: a target for the prevention of neurodegeneration. Annu Rev Pharmacol Toxicol 44:451-474
Witowsky J, Abell A, Johnson NL, Johnson GL, Cuevas BD (2003) MEKK1 is required for inducible urokinase-type plasminogen activator expression. J Biol Chem 278:5941-5946
Witowsky JA, Johnson GL (2003) Ubiquitylation of MEKK1 inhibits its phosphorylation of MKK1 and MKK4 and activation of the ERK1/2 and JNK pathways. J Biol Chem 278:1403-1406
Wong CK, Luo W, Deng Y, Zou H, Ye Z, Lin SC (2004) The DIX domain protein coiled-coil-DIX1 inhibits c-Jun N-terminal kinase activation by Axin and dishevelled through distinct mechanisms. J Biol Chem 279:39366-39373
Xia Y, Kao WW (2004) The signaling pathways in tissue morphogenesis: a lesson from mice with eye-open at birth phenotype. Biochem Pharmacol 68:997-1001
Xia Y, Karin M (2004) The control of cell motility and epithelial morphogenesis by Jun kinases. Trends Cell Biol 14:94-101
Xia Y, Wang J, Xu S, Johnson GL, Hunter T, Lu Z (2007) MEKK1 mediates the ubiquitination and degradation of c-Jun in response to osmotic stress. Mol Cell Biol 27:510-517
Yamashita M, Ying SX, Zhang GM, Li C, Cheng SY, Deng CX, Zhang YE (2005) Ubiquitin ligase Smurf1 controls osteoblast activity and bone homeostasis by targeting MEKK2 for degradation. Cell 121:101-113
Yang DD, Kuan CY, Whitmarsh AJ, Rincon M, Zheng TS, Davis RJ, Rakic P, Flavell RA (1997) Absence of excitotoxicity-induced apoptosis in the hippocampus of mice lacking the Jnk3 gene. Nature 389:865-870
Yujiri T, Sather S, Fanger GR, Johnson GL (1998) Role of MEKK1 in cell survival and activation of JNK and ERK pathways defined by targeted gene disruption. Science 282:1911-1914
Yujiri T, Ware M, Widmann C, Oyer R, Russell D, Chan E, Zaitsu Y, Clarke P, Tyler K, Oka Y, Fanger GR, Henson P, Johnson GL (2000) MEK kinase 1 gene disruption al-

ters cell migration and c-Jun NH2-terminal kinase regulation but does not cause a measurable defect in NF-kappa B activation. Proc Natl Acad Sci USA 97:7272-7277

Zhang L, Deng M, Parthasarathy R, Wang L, Mongan M, Molkentin JD, Zheng Y, Xia Y (2005) MEKK1 transduces activin signals in keratinocytes to induce actin stress fiber formation and migration. Mol Cell Biol 25:60-65

Zhang L, Wang W, Hayashi Y, Jester JV, Birk DE, Gao M, Liu CY, Kao WW, Karin M, Xia Y (2003) A role for MEK kinase 1 in TGF-beta/activin-induced epithelium movement and embryonic eyelid closure. EMBO J 22:4443-4454

Zhang Y, Neo SY, Wang X, Han J, Lin SC (1999) Axin forms a complex with MEKK1 and activates c-Jun NH(2)-terminal kinase/stress-activated protein kinase through domains distinct from Wnt signaling. J Biol Chem 274:35247-35254

Zhang Y, Qiu WJ, Liu DX, Neo SY, He X, Lin SC (2001) Differential molecular assemblies underlie the dual function of Axin in modulating the WNT and JNK pathways. J Biol Chem 276:32152-32159

Zhou T, Raman M, Gao Y, Earnest S, Chen Z, Machius M, Cobb MH, Goldsmith EJ (2004) Crystal structure of the TAO2 kinase domain: activation and specificity of a Ste20p MAP3K. Structure 12:1891-1900

Johnson, Gary L.

Department of Pharmacology, CB# 7365, 1108 Mary Ellen Jones Building, University of North Carolina School of Medicine, Chapel Hill, NC 27599-7365, USA

gary_johnson@med.unc.edu

Stalheim, Lisa

Department of Pharmacology, CB# 7365, 1108 Mary Ellen Jones Building, University of North Carolina School of Medicine, Chapel Hill, NC 27599-7365, USA

Alternative p38 MAPK pathways

Francisco Iñesta-Vaquera, Guadalupe Sabio, Yvonne Kuma, and Ana Cuenda

Abstract

There are four members of the mammalian p38 mitogen-activated protein kinases (MAPKs) family (p38α, p38β, p38γ, and p38δ) which are about 60% identical in their amino acid sequence but differ in their expression patterns, substrate specificities, and sensitivities to chemical inhibitors such as SB203580. Much attention in recent years has been focused on studying the role of the p38α isoform, which is widely referred to as p38 in the literature. However, there are other p38 isoforms (p38β, p38γ, and p38δ) whose roles among the cellular functions and the implication in some of the pathological conditions have not been precisely defined so far. Here, we focus on the emergent roles of the alternative p38γ and p38δ MAPK pathways and their implication in different biological processes. It is now clear that these p38MAPKs show similarities to the classical p38MAPK, but with some differences that challenge the paradigm of the archetypical p38MAPK pathway.

1 Introduction

The mammalian p38 subfamily of MAPK has four members encoded by different genes, which share high sequence homologies and are designated as p38α, p38β, p38γ, and p38δ. Among all p38 MAPK isoforms, p38α is the best characterised and is expressed in most cell types. p38α MAPK was identified in 1994 by several groups as a 38 kDa polypeptide that underwent activation in response to endotoxin treatment, cellular stress or cytokines (Cohen 1997). Two to three years after the identification of p38α, three additional isoforms were described: p38β (also called SAPK2b) (Jiang et al. 1996), p38γ (also called SAPK3 and ERK6) (Mertens et al. 1996; Lechner et al. 1996), and p38δ (also called SAPK4) (Goedert et al. 1997a; Jiang et al. 1997). The initially described p38β (Jiang et al. 1996) was subsequently designated p38β1 to differentiate it from p38β2, cDNA sequences lacks 24 base pairs (resulting in a 8 amino acid deletion) that appears to affect activity of the kinase (Enslen et al. 1998; Stein et al. 1997). The absence of this insertion in the genomic DNA and cDNA sequences suggests that it may be a cloning artefact, and p38β2 is the predominant isoform of p38β (Beardmore et al. 2005).

All p38 MAPKs are strongly activated *in vivo* by environmental stresses and inflammatory cytokines, and less by serum and growth factors. The activation of

Topics in Current Genetics, Vol. 20
F. Posas, A. R. Nebreda (Eds.): Stress-Activated Protein Kinases
DOI 10.1007/4735_2007_0239 / Published online: 1 August 2007

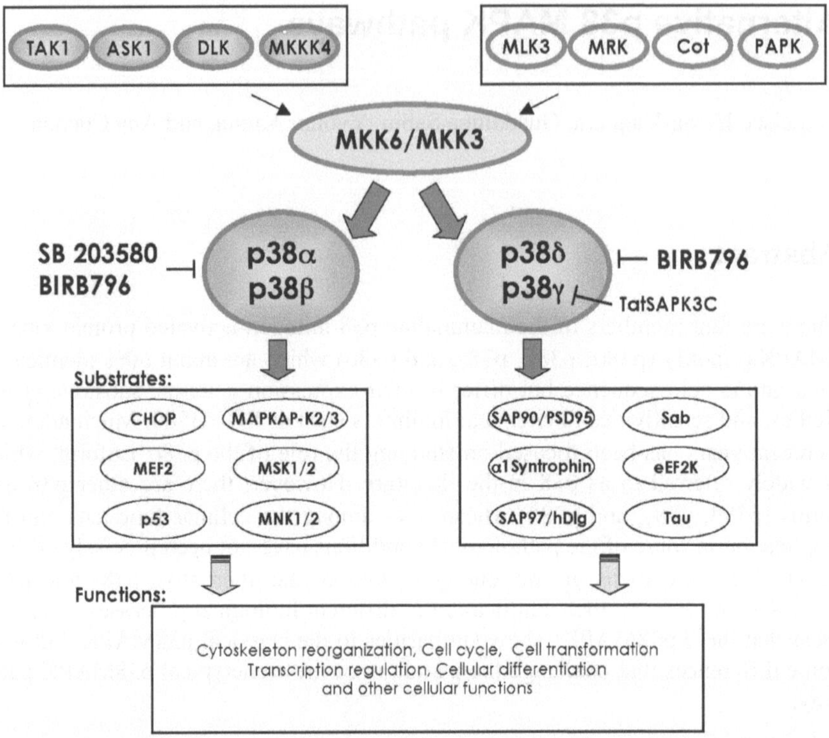

Fig. 1. p38 MAPK signalling pathways. The p38 MAPK subfamily can further be divided into two distinct groups: p38α/p38β and p38γ/p38δ, which differ in their expression patterns, substrate specificities, sensitivities to chemical inhibitors, and biological functions. The activation of p38 MAPKs occurs via dual phosphorylation of their TGY motif by MKK3 and MKK6. Different MKKKs have been described to be specifically upstream of the p38 pathways. It has been described that MLK3, MRK, Cot or PAPK preferentially regulate the p38γ and p38δ MAPK pathways.

p38 MAPKs occurs via dual phosphorylation of their TGY motif, in the activation loop, by MKK3 and MKK6. Although all p38 isoforms are widely expressed, p38α and p38β are ubiquitously expressed, whereas p38γ and p38δ appear to have more restricted expression: p38γ is most significantly abundant in skeletal muscle and p38δ is mainly found in testes, pancreas, kidney, and small intestine (Mertens et al. 1996; Jiang et al. 1997; Goedert et al. 1997a; Court et al. 2002).

The p38 MAPK subfamily can further be divided into two distinct subsets, on the one hand p38α and p38β and on the other, p38γ and p38δ (Fig. 1). This is evident, firstly, from their amino-acid sequence identity; p38α and p38β are 75% identical, whereas p38γ and p38δ are 62% and 61% identical to p38α, respectively. Of note, p38γ and p38δ are more identical (~70%) to each other. Secondly, their susceptibilities to inhibition at low concentrations by the compounds,

SB203580 and SB202190. *In vitro* and *in vivo* assays demonstrated that only p38α and p38β are inhibited by these compounds, whereas p38γ and p38δ were completely unaffected by the drugs (Goedert et al. 1997a; Kuma et al. 2005). The basis of this inhibition was revealed in the crystal structure of p38α complexed with SB203580. Thr106 in the hinge of the p38α and p38β ATP binding pocket interacts with a fluorine atom in the SB203580 structure. This orients the drug to interact with His107 and Leu108 of the pocket preventing ATP binding (Eyers et al. 1998; Gum et al. 1998). p38γ and p38δ possess Met, a large side chain amino acid, at the Thr106 equivalent position in the ATP binding pocket that prevents inhibitor binding (Eyers et al. 1998; Gum et al. 1998). A third difference between these two subgroups of p38MAPKs is with regard to substrate selectivity of these kinases. For example, microtubule-associated protein Tau is a better *in vitro* substrate for p38γ and p38δ than p38α and p38β (Goedert et al. 1997b; Feijoo et al. 2005), and this is also true for the scaffold proteins α1-syntrophin, SAP90/PSD95 and SAP97/hDlg (Hasegawa et al. 1999; Sabio et al. 2004, 2005). Conversely, MAPKAP-K2, MAPKAP-K3 and glycogen synthase are better phosphorylated by p38α and p38β than p38γ and p38δ (Cuenda et al. 1997; Goedert et al. 1997a; Kuma et al. 2004).

2 The p38γ and p38δ MAPK pathways

Upstream activators of p38γ and p38δ are the kinases MKK3 and MKK6, although it has been shown that *in vitro*, MKK4, an activator of JNK, may also phosphorylate and activate them (Cuenda et al. 1997; Goedert et al. 1997a). Using MKK targeted gene disruption and siRNA approaches, it has been shown that in response to most stimuli MKK3 and MKK6 are the main p38α activators but, in some circumstances, such as ultraviolet radiation, MKK4 may contribute to p38 activation (Brancho et al. 2003). Unlike p38α, the activation of the isoforms p38γ and p38δ has not been extensively examined in MKKs knockouts.

The major MKK required for the activation of specific p38 MAPK may be determined by several factors, one is the cell type as the level of expression varies (Tanaka et al. 2002; Wang et al. 2002), another is the nature and also the strength of the stimuli. It has been suggested that the pattern of downstream p38 MAPK activation in the particular response may be determined by the level of MKK6 activity, which activates all p38 isoforms *in vitro* triggered by a given stimulus (Alonso et al. 2000). Moreover, there are two important structural requirements for selective activation of p38 MAPK isoforms by MKKs, common docking sequences in the N-terminal of the MKK and isoform-specific sequences of the p38MAPK isoforms within the activation loop (Enslen et al. 2000).

Different MKKKs has been described to be specifically upstream of the p38γ and/or p38δ pathways (Fig. 1), these include the Mixed lineage kinase 3 (MLK3), which form a complex with the scaffold protein JNK-interacting protein1 (JIP2) and MKKs to activate p38δ (Morrison and Davis 2003). All proteins, polyploidy-associated protein kinase-A (PAPK-A), which is a member of the Ste20/germinal centre kinase family (Nishigaki et al. 2003); MLK related kinase (MRK) (Gross et al. 2002) and the protein Cot, which is a serine/threonine kinase implicated in cellular transformation (Chiariello et al. 2000) activate p38γMAPK pathway. In addition, overexpression of the small G-proteins Rit or RhoA also causes p38γ MAPK activation (Sakabe et al. 2002; Marinissen et al. 2001; Singh et al. 2003).

2.1 p38γ and p38δ substrates

The elucidation of physiological roles and the identification of physiological substrates for p38α and p38β have been facilitated by the availability of specific pyridinyl imidazole inhibitors such as SB203580 or SB202190 (Cuenda et al. 1995; Davies et al. 2000). So far, such specific inhibitors for p38γ and/or p38δ are not available and the delineation of pathways in which these MAPKs are involved has been difficult. However, this problem can be partly solved by the use of p38 knockouts. Recently, knockout mice for p38β, p38γ, and p38δ and double p38γ/p38δ knockout mice have been generated, which are viable and fertile (Sabio et al. 2005; Beardmore et al. 2005). Moreover, the diaryl urea compound BIRB796 (Parguellis et al. 2002) is not only a potent inhibitor of p38α and p38β, but also inhibits p38γ and p38δ at higher concentrations in cell-based assays providing a new tool for identifying physiological roles of these two p38 MAPK isoforms by using varying concentrations of this new compound in combination with the pyridinyl imidazoles (Kuma et al. 2005).

Several physiological substrates for p38γ and p38δ MAPK isoforms have been described in the past years (Fig. 1). A feature that makes p38γ unique among the p38 MAPKs is its short C-terminal sequence –KETXL, an amino acid sequence ideal for binding PDZ domains in proteins. p38γ binds to the PDZ domain of a variety of these proteins, such as α1-syntrophin, SAP90/PSD95 and SAP97/hDlg, and under stress conditions is able to phosphorylate them and may modulate their activity (Hasegawa et al. 1999; Sabio et al. 2004, 2005). One of the tools used in the identification of physiological substrates for p38γ is the cell permeant peptide TatSAPK3C which contains the last nine residues of p38γ fused to the cell-membrane transduction domain of the human immunodeficiency virus-type 1 (HIV-1) Tat protein. This peptide specifically blocks the phosphorylation of PDZ domain-containing proteins by p38γ in intact cells by preventing the association of the kinase with the PDZ domain of the substrate (Sabio et al. 2004, 2005). These PDZ domain containing proteins are scaffold proteins usually targeted to the plasma membrane cytoskeleton at specialised sites such as the neuromuscular junction and gap junctions through protein-protein interactions. In the case of SAP97/hDlg, its phosphorylation by p38γ provided a mechanism of dissociating SAP97/hDlg from the cytoskeleton (Sabio et al. 2005). Another p38γ substrate that does not require PDZ domain binding interactions is the mitochondrial protein Sab (Court et al. 2004).

On the other hand, p38δ possibly plays a role in cytoskeleton regulation as it has been reported to phosphorylate the cytoplasmic protein stathmin, which has been linked to regulation of microtubule dynamics (Parker et al. 1998). Microtubule-associated protein Tau is another protein substrate of p38δ (Goedert et al. 1997b; Feijoo et al. 2005; Yoshida and Goedert 2006). Eukaryotic elongation factor 2 (eEF2) kinase was identified in a screen for substrates for p38δ and later shown to be inhibited upon phosphorylation on Ser359 (Knebel et al. 2001, 2002).

3 Biological roles of p38γ and p38δ MAPK pathways

Evidence from a number of studies carried out during the past few years suggest that many physiological functions of the p38MAPK isoforms may overlap but may not necessarily be redundant and/or identical. Thus, it has been shown that in several human breast cancer cells p38β increases, whereas p38γ and/or p38δ inhibit AP-1-dependent transcription and cell proliferation induced by MKK6 (Pramanick et al. 2003). Additionally, it has been found a selective regulation of p38MAPK isoforms activation in response to a variety of extracellular stimuli. For example: in murine mesangial cells, transforming growth factor-β selectively activated p38α and p38δ (Wang et al. 2002); in PC12 cells, hypoxia selectively activated p38α and p38γ (Conrad et al. 1999); and both the effect of the pro-apoptotic compound LY83583 in renal carcinoma cells and the suppression of antioxidant response element by MKK6 in HepG2 cells, depended on p38γ and p38δ, but not p38α or p38β (Ambrose et al. 2006; Keum et al. 2006).

3.1 Cell cycle and cellular transformation

Most of the work published on cell cycle regulation by p38 MAPK pathway has been focussed on studying the role of the isoforms p38α and β, but in the case of p38γ, several reports indicate its role controlling cell cycle. Thus, it has been claimed that its activation after ionizing radiation could be dependent on ATM and cause G2/M cell cycle arrest (Wang et al. 2000). However, all p38 MAPKs show robust induction by stresses such as ultraviolet radiation, whereas their induction by ionizing radiation is highly infrequent (Liu et al. 1996; Y. Kuma, A. Cuenda, unpublished results). Moreover, overexpression of MRK-β, which is a MKKK that preferentially activates p38γ and JNK, causes G2 arrest (Gross et al. 2002). On the other hand, it has been shown that *Xenopus* p38γ promotes meiotic G2/M transition in *Xenopus* oocytes treated with progesterone and activates XCdc25C by phosphorylating it at Ser205, whereas p38α or p38β have no effect (Perdiguero et al. 2003). Fully grown *Xenopus* oocytes are arrested in G2/prophase of meiosis I and are induced to proceed through meiosis by progesterone stimulation (Perdiguero et al. 2003).

There are a number of recent publications providing evidence for the role of p38γ MAPK pathway in cellular transformation. Overexpression of the active form of Rit, a Ras family member, in NIH3T3 cells, causes transformation and stimulates p38γ, but not other p38MAPKs, ERK1/2, or ERK5. These results suggest that p38γ activation may be requited for the ability of Rit to stimulate gene expression and cellular transformation (Sakabe et al. 2002). Recent data suggest that K-Ras positively regulate the expression of p38γ isoform and that depletion of p38γ suppressed Ras transformation in rat intestinal epithelial cells (Tang et al. 2005) and Ras-increased invasion in breast cancer cells (Qi et al. 2006). The mechanism by which p38γ may promote Ras transformation is not clear, but it has been suggested that may be through a complex formation with ERK proteins (Tang et al. 2005). Additionally, functional p38γ protein was expressed only in K-Ras-mutated human colon cancer cells, and p38γ transcripts were ubiquitously increased in a set of primary human colon cancer tissues (Tang et al. 2005). Moreover, p38γ expression is increased in hepatoma cell line HLE (Liu et al. 2003) and is regulated by the TNF-related apoptosis inducing ligand (TRIAL) and IL-8 in cellular lines from ovarian cancer (Abdollahi et al. 2003). However, in pancreatic cancer cells the levels of p38γ seems to be decreased (Crnogorac-Jurcevic et al. 2001).

3.2 Cellular differentiation

Similar to other systems, p38α and p38β have been suggested to play important roles in mediating keratinocytes responses to cellular stresses and cytokines, whereas p38δ has been suggested to play an important role in inducing keratinocyte differentiation by regulating the expression of involucrin, which is a protein expressed during keratinocyte differentiation (Eckert et al. 2003). Keratinocyte differentiation is a multistage process that is initiated in the proliferative basal layer of the epidermis and proceeds through the metabolically active spinous and granular layers, until the cell is released from the cell surface at the cornified envelope (Eckert et al. 2003). It has been shown that activation of exogenously expressed p38δ by differentiation-inducing agents such as a bioactive green tea polyphenol (EGCG), okadaic acid (OA), or the phorbol ester TPA correlated with increased involucrin promoter activity in keratinocytes via increased activity at AP1, Sp1, and C/EBP sites (Balasubramanian et al. 2002; Efimova et al. 2003). Of note, this occurred in an SB203580-independent manner, moreover, p38γ is not expressed in keratinocytes (Dashti et al. 2001). The mechanisms by which p38δ may regulate keratinocyte differentiation is still unknown. It has been found that in keratinocytes expressing exogenous p38δ this forms a complex with ERK1/2 (Efimova et al. 2003). The complex is constitutively present in both treated and untreated cells. Significant ERK1/2 activity, but little p38δ activity, is observed in untreated cell; but treatment with the above mentioned agents resulted in a reduction of ERK1/2 activity and an increase in the p38δ activity within the complex (Eckert et al. 2004).

More data supporting the idea that p38δ may play a role in keratinocyte differentiation came from a study carried out in lesional psoriasis skin (Johansen et al. 2005). Psoriasis is a chronic inflammatory skin disorder characterised by keratinocytes hyperproliferation and differentiation, as well as by increased expression of inflammatory cytokines such as TNFα and IL-8. It has been shown that the activity of p38α, p38β, and p38δ are augmented in lesional psoriasis skin compared with nonlesional psoriasis skin (Johansen et al. 2005). However, the regulation of keratinocyte differentiation does not seem to be exclusive for p38δ, thus, it has been shown that treatment of keratinocytes with agents to deplete cholesterol, induces the upregulation of involucrin mRNA in a p38α depending manner, but not by p38δ (Jans et al. 2004). Moreover, in the presence of exogenous constitutively active MKK6 or MKK7, a role of p38α has also been identified (Dashti et al. 2001).

On the other hand, it has been also claimed that p38δ may have a dual role in keratinocytes contributing not only to the differentiation process, but also to their apoptosis in a PKCδ dependent manner and in response to OA or H_2O_2 (Efimova et al. 2004; Kraft et al. 2006). It is important to notice that most of the evidences involving p38δ in regulating keratinocyte differentiation or apoptosis are based in overexpression experiments and require verification using other tools to both, inhibit the activity, or the expression of different p38MAPKs.

It has been suggested that p38γ and/or p38δ may also be involved in other differentiation processes such as myogenesis or T-cell development. Most of the work that demonstrates the requirement for p38 MAPK in myogenesis is based on the use of the compound SB203580, which only inhibits p38α and p38β (Lluis et al. 2006). Whether different p38MAPK family members have a role in the differentiation processes remains to be determined. Given that p38γ expression is exceptionally high in skeletal muscle in comparison to other tissues, it is not surprising that it may play a cardinal role in skeletal muscle differentiation. Indeed, Lechner et al. (1996) initially showed that overexpression of p38γ in skeletal muscle cells leads to differentiation from myoblast to myotubes, and that a dominant-negative mutant of p38γ prevented this differentiation process. Moreover, others have reported an increase in endogenous p38γ protein levels as myoblast differentiate into myotubes (Tortorela et al. 2003; Cuenda and Cohen 1999).

Additionally, p38α/β pathway has been implicated in T-cell development through the use of both, specific inhibitors such as SB203580 and the expression of dominant negative or active p38α or MKK6 (Diehl et al. 2000). These results have suggested that the p38α/β activity is required to maintain normal numbers of CD4/CD8 double-negative thymocytes, but that p38 activity inhibits the formation of double positive cells. However, the role of the isoforms p38δ and/or p38γ cannot be rule out since it has been shown that p38α knockout does not result in defects in T and B cell development and proliferation in response to different stimuli, including antigen-receptor ligation (Kim et al. 2005) and p38β knockout does also not result in major problems in thymocyte development (Beardmore et al. 2005). This suggests that either p38α or p38β is dispensable for thymocyte development or that in T cells there is compensation by other p38MAPK isoforms in the single knockouts.

3.3 Cytoskeletal organization

Evidence indicating a role of these p38 MAPK isoforms in modulation of cytoskeletal organization comes from recent studies. Thus, it has been shown that the protein SAP97/hDlg, which is the mammalian homologue of the *Drosophila* tumour suppressor Dlg, is a physiological substrate for the p38γ MAPK. SAP97/hDlg is a scaffold protein that forms multiprotein complexes with a variety of proteins and is targeted to the cytoskeleton by its association with the protein guanylate kinase-associated protein (GKAP), which is associated to the intermediate filaments. The p38γ-catalysed phosphorylation of SAP97/hDlg triggers its dissociation from GKAP and, therefore, releases it from the cytoskeleton (Sabio et al. 2005). This is likely to regulate the integrity of intercellular complexes, cell shape and volume as in adaptative to changes in the environment.

Moreover, p38δ may play a role in cytoskeleton regulation as it has been reported to phosphorylate the cytoplasmic protein stathmin, which is linked to regulation of microtubule dynamics (Parker et al. 1998). Microtubule-associated protein Tau is another protein substrate of p38δ (Goedert et al. 1997b; Feijoo et al. 2005; Yoshida and Goedert 2006). Tau binds to β-tubulin and promotes microtubule assembly playing major regulatory roles in the organization and integrity of the cytoskeleton network under normal physiological conditions. Functionally, Tau is modulated by phosphorylation, since its ability to bind and stabilise microtubules correlates inversely with its phosphorylation which may facilitates its self-assembly. In past few years, it has been shown that Tau is a good *in vitro* substrate for the p38 isoforms p38δ and p38γ, and its phosphorylation by these two enzymes results in a reduction in its ability to promote microtubule assembly (Goedert et al. 1997b; Feijoo et al. 2005). Moreover, overexpression of p38γ in neuroblastoma induces Tau phosphorylation, which correlates with a decrease on Tau associated to the cytoskeleton and an increase of soluble Tau (Jenkins et al. 2000). It has been reported as well that p38δ is the major Tau kinase in neuroblastoma in response to osmotic shock (Feijoo et al. 2005) and that the p38MAPK activator, MKK6, has also been found to be active in neurodegenerative diseases (Zhu et al. 2001). Moreover, oxidant agents implicated in Alzheimer's disease can cause hyperphosphorylation in rat brain and also induce the activation of p38δ, which indicates that this kinase may be involved in Tau phosphorylation (Yin et al. 2006). On the other hand, it has been shown using phosphospecific antibodies that p38MAPKs phosphorylate Tau on residues phosphorylated in a Tau obtained from patients suffering Alzheimer's disease (Goedert et al. 1997b; Feijoo et al. 2005). All these evidences indicate that p38MAPKs can regulate Tau hyperphosphorilation in neurodegenerative disease and could be potentially good therapeutic targets for those diseases.

More evidences suggesting a possible role of p38γ and/or p38δ in citoskeleton remodelling come from the role of some of their upstream activators, Rho-A and PARP-A, have controlling this process (Nishigaki et al. 2003; Singh et al. 2003).

4 More alternative p38MAPK pathways?

In addition to four p38MAPK isoforms, there are some alternative spliced variants of p38α, which regulation, biological functions, and whether or not these differ from the described for p38α, remain to be elucidated. Thus, four alternatively spliced forms of p38α have been described: p38α, CSAID-binding protein 1 (CSBP1), Max-interacting protein 2 (Mxi2), and Exon skip (Exip) (Han et al. 1994; Lee et al. 1994; Zervos et al. 1995; Sudo et al. 2002). p38α is the major splice variant, it is also known as a stress activated protein kinase-2a (SAPK2a) (Cohen 1997) and was identified also as a polypeptide receptor for a class of pyridinyl imidazole anti-inflammatory drugs, called cytokine-suppressive anti-inflammatory drugs (CSAIDs), with SB203580 being the most extensively studied compound (Lee et al. 1994). p38α was also named CSBP2 from CSAID-binding protein 2. CSBP1 was identified in the human monocytic cell line THP.1 as a polypeptide that differs from p38α only in an internal 25-amino acid sequence (Lee et al. 1994), whereas in Exip, the absence of exons 10, 11, and 112 results in the shift of the reading frame at the exon 9–12 junction to produce a unique 53-amino-acid C-terminus. Exip overexpression can induce an early onset of apoptosis in Hela cells (Sudo et al. 2002), and it has been reported that participates in interleukin-1 (IL-1) receptor-proximal complex and in downregulation of the NF-kappaB pathway (Yagasaki et al. 2004). The alternative splicing variant Mxi2 was identified in Hela cells and is abundantly expressed in kidney (Zervos et al. 1995; Faccio et al. 2000). Mxi2 lacks the last 80 residues of p38α, most of the kinase domain XI, and has a distinct C-terminal sequence of 17 amino acids (Zervos et al. 1995). Mxi2 is not activated in response to stress stimuli or in the presence of MKK3/MKK6, as judged using ATF2 as a substrate, but deletion of the unique C-terminal amino acid generates an active kinase (Faccio et al. 2000). Conversely, Mxi2 interacts with ERK1/2 sustaining ERK phosphorylation levels, therefore, prolonging its signal (Sanz-Moreno et al. 2003). It is likely that the C-terminus of Mxi2 has a role in mediating other protein-protein interactions since it is reported to bind to the serine protease Omi (Faccio et al. 2000). Mxi2 also interacts with and phosphorylates the c-Myc partner protein Max (Zervos et al. 1995).

5 Concluding remarks

Studies carried out during the last few years tried to answer the question of why a variety of p38MAPK isoforms and alternative splice variants are needed in mammalian cells. It should be noted that most of the studies to date have focused on the role of the p38α isoform, which is widely referred as p38 in the literature. However, there are three other alternative p38 isoforms (p38β, p38γ, and p38δ) whose roles among the cellular functions and the implication in some of the pathological conditions described here have not been precisely defined so far. One important question that remains to be answered is whether these p38MAPK isoforms are differentially activated by certain stimuli to mediate specific signals. For example, in the implication of p38MAPK in cancer, some isoforms may play a pro-oncogenic role whereas other p38 isoforms act as tumour suppressors, although, not much evidence exist so far. It is important to notice that the cell culture and overexpression studies outlined here require verification using *in vivo* models, including the generation of mice with tissue-specific inactivation of the individual p38MAPK family members. These mice, in combination with both, the use of siRNA techniques and new kinase inhibitors, such as the compound BIRB0796, should provide powerful biological models and tools to address the specific roles of each p38MAPK isoform.

Acknowledgments

The work in the authors' laboratory is supported by grants from the Medical Research Council UK, pharmaceutical companies that support the Division of Signal Transduction Therapy (Astra-Zeneca, Boehringer-Ingelheim, GlaxoSmithKline, Merck & Co. Inc, Merck KgaA and Pfizer), Ministerio de Educacion y Ciencia (MEC) Spain (SAF2004-1933). FIV was supported by fellowships from the Spanish Government (Becas FPU, Ministerio de Educacion y Ciencia).

References

Abdollahi T, Robertson NM, Abdollahi A, Litwack G (2003) Identification of interleukin 8 as an inhibitor of tumor necrosis factor-related apoptosis-inducing ligand-induced apoptosis in the ovarian carcinoma cell line OVCAR3. Cancer Res 63:4521-4526

Alonso G, Ambrosino C, Jones M, Nebreda AR (2000) Differential activation of p38 mitogen-activated protein kinase isoforms depending on signal strength. J Biol Chem 275:40641-40648

Ambrose M, Ryan A, O'Sullivan GC, Dunne C, Barry OP (2006) Induction of apoptosis in renal cell carcinoma by reactive oxygen species: involvement of extracellular signal-regulated kinase 1/2, p38delta/gamma, cyclooxygenase-2 down-regulation, and translocation of apoptosis-inducing factor. Mol Pharmacol 69:1879-1890

Balasubramanian S, Efimova T, Eckert RL (2002) Green tea polyphenol stimulates a Ras, MEKK1, MEK3, and p38 cascade to increase activator protein 1 factor-dependent involucrin gene expression in normal human keratinocytes. J Biol Chem 277:1828-1836

Beardmore VA, Hinton HJ, Eftychi C, Apostolaki M, Armaka M, Darragh J, McIlrath J, Carr JM, Armit LJ, Clacher C, Malone L, Kollias G, Arthur JS (2005) Generation and characterization of p38beta (MAPK11) gene-targeted mice. Mol Cell Biol 25:10454-10464

Brancho D, Tanaka N, Jaeschke A, Ventura JJ, Kelkar N, Tanaka Y, Kyuuma M, Takeshita T, Flavell RA, Davis RJ (2003) Mechanism of p38 MAP kinase activation *in vivo.* Genes Dev 17:1969-1978

Chiariello M, Marinissen MJ, Gutkind JS (2000) Multiple mitogen-activated protein kinase signaling pathways connect the cot oncoprotein to the c-Jun promoter and to cellular transformation. Mol Cell Biol 20:1747-1758

Cohen P (1997) The search for physiological substrate of MAP and SAP kinases in mammalian cells. Trends Cell Biol 7:353-361

Conrad PW, Rust RT, Han J, Millhorn DE, Beitner-Johnson D (1999) Selective activation of p38alpha and p38gamma by hypoxia. Role in regulation of cyclin D1 by hypoxia in PC12 cells. J Biol Chem 274:23570-23576

Court NW, dos Remedios CG, Cordell J, Bogoyevitch MA (2002) Cardiac expression and subcellular localization of the p38 mitogen-activated protein kinase member, stress-activated protein kinase-3 (SAPK3). J Mol Cell Cardiol 34:413-426

Court NW, Kuo I, Quigley O, Bogoyevitch MA (2004) Phosphorylation of the mitochondrial protein Sab by stress-activated protein kinase 3. Biochem Biophys Res Commun 319:130-137

Crnogorac-Jurcevic T, Efthimiou E, Capelli P, Blaveri E, Baron A, Terris B, Jones M, Tyson K, Bassi C, Scarpa A, Lemoine NR (2001) Gene expression profiles of pancreatic cancer and stromal desmoplasia. Oncogene 20:7437-7446

Cuenda A, Cohen P (1999) Stress-activated protein kinase-2/p38 and a rapamycin-sensitive pathway are required for C2C12 myogenesis, J Biol Chem 274:4341–4346

Cuenda A, Cohen P, Buee-Scherrer V, Goedert M (1997) Activation of stress-activated protein kinase-3 (SAPK3) by cytokines and cellular stresses is mediated via SAPKK3 (MKK6); comparison of the specificities of SAPK3 and SAPK2 (RK/p38). EMBO J 16:295-305

Cuenda A, Rouse J, Doza YN, Meier R, Cohen P, Gallagher TF, Young PR, Lee JC (1995) SB 203580 is a specific inhibitor of a MAP kinase homologue which is stimulated by cellular stresses and interleukin-1. FEBS Lett 364:229-233

Dashti SR, Efimova T, Eckert RL (2001) MEK6 regulates human involucrin gene expression via a p38alpha - and p38delta -dependent mechanism. J Biol Chem 276:27214-27220

Davies SP, Reddy H, Caivano M, Cohen P (2000) Specificity and mechanism of action of some commonly used protein kinase inhibitors. Biochem J 351:95-105

Diehl NL, Enslen H, Fortner KA, Merritt C, Stetson N, Charland C, Flavell RA, Davis RJ, Rincon M (2000) Activation of the p38 mitogen-activated protein kinase pathway arrests cell cycle progression and differentiation of immature thymocytes *in vivo.* J Exp Med 191:321-34

Eckert RL, Crish JF, Efimova T, Dashti SR, Deucher A, Bone F, Adhikary G, Huang G, Gopalakrishnan R, Balasubramanian S (2004) Regulation of involucrin gene expression. J Invest Dermatol 123:13-22

Eckert RL, Efimova T, Balasubramanian S, Crish JF, Bone F, Dashti S (2003) p38 Mitogen-activated protein kinases on the body surface--a function for p38 delta. J Invest Dermatol 120:823-828

Efimova T, Broome AM, Eckert RL (2003) A regulatory role for p38 delta MAPK in keratinocyte differentiation. Evidence for p38 delta-ERK1/2 complex formation. J Biol Chem 278:34277-34285

Efimova T, Broome AM, Eckert RL (2004) Protein kinase Cdelta regulates keratinocyte death and survival by regulating activity and subcellular localization of a p38delta-extracellular signal-regulated kinase 1/2 complex. Mol Cell Biol 24:8167-8183

Enslen H, Brancho DM, Davis RJ (2000) Molecular determinants that mediate selective activation of p38 MAP kinase isoforms. EMBO J 19:1301-1311

Enslen H, Raingeaud J, Davis RJ (1998) Selective activation of p38 mitogen-activated protein (MAP) kinase isoforms by the MAP kinase kinases MKK3 and MKK6. J Biol Chem 273:1741-1748

Eyers PA, Craxton M, Morrice N, Cohen P, Goedert M (1998) Conversion of SB 203580-insensitive MAP kinase family members to drug-sensitive forms by a single amino-acid substitution. Chem Biol 5:321-328

Faccio L, Chen A, Fusco C, Martinotti S, Bonventre JV, Zervos AS (2000) Mxi2, a splice variant of p38 stress-activated kinase, is a distal nephron protein regulated with kidney ischemia. Am J Physiol Cell Physiol 278:C781-C790

Feijoo C, Campbell DG, Jakes R, Goedert M, Cuenda A (2005) Evidence that phosphorylation of the microtubule-associated protein Tau by SAPK4/p38delta at Thr50 promotes microtubule assembly. J Cell Sci 118:397-408

Goedert M, Cuenda A, Craxton M, Jakes R, Cohen P (1997a) Activation of the novel stress-activated protein kinase SAPK4 by cytokines and cellular stresses is mediated by SKK3 (MKK6); comparison of its substrate specificity with that of other SAP kinases. EMBO J 16:3563-3571

Goedert M, Hasegawa M, Jakes R, Lawler S, Cuenda A, Cohen P (1997b) Phosphorylation of microtubule-associated protein tau by stress-activated protein kinases. FEBS Lett 409:57-62

Gross EA, Callow MG, Waldbaum L, Thomas S, Ruggieri R (2002) MRK, a mixed lineage kinase-related molecule that plays a role in gamma-radiation-induced cell cycle arrest. J Biol Chem 277:13873-82

Gum RJ, McLaughlin MM, Kumar S, Wang Z, Bower MJ, Lee JC, Adams JL, Livi GP, Goldsmith EJ, Young PR (1998) Acquisition of sensitivity of stress-activated protein kinases to the p38 inhibitor, SB 203580, by alteration of one or more amino acids within the ATP binding pocket. J Biol Chem 273:15605-15610

Han J, Lee JD, Bibbs L, Ulevitch RJA (1994) MAP kinase targeted by endotoxin and hyperosmolarity in mammalian cells. Science 265:808-811

Hasegawa M, Cuenda A, Spillantini MG, Thomas GM, Buee-Scherrer V, Cohen P, Goedert M (1999) Stress-activated protein kinase-3 interacts with the PDZ domain of alpha1-syntrophin. A mechanism for specific substrate recognition. J Biol Chem 274:12626-12631

Jans R, Atanasova G, Jadot M, Poumay Y (2004) Cholesterol depletion upregulates involucrin expression in epidermal keratinocytes through activation of p38. J Invest Dermatol 123:564-573

Jenkins SM, Zinnerman M, Garner C, Johnson GV (2000) Modulation of Tau phosphorylation and intracellular localization by cellular stress. Biochem J 345:263-270

Jiang Y, Chen C, Li Z, Guo W, Gegner JA, Lin S, Han J (1996) Characterization of the structure and function of a new mitogen-activated protein kinase (p38beta). J Biol Chem 271:17920-17926

Jiang Y, Gram H, Zhao M, New L, Gu J, Feng L, Di Padova F, Ulevitch RJ, Han J (1997) Characterization of the structure and function of the fourth member of p38 group mitogen-activated protein kinases, p38delta. J Biol Chem 272:30122-30128

Johansen C, Kragballe K, Westergaard M, Henningsen J, Kristiansen K, Iversen L (2005) The mitogen-activated protein kinases p38 and ERK1/2 are increased in lesional psoriatic skin. Br J Dermatol 152:37-42

Keum YS, Yu S, Chang PP, Yuan X, Kim JH, Xu C, Han J, Agarwal A, Kong AN (2006) Mechanism of action of sulforaphane: inhibition of p38 mitogen-activated protein kinase isoforms contributing to the induction of antioxidant response element-mediated heme oxygenase-1 in human hepatoma HepG2 cells. Cancer Res 66:8804-8813

Kim JM, White JM, Shaw AS, Sleckman BP (2005) MAPK p38 alpha is dispensable for lymphocyte development and proliferation. J Immunol 174:1239-1244

Knebel A, Haydon CE, Morrice N, Cohen P (2002) Stress-induced regulation of eukaryotic elongation factor 2 kinase by SB 203580-sensitive and -insensitive pathways. Biochem J 367:525-532

Knebel A, Morrice N, Cohen P (2001) A novel method to identify protein kinase substrates: eEF2 kinase is phosphorylated and inhibited by SAPK4/p38delta. EMBO J 20:4360-4369

Kraft CA, Efimova T, Eckert RL (2007) Activation of PKCdelta and p38delta MAPK during okadaic acid dependent keratinocyte apoptosis. Arch Dermatol Res 2007 Jan 26: [Epub ahead of print]

Kuma Y, Campbell DG, Cuenda A (2004) Identification of glycogen synthase as a new substrate for stress-activated protein kinase 2b/p38beta. Biochem J 379:133-139

Kuma Y, Sabio G, Bain J, Shpiro N, Marquez R, Cuenda A (2005) BIRB796 inhibits all p38 MAPK isoforms *in vitro* and *in vivo*. J Biol Chem 280:19472-19479

Lechner C, Zahalka MA, Giot JF, Moller NP, Ullrich A (1996) ERK6, a mitogen-activated protein kinase involved in C2C12 myoblast differentiation. Proc Natl Acad Sci USA 93:4355-4359

Lee JC, Laydon JT, McDonnell PC, Gallagher TF, Kumar S, Green D, McNulty D, Blumenthal MJ, Heys JR, Landvatter SW, Strickler JE, McLaughlin MM, Siemens IR, Fisher SM, Livi GP, White JR, Adams JL, Young PR (1994) A protein kinase involved in the regulation of inflammatory cytokine biosynthesis. Nature 372:739-746

Liu ZG, Baskaran R, Lea-Chou ET, Wood LD, Chen Y, Karin M, Wang JY (1996) Three distinct signalling responses by murine fibroblasts to genotoxic stress. Nature 384:273-276

Liu LX, Liu ZH, Jiang HC, Zhang WH, Qi SY, Hu J, Wang XQ, Wu M (2003) Gene expression profiles of hepatoma cell line HLE. World J Gastroenterol 9:683-687

Lluis F, Perdiguero E, Nebreda AR, Munoz-Canoves P (2006) Regulation of skeletal muscle gene expression by p38 MAP kinases. Trends Cell Biol 16:36-44

Marinissen MJ, Chiariello M, Gutkind JS (2001) Regulation of gene expression by the small GTPase Rho through the ERK6 (p38 gamma) MAP kinase pathway. Genes Dev 15:535-553

Mertens S, Craxton M, Goedert M (1996) SAP kinase-3, a new member of the family of mammalian stress-activated protein kinases. FEBS Lett 383:273-276

Morrison DK, Davis RJ (2003) Regulation of MAP kinase signaling modules by scaffold proteins in mammals. Annu Rev Cell Dev Biol 19:91-118

Nishigaki K, Thompson D, Yugawa T, Rulli K, Hanson C, Cmarik J, Gutkind JS, Teramoto H, Ruscetti S (2003) Identification and characterization of a novel Ste20/germinal center kinase-related kinase, polyploidy-associated protein kinase. J Biol Chem 278:13520-13530

Pargellis C, Tong L, Churchill L, Cirillo PF, Gilmore T, Graham AG, Grob PM, Hickey ER, Moss N, Pav S, Regan J (2002) Inhibition of p38 MAP kinase by utilizing a novel allosteric binding site. Nat Struct Biol 9:268-272

Parker CG, Hunt J, Diener K, McGinley M, Soriano B, Keesler GA, Bray J, Yao Z, Wang XS, Kohno T, Lichenstein HS (1998) Identification of stathmin as a novel substrate for p38 delta. Biochem Biophys Res Commun 249:791-796

Perdiguero E, Pillaire MJ, Bodart JF, Hennersdorf F, Frodin M, Duesbery NS, Alonso G, Nebreda AR (2003) Xp38gamma/SAPK3 promotes meiotic G(2)/M transition in *Xenopus* oocytes and activates Cdc25C. EMBO J 22:5746-5756

Pramanik R, Qi X, Borowicz S, Choubey D, Schultz RM, Han J, Chen G (2003) p38 isoforms have opposite effects on AP-1-dependent transcription through regulation of c-Jun. The determinant roles of the isoforms in the p38 MAPK signal specificity. J Biol Chem 278:4831-4839

Qi X, Tang J, Loesch M, Pohl N, Alkan S, Chen G (2006) p38gamma mitogen-activated protein kinase integrates signaling crosstalk between Ras and estrogen receptor to increase breast cancer invasion. Cancer Res 66:7540-7547

Sabio G, Arthur JS, Kuma Y, Peggie M, Carr J, Murray-Tait V, Centeno F, Goedert M, Morrice NA, Cuenda A (2005) p38gamma regulates the localisation of SAP97 in the cytoskeleton by modulating its interaction with GKAP. EMBO J 24:1134-1145

Sabio G, Reuver S, Feijoo C, Hasegawa M, Thomas GM, Centeno F, Kuhlendahl S, Leal-Ortiz S, Goedert M, Garner C, Cuenda A (2004) Stress- and mitogen-induced phosphorylation of the synapse-associated protein SAP90/PSD-95 by activation of SAPK3/p38gamma and ERK1/ERK2. Biochem J 380:19-30

Sakabe K, Teramoto H, Zohar M, Behbahani B, Miyazaki H, Chikumi H, Gutkind JS (2002) Potent transforming activity of the small GTP-binding protein Rit in NIH 3T3 cells: evidence for a role of a p38gamma-dependent signaling pathway. FEBS Lett 511:15-20

Sanz-Moreno V, Casar B, Crespo P (2003) p38alpha isoform Mxi2 binds to extracellular signal-regulated kinase 1 and 2 mitogen-activated protein kinase and regulates its nuclear activity by sustaining its phosphorylation levels. Mol Cell Biol 23:3079-3090

Singh US, Pan J, Kao YL, Joshi S, Young KL, Baker KM (2003) Tissue transglutaminase mediates activation of RhoA and MAP kinase pathways during retinoic acid-induced neuronal differentiation of SH-SY5Y cells. J Biol Chem. 278:391-399

Stein B, Yang MX, Young DB, Janknecht R, Hunter T, Murray BW, Barbosa MS (1997) p38-2, a novel mitogen-activated protein kinase with distinct properties. J Biol Chem 272:19509-19517

Sudo T, Yagasaki Y, Hama H, Watanabe N, Osada H (2002) Exip, a new alternative splicing variant of p38 alpha, can induce an earlier onset of apoptosis in HeLa cells. Biochem Biophys Res Commun 291:838-843

Tanaka N, Kamanaka M, Enslen H, Dong C, Wysk M, Davis RJ, Flavell RA (2002) Differential involvement of p38 mitogen-activated protein kinase kinases MKK3 and MKK6 in T-cell apoptosis. EMBO Rep 3:785-791

Tang J, Qi X, Mercola D, Han J, Chen G (2005) Essential role of p38gamma in K-Ras transformation independent of phosphorylation. J Biol Chem 280:23910-23917

Tortorella LL, Lin CB, Pilch PF (2003) ERK6 is expressed in a developmentally regulated manner in rodent skeletal muscle. Biochem Biophys Res Commun 306:163-168

Wang L, Ma R, Flavell RA, Choi ME (2002) Requirement of mitogen-activated protein kinase kinase 3 (MKK3) for activation of p38alpha and p38delta MAPK isoforms by TGF-beta 1 in murine mesangial cells. J Biol Chem 277:47257-47262

Wang X, McGowan CH, Zhao M, He L, Downey JS, Fearns C, Wang Y, Huang S, Han J (2000) Involvement of the MKK6-p38gamma cascade in gamma-radiation-induced cell cycle arrest. Mol Cell Biol 20:4543-4552

Yagasaki Y, Sudo T, Osada H (2004) Exip, a splicing variant of p38alpha, participates in interleukin-1 receptor proximal complex and downregulates NF-kappaB pathway. FEBS Lett 575:136-140

Yin J, Liu YH, Xu YF, Zhang YJ, Chen JG, Shu BH, Wang JZ (2006) Melatonin arrests peroxynitrite-induced tau hyperphosphorylation and the overactivation of protein kinases in rat brain. J Pineal Res 41:124-129

Yoshida H, Goedert M (2006) Sequential phosphorylation of tau protein by cAMP-dependent protein kinase and SAPK4/p38delta or JNK2 in the presence of heparin generates the AT100 epitope. J Neurochem 99:154-164

Zervos AS, Faccio L, Gatto JP, Kyriakis JM, Brent R (1995) Mxi2, a mitogen-activated protein kinase that recognizes and phosphorylates Max protein. Proc Natl Acad Sci USA 92:10531-10534

Zhang QS, Maddock DA, Chen JP, Heo S, Chiu C, Lai D, Souza K, Mehta S, Wan YS (2001) Cytokine-induced p38 activation feedback regulates the prolonged activation of AKT cell survival pathway initiated by reactive oxygen species in response to UV irradiation in human keratinocytes. Int J Oncol 19:1057-1061

Zhu X, Rottkamp CA, Hartzler A, Sun Z, Takeda A, Boux H, Shimohama S, Perry G, Smith MA (2001) Activation of MKK6, an upstream activator of p38, in Alzheimer's disease. J Neurochem 79:311-318

Cuenda, Ana

Departamento de Inmunología y Oncología, Centro Nacional de Biotecnología-CSIC, Campus de Cantoblanco-UAM, 28049-Madrid, Spain and MRC Protein Phosphorylation Unit, School of life Sciences, University of Dundee, Dundee DD1 5EH, Scotland, UK

a.i.cuenda@dundee.ac.uk

Iñesta-Vaquera, Francisco

Departamento de Inmunología y Oncología, Centro Nacional de Biotecnología-CSIC, Campus de Cantoblanco-UAM, 28049-Madrid, Spain

Kuma, Yvonne

MRC Protein Phosphorylation Unit, School of life Sciences, University of Dundee, Dundee DD1 5EH, Scotland, UK

Sabio, Guadalupe
MRC Protein Phosphorylation Unit, School of life Sciences, University of Dundee, Dundee DD1 5EH, Scotland, UK

The regulation of stress-activated MAP kinase signalling by protein phosphatases

Stephen M. Keyse

Abstract

The regulated dephosphorylation of mitogen-activated protein kinases (MAPKs) is a key determinant of the biological outcome of signalling. Because MAPKs require phosphorylation on both threonine and tyrosine residues for activation, they are potential substrates for all three major classes of protein phosphatase. These include serine/threonine phosphatases, protein tyrosine phosphatases (PTPs) and a dedicated subfamily of dual-specificity thr/tyr MAPK phosphatases (MKPs). This review summarises genetic and biochemical studies in model organisms including yeasts, *Drosophila*, and *C. elegans*, which have provided evidence for a complex interplay between upstream activators and these enzymes in regulating stress responsive MAPK pathways. Such studies have provided important insights into the regulation of the stress-responsive JNK and p38 MAPK pathways in mammals, where genetic experiments are beginning to reveal important roles for dual-specificity MKPs in regulating diverse physiological endpoints. These include resistance to environmental stress, immune function, and metabolic homeostasis.

1 Introduction

Mitogen-activated protein kinase (MAPK) signalling pathways are conserved from yeast to man and regulate numerous physiological processes including proliferation, differentiation, development, immune function, stress responses, and apoptosis (Chang and Karin 2001; Pearson et al. 2001; Johnson and Lapadat 2002; Wada and Penninger 2004; Qi and Elion 2005). Prominent amongst these pathways are MAPK isoforms, which respond to various physical and chemical stresses. In *Saccharomyces cerevisisiae* these include the high osmolarity glycerol (Hog1) MAPK which regulates the cellular response to osmotic stress and Mpk1 which regulates cell integrity (Qi and Elion 2005). In mammalian cells there are two major stress-responsive MAPK pathways. Firstly, the c-Jun amino-terminal kinases (JNKs 1-3) that regulate responses to radiation and chemical stress and mediate inflammation and apoptosis (Davis 2000). Secondly, the p38 MAPKs (α,β,δ, and γ), which are most similar to the Hog1 MAPK of budding yeast and are activated in response to stress and pro-inflammatory cytokines. (Kumar et al. 2003; Schieven 2005).

Topics in Current Genetics, Vol. 20
F. Posas, A. R. Nebreda (Eds.): Stress-Activated Protein Kinases
DOI 10.1007/4735_2007_0240 / Published online: 4 July 2007

MAPK signalling cascades consist of a conserved module in which a MAP kinase kinase kinase (MKKK or MEKK) phosphorylates and activates a MAP kinase kinase (MKK or MEK). The latter enzyme is a dual-specificity protein kinase that activates the MAPK by phosphorylating both threonine and tyrosine residues within the conserved signature sequence T-X-Y (Marshall 1994). The component kinases of the MAPK module may interact sequentially to propagate a signal, but can also be further organised into specific signalling complexes through interactions with scaffold proteins such as the JNK-interacting proteins (JIPs) (Garrington and Johnson 1999; Morrison and Davis 2003).

One highly conserved property of MAPK cascades is that biological outcome is determined by the magnitude and duration of signalling (Marshall 1995; Sabbagh et al. 2001). This highlights the potential importance of negative regulatory mechanisms in determining the output of MAPK pathways, a concept reinforced by mathematical modelling of MAPK signalling behaviour (Asthagiri and Lauffenburger 2001; Bhalla et al. 2002). One major point of control occurs at the level of the MAPK itself, via the intervention of specific protein phosphatases. As MAPK activation requires phosphorylation of both a threonine residue and a tyrosine residue, dephosphorylation of either can inactivate these enzymes. This can be achieved by serine-threonine phosphatases, tyrosine-specific phosphatases, or dual-specificity (Thr/Tyr) phosphatases. With respect to the regulation of stress-activated MAPK pathways, work in a variety of model systems including yeasts, *drosophila, C. elegans* and mammals has demonstrated that all three classes of protein phosphatase act to regulate these pathways *in vivo* (Keyse 2000; Saxena and Mustelin 2000).

2 Regulation of stress-activated MAPKs by protein phosphatases in *S. cerevisiae*

A combination of biochemical and genetic studies have revealed considerable complexity in the regulation of stress-activated MAPK signalling by protein phosphatases in *S. cerevisiae* and these studies may provide important paradigms for the regulation of their mammalian counterparts. The Hog1 MAPK is activated in response to osmotic stress and mediates an adaptive response by increasing glycerol production and mediating transient inhibition of both protein synthesis and progression through the cell cycle (O'Rourke et al. 2002). In common with other MAPK pathways, signalling must be returned to basal levels so that the cell can then respond to further osmotic challenge. In addition, constitutive activation of Hog1 is lethal. Two tyrosine-specific protein phosphatases Ptp2 and Ptp3 are responsible for inactivation of Hog1 (Fig. 1). Of these, Ptp2, which is a nuclear phosphatase, is the major regulator of Hog1, with Ptp3 playing a lesser role (Jacoby et al. 1997; Wurgler-Murphy et al. 1997). Interestingly, in addition to regulating the activity of Hog1, these two phosphatases also play a role in determining the subcellular localisation of this MAPK (Mattison and Ota 2000). It was noted that the nuclear retention of Hog1 was impaired in a strain lacking Ptp2,

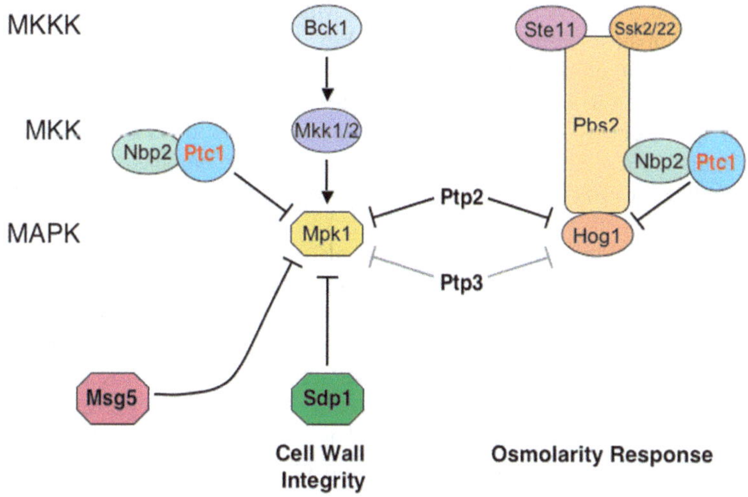

Fig. 1. A complex network of protein phosphatases is responsible for the regulation of stress-activated MAPK signalling in *S. cerevisiae*. Positive interactions are indicated by arrows, while negative regulation by protein phosphatases is indicated by bars. Different line weights indicate the relative potency of activities. For details see text.

while in a strain lacking Ptp3, Hog1 accumulated in the cell nucleus. This suggested that these two proteins acted as a nuclear anchor and cytoplasmic tether for Hog1 respectively. In agreement with this idea, overexpression of Ptp2 caused nuclear sequestration of Hog1 while overexpressing Ptp3 caused cytoplasmic retention of the MAPK.

In addition to these tyrosine-specific phosphatases, three type-2C serine/threonine phosphatases (PP2C) known as Ptc1, Ptc2, and Ptc3 in yeast are also involved in regulation of Hog1 activity (Fig. 1) (Maeda et al. 1994; Warmka et al. 2001; Young et al. 2002). Ptc1 suppresses basal Hog1 activity and inactivates Hog1 during adaptation to osmotic stress (Warmka et al. 2001), while both Ptc2 and 3 seem to set a maximum limit on the levels of Hog1 activity which can be achieved in response to stress (Young et al. 2002). The mechanism by which PP2Cs are able to target components of the MAPK pathways is unclear. However global protein-protein interaction studies in *S. cerevisiae* using 2-hybrid screening identified a Nap-1 binding protein (Nbp2) as a binding partner for Ptc1. This 237aa SH3 domain containing protein was then found to bind specifically to Ptc1 and to negatively regulate the Hog1 kinase (Mapes and Ota 2004). The Pbs2 protein kinase acts as a MKK for Hog1 and also as a scaffold to assemble Hog1 and upstream activators including the MKKKs Ste11 and Ssk2/22. Nbp2 binds Pbs2 via its SH3 domain, thus acting as an adaptor protein to bring Ptc1 to the scaffold complex where it exerts its phosphatase activity towards Hog1 (Mapes and Ota 2004).

A second stress-responsive MAPK in budding yeast is Mpk1, which acts in a signalling pathway to maintain cell integrity and can be activated in response to heat shock and oxidative stress. Like Hog1, Mpk1 is also targeted by the tyrosine-specific phosphatases Ptp2 and Ptp3, with Ptp2 the major regulator of this MAPK (Fig. 1) (Mattison et al. 1999). Interestingly, Ptp2 and Ptp3 also target the pheromone-responsive Fus3 MAPK which regulates the mating response, but with a reciprocal specificity when compared with their activities towards Hog1 and Mpk1 (Zhan et al. 1997; Zhan and Guan 1999). This serves to emphasise the complexity and potential for cross-regulation of MAPK pathways by protein phosphatases *in vivo.*

In addition to Ptp2 and Ptp3, Mpk1 is also regulated by two dual-specificity protein phosphatases Msg5 and Sdp1 (Fig. 1). Msg5 was first identified as a regulator of the Fus3 MAPK, where it acts as part of a negative feedback loop to control the adaptive response to pheromone (Doi et al. 1994). With respect to Mpk1, Msg5 appears to be the major phosphatase acting on this MAPK at least in unstressed cells (Flandez et al. 2004; Martin et al. 2005). Sdp1, a phosphatase which is most similar in sequence to Msg5, also has activity towards Mpk1. Sdp1 interacts with Mpk1 and regulates the activation of this MAPK in response to heat stress (Collister et al. 2002; Hahn and Thiele 2002). Unlike Msg5, which also regulates Fus3, the binding of Sdp1 appears to be specific for Mpk1 indicating that this phosphatase is dedicated solely to the regulation of this MAPK pathway. Finally, the type-2C serine/threonine phosphatase Ptc1 has also been implicated in regulation of Mpk1 in the context of inheritance of endoplasmic reticulum (ER). Both Ptc1 and its binding partner Nbp1 are required for the delivery of ER tubules to the periphery of daughter cells and loss of either protein results in increased levels of Mpk1 phosphorylation (Du et al. 2006). However, it is as yet unclear how Ptc1 is targeted to Mpk1 in this process (Fig. 1).

In conclusion, the complex negative regulatory network of enzymes, which regulate stress-activated MAPK signalling in budding yeast encompasses members of all three major classes of protein phosphatase. Within this network, a single MAPK can be regulated by members of all three major classes of protein phosphatase, as is the case for Mpk1, or like Hog1 by a combination of tyrosine-specific and serine/threonine phosphatases. The fact that both Ptp2/Ptp3 and Ptc1 are involved in the regulation of the Hog1 and Mpk1 pathway adds the possibility of signal integration at the level of MAPK dephosphorylation. Finally, Ptp2 is regulated at the transcriptional level by activation of both Hog1 and Mpk1 indicating that it functions as part of a negative feedback control in regulating these MAPKs. In contrast, although Sdp1 is inducible in response to activators of the Mpk1 pathway, such as heat and oxidative stress, the Sdp1 gene is transcriptionally regulated downstream of protein kinase A (PKA) (Hahn and Thiele 2002), thus providing a mechanism of regulated crosstalk between these distinct signalling pathways.

3 Regulation of JNK and p38 MAPKs in *Drosophila* and *C. elegans*

3.1 The *Drosophila* phosphatase *puckered* plays key roles in development, stress responses and ageing

The first genetic evidence linking a dual-specificity MKP to the physiological regulation of a stress activated MAPK in a higher eukaryote came from studies of the *puckered (puc)* phosphatase in *Drosophila*. *Basket (bsk)*, the Drosophila homologue of JNK plays a key role in dorsal closure during embryogenesis. This process involves morphological change and migration of lateral epithelial cells, which eventually cover the dorsal region of the embryo. Mutations in *puc* affect this process (Ring and Martinez Arias 1993) and *puc* was later found to encode a dual-specificity protein phosphatase with significant homology to DUSP1/MKP-1 a mammalian dual-specificity MAPK phosphatase (MKP) (Martin-Blanco et al. 1998). Furthermore, *puc* expression is dependent on the activity of either *Bsk* or its upstream activator *hemipterous*, indicating that, like certain of the phosphatases which regulate MAPK signalling in yeast, *puc* functions as part of a negative feedback loop to regulate JNK activity (Martin-Blanco et al. 1998). Interestingly, flies which are heterozygous for loss of function mutations in *puc* and thus display increased JNK signalling activity are both resistant to oxidative stress and showed dramatic increases in median and maximum life expectancy (Wang et al. 2003). More recently it has also been shown that *puc* is a key regulator of apoptosis in epithelial cells, where it acts to restrain basal JNK activity. Furthermore, *puc* also regulates γ-radiation-induced apoptosis where radiation both upregulates JNK activity and induces *puc* expression in a p53 dependent manner (McEwen and Peifer 2005). Overall, this work demonstrates a central role for *puckered* in regulating the JNK pathway both during development and also in response to cellular stresses which affect cell viability and the ageing process and underlines a key role for a dual-specificity MKP in setting the balance between cell proliferation differentiation and apoptosis.

3.2 The *vhp-1* phosphatase plays key roles in regulating stress responses and immunity in *C. elegans*

Despite sequence similarity between *puckered* and mammalian MKPs within its catalytic domain, this enzyme does not contain an amino-terminal rhodanese- homology domain and the relationship between *puc* and mammalian dual-specificity MKPs remains unclear. However, recent work in *C. elegans* has now identified *vhp-1*, a phosphatase which is most similar to DUSP16/MKP-7, a mammalian MKP which selectively inactivates the JNK and p38 MAPKs (Mizuno et al. 2004). Unlike *puc*, *vhp-1* does possess an N-terminal rhodanese-homology domain and this also contains a conserved MAPK interaction motif (KIM). Loss of *vhp-1* function causes larval arrest and this phenotype is suppressed by deletion of the

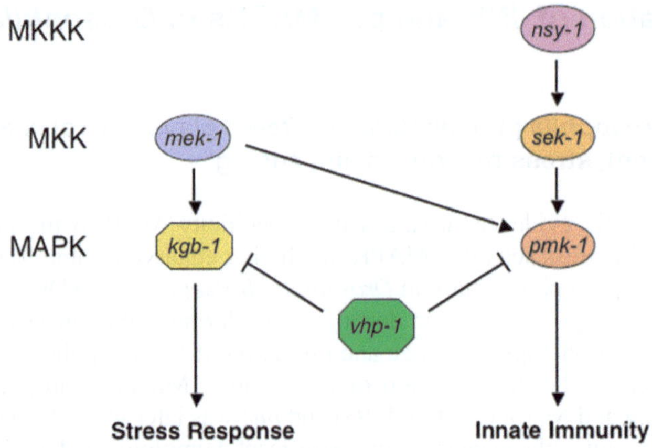

Fig. 2. The integration of signalling through the *pmk-1* (p38) and *kgb-1* (JNK) MAPK pathways by the dual-specificity MAPK phosphatase vhp-1 in *C. elegans*. Positive interactions are indicated by arrows, while negative regulation by protein phosphatases is indicated by bars. For details see text.

C. elegans JNK homologue *kgb-1*. Interestingly, deletion of two other *C. elegans* JNK homologues *jnk-1* and *kgb-2* did not rescue the larval arrest caused by loss of *vhp-1,* indicating that this phosphatase is specific for one individual JNK isoform (Mizuno et al. 2004). *Mek-1*, the upstream activator of *kgb-1,* had previously been implicated in stress responses triggered by exposure to metal ions (Koga et al. 2000). Consistent with a role for *vhp-1* in regulating the response to heavy metal stress, loss of *vhp-1* partially suppressed the Cu^{2+} sensitive phenotypes of *mek-1* and *kgb-1* mutants. Furthermore, the latter result suggested that *kgb-1* was not the only MAPK target of *vhp-1* involved in regulating metal resistance. As DUSP16/MKP-7 can also inactivate p38 MAPK in mammalian cells, the *C. elegans* p38 homologue *pmk-1* was investigated as a possible target for *vhp-1*. Mutation of *pmk-1* caused a modest sensitivity to Cu^{2+} and enhanced the metal sensitivity of the *kgb-1* mutant but, most importantly, it also suppressed the resistance of the *kgb-1,vhp-1* double mutant. Taken together these data suggest that the *pmk-1* and *kgb-1* MAPK pathways serve mutually redundant roles in the response to heavy metals and that *vhp-1* negatively regulates both pathways (Fig. 2) (Mizuno et al. 2004).

Finally, both p38 and JNK serve evolutionarily conserved roles in mediating immune function and, in *C. elegans,* the *pmk-1* pathway mediates innate immunity against bacterial pathogens (Kim et al. 2002). Subsequent work showed that *vhp-1* also regulates pathogen resistance by inactivating *pmk-1*. The activity of *pmk-1* in this pathway was mediated by *mek-1* the upstream activator of *kgb-1*(Kim et al. 2004). Thus, the regulation of p38 and JNK pathways in mediating immunity and stress responses by common components suggests critical roles for *mek-1* and *vhp-1* in the integration of diverse signals contributing to pathogen resistance in *C. elegans* (Fig. 2). Furthermore, this work suggests a mechanism by which signals transduced by distinct MAPK pathways may be subject to physiological integration at the level of regulation by MKPs.

4 Mammalian MKPs play essential roles in the regulation of both JNK and p38 MAP kinase signalling

4.1 A subset of mammalian MKPs can regulate stress-activated MAPK pathways

The dual-specificity MKPs constitute a family of ten proteins which share a common structure comprising a C-terminal catalytic domain with sequence similarity to the prototypic dual-specificity protein phosphatase VH-1 of vaccinia virus and an amino-terminal domain containing two regions of sequence similarity with the cell cycle regulatory phosphatase Cdc25 (Camps et al. 2000; Keyse 2000; Theodosiou and Ashworth 2002). This latter region has structural homology with the rhodanese family of sulphotransferases (Bordo and Bork 2002) and also contains a conserved kinase interaction motif (KIM) which is responsible for MAPK substrate recognition and binding. Of these proteins, six MKPs have been identified as potential regulators of the stress-activated MAPK pathways in mammalian cells (Table 1). Three of these, DUSP1/MKP-1, DUSP2/PAC1, and DUSP4/MKP-2 are inducible nuclear phosphatases which are able to interact with and inactivate all three classes of MAPK, namely ERKs, JNKs, and p38 MAPK, albeit with varying specificities (Chu et al. 1996). A second group of three MKPs comprising DUSP8, DUSP10/MKP-5 and DUSP16/MKP-7 are highly specific for the JNK and p38 MAPKs having little or no activity towards the "classical" Ras/ERK MAPK pathway (Muda et al. 1996; Tanoue et al. 1999, 2001; Theodosiou et al. 1999; Masuda et al. 2001).

4.2 DUSP10/MKP-5 plays a key role in adaptive and innate immunity

DUSP10/MKP-5 is the first of three MKPs that specifically target the stress-activated MAPKs to be the subject of a mouse knockout experiment (Zhang et al. 2004). DUSP10/MKP-5 is inducible in a mouse macrophage cell line in response to bacterial lipopolysaccharide (LPS) and also expressed in the T cell compartment suggesting an involvement in immune regulation. Deletion of the murine DUSP10MKP-5 gene does not cause gross developmental abnormalities, nor is DUSP10/MKP-5 essential for the development of the immune system. However, mice lacking DUSP10/MKP-5 exhibit higher JNK (but not p38 MAPK) activity in both the Th1 and Th2 subsets of T helper cells and also in LPS treated macrophages, suggesting that DUSP10/MKP-5 might regulate innate or adaptive immune responses.

In support of this, LPS-treated peritoneal macrophages from DUSP10/MKP5$^{-/-}$ animals produce increased levels of the pro-inflammatory cytokines interleukin-6

Table 1. Nomenclature and key properties for dual-specificity MKPs

Gene	MKP	Trivial names	Chromosomal localisation	Subcellular localisation	Substrate preference
DUSP1	MKP-1	CL100, erp, 3CH134, hVH1	5q34	Nuclear	JNK, p38, ERK
DUSP4	MKP-2	Typ1, Sty8, hVH2	8p12-p11	Nuclear	JNK, p38, ERK
DUSP2		PAC-1	2q11	Nuclear	ERK, p38
DUSP5		hVH-3, B23	10q25	Nuclear	ERK
DUSP6	MKP-3	Pyst1, rVH6	12q22-q23	Cytoplasmic	ERK
DUSP7	MKP-X	Pyst2, B59	3p21	Cytoplasmic	ERK
DUSP9	MKP-4	Pyst3	Xq28	Cytoplasmic	ERK>p38
DUSP8		M3/6, hVH5, HB5	11p15.5	Cytoplasmic/nuclear	JNK, p38
DUSP10	MKP-5		1q41	Cytoplasmic/nuclear	JNK, p38
DUSP16	MKP-7	MKP-M	12p12	Cytoplasmic/nuclear	JNK, p38

(IL-6) and tumour necrosis factor (TNF) when compared to wild type cells and injection of LPS into DUSP10/MKP-5$^{-/-}$ mice leads to increased serum TNF production (Zhang et al. 2004). The T-cell priming activity of LPS-treated antigen presenting cells (APC) lacking DUSP10/MKP-5 was also greatly enhanced indicating that MKP-5 is a negative regulator of innate immunity (Zhang et al. 2004). To examine the role of DUSP10/MKP-5 in effector T-cell function, CD4 T cells were differentiated into Th1 and Th2 cells. These cells were found to produce increased levels of IFN-γ and IL-4 respectively when compared to wild type. DUSP10/MKP-5$^{-/-}$ CD8 T-cells also produced more IFN-γ and TNF *in vitro*. These results indicate that cytokine expression in effector T-cells is also subject to negative regulation by DUSP10/MKP-5.

Finally, DUSP10/MKP-5 $^{-/-}$ mice were characterised using *in vivo* immune and autoimmune models. Firstly, mice were immunised with myelin oligodendrocyte glycoprotein (MOG) peptide to induce experimental autoimmune encephalomyelitis (EAE). DUSP10/MKP-5 $^{-/-}$ animals exhibited a reduced incidence and severity of disease suggesting that MKP-5 plays a crucial role in the generation and/or expansion of autoreactive T cells in EAE (Zhang et al. 2004). To examine T-cell mediated immunity to infection, animals were infected with lymphocytic choriomeningitis virus (LCMV). MKP-5 $^{-/-}$ mice cleared virus and exhibited little difference in primary T-cell response when compared with wild type mice. However, they responded to a second viral challenge by producing markedly elevated levels of serum TNF and this is probably responsible for the immune-mediated death of these animals two to four days later. Overall this work demonstrates that DUSP10/MKP-5 negatively regulates the JNK signalling pathway and serves an important role as a negative regulator of innate and adaptive immunity. The role of DUSP10/MKP-5 in regulating p38 MAPK activity is not yet clear nor is the extent to which this MKP may be involved in regulating mammalian stress responses. Further study of these knockout mice and cells derived from them will be required to clarify these issues. In addition, it will be crucial to generate mouse knockouts for DUSP8 and DUSP16/MKP-7, which also encode MKPs that selectively inactivate the stress-activated MAPKs.

5 Inducible nuclear MKPs play key roles in stress resistance, immune function, and metabolic homeostasis

The inducible nuclear MKPs constitute a subfamily of four enzymes DUSP1/MKP-1, DUSP2/PAC-1, DUSP4/MKP-2, and DUSP5/hVH3. Initial studies of mice lacking DUSP1/MKP-1 failed to uncover any significant phenotype (Dorfman et al. 1996). However, more extensive and detailed characterisation of this animal model has now provided compelling evidence for key physiological roles for DUSP1/MKP-1 in regulating signalling through the ERK, JNK, and p38 MAPK pathways.

5.1 DUSP1/MKP-1 is a key regulator of stress resistance

Mouse embryo fibroblasts (MEFs) obtained from DUSP1/MKP-1 null animals show increased JNK and p38 activities in response to a range of activating stimuli when compared to wild type cells (Wu and Bennett 2005). Furthermore, MEFs lacking DUSP1MKP-1 display lower growth rates and are more sensitive to anisomycin-induced apoptosis when compared with wild type cells. This was the first indication that DUSP1/MKP-1 might promote cell survival by attenuating signalling through the stress-activated MAPK pathways (Wu and Bennett 2005). Subsequent studies have now demonstrated that MEFs lacking DUSP1/MKP-1 exhibit elevated JNK and p38 activities when exposed to either hydrogen peroxide or cisplatin and are more sensitive to killing by these agents (Wang et al. 2006; Zhou et al. 2006). Interestingly, DUSP1/MKP-1was first identified as a hydrogen peroxide inducible gene in human skin fibroblasts (Keyse and Emslie 1992) and is also inducible by cisplatin (Sanchez-Perez et al. 2000). This work links induction of DUSP1/MKP-1 to a protective role against oxidative stress and DNA damaging agents and suggests that pharmacological inhibition of DUSP1/MKP-1 activity might be a useful strategy to modify tumour response to cisplatin.

5.2 DUSP1/MKP-1 regulates both immune and metabolic function

Loss of DUSP1/MKP-1 increases the duration of p38 activation in mouse macrophages exposed to bacterial lipopolysaccharide (LPS), suggesting that this phosphatase might regulate the innate immune response (Zhao et al. 2005). Recent work has now confirmed such a role as DUSP-1/MKP-1$^{-/-}$ mice respond to LPS challenge by overproducing certain cytokines, including TNFα, IL6, and IL10, are acutely sensitive to lethal endotoxic shock and also exhibit a marked increase in both the incidence and severity of autoimmune arthritis following injection of chicken type-II collagen (Chi et al. 2006; Hammer et al. 2006; Salojin et al. 2006; Zhao et al. 2006). Thus, DUSP1/MKP-1 controls the levels of both pro-inflammatory (TNFα) and anti-inflammatory (IL10) cytokines in response to LPS challenge. The overproduction of TNF in MKP$^{-/-}$ macrophages was seen only at early times after exposure to LPS and was suppressed as production of IL10 increased at later times. This indicates that the regulation of immune responses by DUSP1/MKP-1 is complex and reflects a dynamic balance between MAPK activation and the activity of MKP-1 in the temporal regulation of both pro-and anti-inflammatory mediators during the innate immune response (Chi et al. 2006).

It is interesting to compare the immune defects seen in the DUSP1/MKP-1$^{-/-}$ animals with those reported for the DUSP10/MKP-5 knockout (Zhang et al. 2004) as both phosphatases are able to target the same subset of MAPKs. However, loss of DUSP10/MKP-5 appears to have a greater effect on signalling through the JNK pathway while the effects of knocking out DUSP1/MKP-1 in cells of the immune system are more consistent with increased signalling through the p38 MAPK pathway (Hammer et al. 2006).

DUSP1/MKP-1 is widely expressed in mouse tissues (Carrasco and Bravo 1993; Kwak et al. 1994), and therefore its function is unlikely to be restricted to immune regulation. A recent study, again using the DUSP1/MKP$^{-/-}$ mice, reveals that this enzyme also plays a key role in metabolic homeostasis. Increased activation of JNK and p38 is seen in white adipose tissue (WAT), skeletal muscle and liver in DUSP1/MKP-1 null mice, while ERK activities are only elevated in WAT and muscle (Wu et al. 2006). It was noticed that the null animals gained significantly less weight than wild type animals following weaning onto a chow diet and were also resistant to obesity induced by a high-fat diet. However, despite increased MAPK activities in insulin-responsive tissues, DUSP1/MKP$^{-/-}$ mice fed a high fat diet still developed glucose intolerance and hyperinsulinemia. Thus, DUSP1/MKP-1 must regulate body mass independently of glucose homeostasis. This was unexpected as elevated JNK activity has been associated with obesity and insulin resistance, while mice lacking JNK1 are resistant to diet induced obesity and also exhibit insulin sensitivity.

The reasons for this apparent discrepancy could lie in two key properties of DUSP1/MKP-1. First, DUSP1/MKP-1 targets all three major MAPK pathways in insulin responsive tissues and each of these may play an independent role in metabolic regulation. For instance, p38 MAPK can regulate energy expenditure via activation of peroxisome proliferator-activated receptor gamma (PPARgamma) coactivator 1alpha (PGC-1alpha)(Fan et al. 2004), while mice lacking the ERK1 MAPK also display resistance to diet induced obesity coupled with protection from the development of insulin resistance (Bost et al. 2005). Thus, it is possible that the phenotype of the DUSP1/MKP-1$^{/-}$ mice reflects complex changes in the activities of all three major classes of MAPK in target tissues rather than just the hyperactivation of JNK. Second, DUSP1/MKP-1 is a nuclear phosphatase and its inability to regulate cytosolic pools of MAPKs such as JNK may underpin some of the phenotypic changes observed. For instance while nuclear accumulation of phospho c-Jun is seen in the livers of DUSP1/MKP-1$^{/-}$ mice, phosphorylation of insulin receptor substrate-1 (IRS-1), a cytoplasmic target for JNK, was unchanged (Wu et al. 2006)

5.3 DUSP2/PAC-1 is a positive regulator of certain inflammatory responses

DUSP2/PAC-1 was originally identified as a mitogen-inducible MKP of human T cells and can inactivate ERK1/2 and p38 MAP kinases *in vitro* and *in vivo* (Rohan et al. 1993; Ward et al. 1994; Chu et al. 1996). The DUSP2/PAC-1 gene is also a transcriptional target of p53 in cells following either growth factor withdrawal or oxidative stress and is implicated in p53-mediated apoptosis (Yin et al. 2003). The results of deleting the murine DUSP2/PAC-1 gene have only recently been reported (Jeffrey et al. 2006). Animals lacking DUSP2/PAC-1 have a complex phenotype with respect to the regulation of MAP kinase activities in immune effector cells and this has revealed an unexpected positive function for DUSP2/PAC-1 in the induction and maintenance of certain inflammatory responses.

Micro-array analysis of gene expression in wild type versus DUSP2/PAC-1 null macrophages following LPS stimulation reveals lower levels of pro-inflammatory mediators and cytokines including IL-6, IL12α, cyclooxygenase-2, IL-1β, and inflammatory chemokines. Furthermore, this deficit was rescued by re-expression of wild type but not a catalytically inactive mutant of DUSP2/PAC-1(Jeffrey et al. 2006). DUSP2/PAC-1 mice were then tested in an autoimmune (K/BxN serum-induced) model of inflammatory arthritis in which pathology is driven by effector leukocytes (mast cells, neutrophils, and macrophages). Wild type mice injected with arthritogenic K/BxN serum developed inflammatory arthritis within two days and showed eventual joint destruction. However, DUSP2/PAC-1 null littermates were protected and showed greatly diminished pathology.

Biochemical evidence links DUSP2/PAC-1 to negative regulation of ERK and p38 but not JNK *in vitro* and *in vivo* (Chu et al. 1996). However, both mast cells and macrophages derived from animals lacking DUSP2/PAC-1 display increased activation of JNK while ERK and p38 activities are actually reduced when compared to wild type controls (Jeffrey et al. 2006). In addition, pharmacological inhibition of JNK activity in these cells led to a significant increase in ERK activity suggesting the elevated JNK activity in DUSP2/PAC-1 null cells is responsible for the decrease in ERK activation.

In conclusion, recent knockout experiments have provided contrasting and complex immune system phenotypes. Loss of DUSP10/MKP-5 and DUSP1/MKP-1 reveals that these phosphatases are negative regulators of cytokine synthesis and play essential non-redundant functions in innate and adaptive immunity. In contrast, the combination of increased JNK activation and reduced ERK and p38 activities seen in DUSP2/PAC-1 null animals translates into a quite different outcome, with a deficit in cytokine gene expression accompanied by protection from inflammatory arthritis. These studies demonstrate that MKP function may be complex and that the ability of these enzymes to differentially modulate multiple MAP kinase pathways can give rise to quite distinct biological outcomes.

6 Summary

Studies of the functional interactions between stress-activated MAPK pathways and protein phosphatases in yeasts, *Drosophila* and *C. elegans* have revealed important regulatory principles that extend beyond simple dephosphorylation/inactivation and are often conserved in mammalian systems. Studies in *S. cerevisiae* were the first to invoke the regulation of a single MAPK by more than one class of protein phosphatase and to suggest that these enzymes could also provide a mechanism for crosstalk between distinct signalling pathways. In addition, the concept that protein phosphatases act to determine not only the activity but also the subcellular localisation of their MAPK substrates was first demonstrated in *S. cerevisiae*. In *Drosophila*, functional studies of the *puckered* phosphatase reinforced the concept that MKPs can participate as components of negative feedback loops to control MAPK activity, while the observation that the *vhp-1* phosphatase regulates both JNK and p38 MAPKs in *C. elegans* suggested that MKPs could serve to integrate signalling through multiple MAPK pathways. Recent genetic studies of the regulation of stress-activated MAPK pathways in mice lacking single MKPs such as DUSP1/MKP-1 and DUSP2/PAC1 have revealed that these enzymes control the activities of two or more MAPK pathways in certain tissues and that the resulting integration of these distinct signalling pathways is crucial to the regulation of important physiological endpoints such as immune function and metabolic control. Certain issues remain to be addressed with respect to these knockout models. These include the apparent discordance between DUSP2/PAC1 substrate specificity *in vitro* and *in vivo* and the changes in individual MAPK activities on deletion of this gene. More detailed studies of existing mouse mutants coupled with the generation of additional single and compound null mice will be essential in revealing the full extent to which the mammalian MKPs exert regulatory control over the physiological outputs of MAPK signalling.

Acknowledgements

Work in the authors' laboratory is supported by Cancer Research UK.

References

Asthagiri AR, Lauffenburger DA (2001) A computational study of feedback effects on signal dynamics in a mitogen-activated protein kinase (MAPK) pathway model. Biotechnol Prog 17:227-239

Bhalla US, Ram PT, Iyengar R (2002) MAP kinase phosphatase as a locus of flexibility in a mitogen-activated protein kinase signaling network. Science 297:1018-1023

Bordo D, Bork P (2002) The rhodanese/Cdc25 phosphatase superfamily. Sequence-structure-function relations. EMBO Rep 3:741-746

Bost F, Aouadi M, Caron L, Even P, Belmonte N, Prot M, Dani C, Hofman P, Pages G, Pouyssegur J, Le Marchand-Brustel Y, Binetruy B (2005) The extracellular signal-regulated kinase isoform ERK1 is specifically required for *in vitro* and *in vivo* adipogenesis. Diabetes 54:402-411

Camps M, Nichols A, Arkinstall S (2000) Dual specificity phosphatases: a gene family for control of MAP kinase function. Faseb J 14:6-16

Carrasco D, Bravo R (1993) Expression of the nontransmembrane tyrosine phosphatase gene erp during mouse organogenesis. Cell Growth Differ 4:849-859

Chang L, Karin M (2001) Mammalian MAP kinase signalling cascades. Nature 410:37-40

Chi H, Barry SP, Roth RJ, Wu JJ, Jones EA, Bennett AM, Flavell RA (2006) Dynamic regulation of pro- and anti-inflammatory cytokines by MAPK phosphatase 1 (MKP-1) in innate immune responses. Proc Natl Acad Sci USA 103:2274-2279

Chu Y, Solski PA, Khosravi-Far R, Der CJ, Kelly K (1996) The mitogen-activated protein kinase phosphatases PAC1, MKP-1, and MKP-2 have unique substrate specificities and reduced activity *in vivo* toward the ERK2 sevenmaker mutation. J Biol Chem 271:6497-6501

Collister M, Didmon MP, MacIsaac F, Stark MJ, McDonald NQ, Keyse SM (2002) YIL113w encodes a functional dual-specificity protein phosphatase which specifically interacts with and inactivates the Slt2/Mpk1p MAP kinase in *S. cerevisiae*. FEBS Lett 527:186-192

Davis RJ (2000) Signal transduction by the JNK group of MAP kinases. Cell 103:239-252

Doi K, Gartner A, Ammerer G, Errede B, Shinkawa H, Sugimoto K, Matsumoto K (1994) MSG5, a novel protein phosphatase promotes adaptation to pheromone response in *S. cerevisiae*. Embo J 13:61-70

Dorfman K, Carrasco D, Gruda M, Ryan C, Lira SA, Bravo R (1996) Disruption of the erp/mkp-1 gene does not affect mouse development: normal MAP kinase activity in ERP/MKP-1-deficient fibroblasts. Oncogene 13:925-931

Du Y, Walker L, Novick P, Ferro-Novick S (2006) Ptc1p regulates cortical ER inheritance via Slt2p. Embo J 25:4413-4422

Fan M, Rhee J, St-Pierre J, Handschin C, Puigserver P, Lin J, Jaeger S, Erdjument-Bromage H, Tempst P, Spiegelman BM (2004) Suppression of mitochondrial respiration through recruitment of p160 myb binding protein to PGC-1alpha: modulation by p38 MAPK. Genes Dev 18:278-289

Flandez M, Cosano IC, Nombela C, Martin H, Molina M (2004) Reciprocal regulation between Slt2 MAPK and isoforms of Msg5 dual-specificity protein phosphatase modulates the yeast cell integrity pathway. J Biol Chem 279:11027-11034

Garrington TP, Johnson GL (1999) Organization and regulation of mitogen-activated protein kinase signaling pathways. Curr Opin Cell Biol 11:211-218

Hahn JS, Thiele DJ (2002) Regulation of the *Saccharomyces cerevisiae* Slt2 kinase pathway by the stress-inducible Sdp1 dual specificity phosphatase. J Biol Chem 277:21278-21284

Hammer M, Mages J, Dietrich H, Servatius A, Howells N, Cato AC, Lang R (2006) Dual specificity phosphatase 1 (DUSP1) regulates a subset of LPS-induced genes and protects mice from lethal endotoxin shock. J Exp Med 203:15-20

Jacoby T, Flanagan H, Faykin A, Seto AG, Mattison C, Ota I (1997) Two protein-tyrosine phosphatases inactivate the osmotic stress response pathway in yeast by targeting the mitogen-activated protein kinase, Hog1. J Biol Chem 272:17749-17755

Jeffrey KL, Brummer T, Rolph MS, Liu SM, Callejas NA, Grumont RJ, Gillieron C, Mackay F, Grey S, Camps M, Rommel C, Gerondakis SD, Mackay CR (2006) Positive regulation of immune cell function and inflammatory responses by phosphatase PAC-1. Nat Immunol 7:274-283

Johnson GL, Lapadat R (2002) Mitogen-activated protein kinase pathways mediated by ERK, JNK, and p38 protein kinases. Science 298:1911-1912

Keyse SM (2000) Protein phosphatases and the regulation of mitogen-activated protein kinase signalling. Curr Opin Cell Biol 12:186-192

Keyse SM, Emslie EA (1992) Oxidative stress and heat shock induce a human gene encoding a protein-tyrosine phosphatase. Nature 359:644-647

Kim DH, Feinbaum R, Alloing G, Emerson FE, Garsin DA, Inoue H, Tanaka-Hino M, Hisamoto N, Matsumoto K, Tan MW, Ausubel FM (2002) A conserved p38 MAP kinase pathway in *Caenorhabditis elegans* innate immunity. Science 297:623-626

Kim DH, Liberati NT, Mizuno T, Inoue H, Hisamoto N, Matsumoto K, Ausubel FM (2004) Integration of *Caenorhabditis elegans* MAPK pathways mediating immunity and stress resistance by MEK-1 MAPK kinase and VHP-1 MAPK phosphatase. Proc Natl Acad Sci USA 101:10990-10994

Koga M, Zwaal R, Guan KL, Avery L, Ohshima Y (2000) A *Caenorhabditis elegans* MAP kinase kinase, MEK-1, is involved in stress responses. Embo J 19:5148-5156

Kumar S, Boehm J, Lee JC (2003) p38 MAP kinases: key signalling molecules as therapeutic targets for inflammatory diseases. Nat Rev Drug Discov 2:717-726

Kwak SP, Hakes DJ, Martell KJ, Dixon JE (1994) Isolation and characterization of a human dual specificity protein-tyrosine phosphatase gene. J Biol Chem 269:3596-3604

Maeda T, Wurgler-Murphy SM, Saito H (1994) A two-component system that regulates an osmosensing MAP kinase cascade in yeast. Nature 369:242-245

Mapes J, Ota IM (2004) Nbp2 targets the Ptc1-type 2C Ser/Thr phosphatase to the HOG MAPK pathway. Embo J 23:302-311

Marshall CJ (1994) MAP kinase kinase kinase, MAP kinase kinase and MAP kinase. Curr Opin Genet Dev 4:82-89

Marshall CJ (1995) Specificity of receptor tyrosine kinase signaling: transient versus sustained extracellular signal-regulated kinase activation. Cell 80:179-185

Martin H, Flandez M, Nombela C, Molina M (2005) Protein phosphatases in MAPK signalling: we keep learning from yeast. Mol Microbiol 58:6-16

Martin-Blanco E, Gampel A, Ring J, Virdee K, Kirov N, Tolkovsky AM, Martinez-Arias A (1998) puckered encodes a phosphatase that mediates a feedback loop regulating JNK activity during dorsal closure in *Drosophila*. Genes Dev 12:557-570

Masuda K, Shima H, Watanabe M, Kikuchi K (2001) MKP-7, a novel mitogen-activated protein kinase phosphatase, functions as a shuttle protein. J Biol Chem 276:39002-39011

Mattison CP, Ota IM (2000) Two protein tyrosine phosphatases, Ptp2 and Ptp3, modulate the subcellular localization of the Hogl MAP kinase in yeast. Genes Dev 14:1229-1235

Mattison CP, Spencer SS, Kresge KA, Lee J, Ota IM (1999) Differential regulation of the cell wall integrity mitogen-activated protein kinase pathway in budding yeast by the protein tyrosine phosphatases Ptp2 and Ptp3. Mol Cell Biol 19:7651-7660

McEwen DG, Peifer M (2005) Puckered, a *Drosophila* MAPK phosphatase, ensures cell viability by antagonizing JNK-induced apoptosis. Development 132:3935-3946

Mizuno T, Hisamoto N, Terada T, Kondo T, Adachi M, Nishida E, Kim DH, Ausubel FM, Matsumoto K (2004) The *Caenorhabditis elegans* MAPK phosphatase VHP-1 mediates a novel JNK-like signaling pathway in stress response. Embo J 23:2226-2234

Morrison DK, Davis RJ (2003) Regulation of MAP kinase signaling modules by scaffold proteins in mammals. Annu Rev Cell Dev Biol 19:91-118

Muda M, Theodosiou A, Rodrigues N, Boschert U, Camps M, Gillieron C, Davies K, Ashworth A, Arkinstall S (1996) The dual specificity phosphatases M3/6 and MKP-3 are highly selective for inactivation of distinct mitogen-activated protein kinases. J Biol Chem 271:27205-27208

O'Rourke SM, Herskowitz I, O'Shea EK (2002) Yeast go the whole HOG for the hyperosmotic response. Trends Genet 18:405-412

Pearson G, Robinson F, Beers Gibson T, Xu BE, Karandikar M, Berman K, Cobb MH (2001) Mitogen-activated protein (MAP) kinase pathways: regulation and physiological functions. Endocr Rev 22:153-183

Qi M, Elion EA (2005) MAP kinase pathways. J Cell Sci 118:3569-3572

Ring JM, Martinez Arias A (1993) puckered, a gene involved in position-specific cell differentiation in the dorsal epidermis of the *Drosophila* larva. Dev Suppl:251-259

Rohan PJ, Davis P, Moskaluk CA, Kearns M, Krutzsch H, Siebenlist U, Kelly K (1993) PAC-1: a mitogen-induced nuclear protein tyrosine phosphatase. Science 259:1763-1766

Sabbagh W Jr, Flatauer LJ, Bardwell AJ, Bardwell L (2001) Specificity of MAP kinase signaling in yeast differentiation involves transient versus sustained MAPK activation. Mol Cell 8:683-691

Salojin KV, Owusu IB, Millerchip KA, Potter M, Platt KA, Oravecz T (2006) Essential role of MAPK phosphatase-1 in the negative control of innate immune responses. J Immunol 176:1899-1907

Sanchez-Perez I, Martinez-Gomariz M, Williams D, Keyse SM, Perona R (2000) CL100/MKP-1 modulates JNK activation and apoptosis in response to cisplatin. Oncogene 19:5142-5152

Saxena M, Mustelin T (2000) Extracellular signals and scores of phosphatases: all roads lead to MAP kinase. Semin Immunol 12:387-396

Schieven GL (2005) The biology of p38 kinase: a central role in inflammation. Curr Top Med Chem 5:921-928

Tanoue T, Moriguchi T, Nishida E (1999) Molecular cloning and characterization of a novel dual specificity phosphatase, MKP-5. J Biol Chem 274:19949-19956

Tanoue T, Yamamoto T, Maeda R, Nishida E (2001) A Novel MAPK phosphatase MKP-7 acts preferentially on JNK/SAPK and p38 alpha and beta MAPKs. J Biol Chem 276:26629-26639

Theodosiou A, Ashworth A (2002) MAP kinase phosphatases. Genome Biol 3:REVIEWS3009

Theodosiou A, Smith A, Gillieron C, Arkinstall S, Ashworth A (1999) MKP5, a new member of the MAP kinase phosphatase family, which selectively dephosphorylates stress-activated kinases. Oncogene 18:6981-6988

Wada T, Penninger JM (2004) Mitogen-activated protein kinases in apoptosis regulation. Oncogene 23:2838-2849

Wang MC, Bohmann D, Jasper H (2003) JNK signaling confers tolerance to oxidative stress and extends lifespan in *Drosophila.* Dev Cell 5:811-816

Wang Z, Xu J, Zhou JY, Liu Y, Wu GS (2006) Mitogen-activated protein kinase phosphatase-1 is required for cisplatin resistance. Cancer Res 66:8870-8877

Ward Y, Gupta S, Jensen P, Wartmann M, Davis RJ, Kelly K (1994) Control of MAP kinase activation by the mitogen-induced threonine/tyrosine phosphatase PAC1. Nature 367:651-654

Warmka J, Hanneman J, Lee J, Amin D, Ota I (2001) Ptc1, a type 2C Ser/Thr phosphatase, inactivates the HOG pathway by dephosphorylating the mitogen-activated protein kinase Hog1. Mol Cell Biol 21:51-60

Wu JJ, Bennett AM (2005) Essential role for mitogen-activated protein (MAP) kinase phosphatase-1 in stress-responsive MAP kinase and cell survival signaling. J Biol Chem 280:16461-16466

Wu JJ, Roth RJ, Anderson EJ, Hong EG, Lee MK, Choi CS, Neufer PD, Shulman GI, Kim JK, Bennett AM (2006) Mice lacking MAP kinase phosphatase-1 have enhanced MAP kinase activity and resistance to diet-induced obesity. Cell Metab 4:61-73

Wurgler-Murphy SM, Maeda T, Witten EA, Saito H (1997) Regulation of the *Saccharomyces cerevisiae* HOG1 mitogen-activated protein kinase by the PTP2 and PTP3 protein tyrosine phosphatases. Mol Cell Biol 17:1289-1297

Yin Y, Liu YX, Jin YJ, Hall EJ, Barrett JC (2003) PAC1 phosphatase is a transcription target of p53 in signalling apoptosis and growth suppression. Nature 422:527-531

Young C, Mapes J, Hanneman J, Al-Zarban S, Ota I (2002) Role of Ptc2 type 2C Ser/Thr phosphatase in yeast high-osmolarity glycerol pathway inactivation. Eukaryot Cell 1:1032-1040

Zhan XL, Deschenes RJ, Guan KL (1997) Differential regulation of FUS3 MAP kinase by tyrosine-specific phosphatases PTP2/PTP3 and dual-specificity phosphatase MSG5 in *Saccharomyces cerevisiae*. Genes Dev 11:1690-1702

Zhan XL, Guan KL (1999) A specific protein-protein interaction accounts for the *in vivo* substrate selectivity of Ptp3 towards the Fus3 MAP kinase. Genes Dev 13:2811-2827

Zhang Y, Blattman JN, Kennedy NJ, Duong J, Nguyen T, Wang Y, Davis RJ, Greenberg PD, Flavell RA, Dong C (2004) Regulation of innate and adaptive immune responses by MAP kinase phosphatase 5. Nature 430:793-797

Zhao Q, Shepherd EG, Manson ME, Nelin LD, Sorokin A, Liu Y (2005) The role of mitogen-activated protein kinase phosphatase-1 in the response of alveolar macrophages to lipopolysaccharide: attenuation of proinflammatory cytokine biosynthesis via feedback control of p38. J Biol Chem 280:8101-8108

Zhao Q, Wang X, Nelin LD, Yao Y, Matta R, Manson ME, Baliga RS, Meng X, Smith CV, Bauer JA, Chang CH, Liu Y (2006) MAP kinase phosphatase 1 controls innate immune responses and suppresses endotoxic shock. J Exp Med 203:131-140

Zhou JY, Liu Y, Wu GS (2006) The role of mitogen-activated protein kinase phosphatase-1 in oxidative damage-induced cell death. Cancer Res 66:4888-4894

Keyse, Stephen M.

Cancer Research UK Stress Response Laboratory, Ninewells Hospital and Medical School, Dundee DD1 9SY, United Kingdom
Stephen.Keyse@cancer.org.uk

Transcriptional regulation by the p38 MAPK signaling pathway in mammalian cells

Eusebio Perdiguero and Pura Muñoz-Cánoves

Abstract

The p38 MAPK signaling pathway plays a key role in numerous physiopathological processes, as inferred from the effects of specific pharmacological inhibitors and of catalytically inactive and constitutively active forms of different pathway components, both in cell culture and in genetically modified mouse models. The p38 pathway is clearly involved in coupling extracellular stimuli to the gene transcription machinery in different mammalian cells. Beyond the classical involvement in inflammation and stress responses, p38 MAPK has been shown to control the growth and differentiation programs of particular cell types. In this review, we first highlight the characteristics of the p38 pathway components and of the mammalian transcription machinery. Secondly, we discuss the mechanisms by which p38 MAPK controls skeletal myocyte, adipocyte, and cardiomyocyte growth and differentiation, by regulating the sequential activation of ubiquitous and tissue-specific transcription factors and transcriptional coactivators, including chromatin-associated activities.

1 Transcription in mammalian cells

Transcription is the primary process that is used by cells to regulate the gene expression program, the development of organs and tissues, and also for the progression of abnormal events like tumors and cancer (Levine and Tjian 2003). The control of transcription in mammalian cells is beautifully and utterly complex and has not yet been fully unraveled. Transcription initiation requires interactions between sequence-specific transcription factors and their cognate regulatory sequences on the promoters of specific genes, as well as remodeling of local chromatin structures. In addition, chromatin histones are subject to a plethora of different modifications. All those steps are controlled in time and space by multiple intracellular signaling pathways which modify the activity of the transcription factors, direct the function of chromatin remodeling activities and RNA polymerases, and therefore control which genes are transcribed, and which are not, along the various stages of the cell cycle (Orphanides and Reinberg 2002; Levine and Tjian 2003).

Topics in Current Genetics, Vol. 20
F. Posas, A. R. Nebreda (Eds.): Stress-Activated Protein Kinases
DOI 10.1007/4735_2007_0256 / Published online: 9 August 2007

1.1 Regulatory DNA elements

A typical mammalian gene contains a compact core promoter upstream the transcription start site. In general, promoters are selected for expression by the binding of TATA-binding protein (TBP) to the TATA element and many associated factors that form generic transcription factor complexes (as TFIID) and promote the formation of the preinitiation complex that recruits and activates the RNA polymerase II. In addition to the core promoter, there are binding sites for proximal regulatory factors and enhancer sequences. Each enhancer is responsible for a subset of the total gene expression pattern, mediating the expression within a specific cell type or developmental stage, and contains binding sites for different sequence-specific transcription factors (Orphanides and Reinberg 2002; Levine and Tjian 2003).

1.2 Chromatin modifying activities

Two main enzymatic activities induce chromatin modifications and regulate accessibility of promoter binding sites to transcription factors: chromatin modifying complexes and chromatin remodeling complexes (Narlikar et al. 2002; Kadam and Emerson 2003). Core histones are acetylated by histone acetyltransferase (HAT) enzymes, which weaken the histone–DNA interaction and have been therefore associated with transcriptional activation. Acetylation, however, is a reversible process as deacetylation is regulated by histone deacetylases (HDACs), being generally associated with transcriptional repression (McKinsey et al. 2002b; Narlikar et al. 2002). On the other hand, ATP-dependent chromatin remodeling enzymes, like those of the SWI/SNF complex, loosen DNA-histone contacts and thus facilitate the movement of the nucleosomes along a particular DNA sequence (Kadam and Emerson 2003).

1.3 Transcription factors

Transcription factors are proteins responsible for the coordinated expression of genes through specific binding to gene promoter and enhancer sites (Orphanides and Reinberg 2002). Their function can be directly or indirectly regulated by protein phosphorylation and dephosphorylation through different mechanisms: by controlling the time that transcription factors spend in the nucleus, by targeting transcription factors or their coregulators for proteolytic degradation, by modulating protein-protein interactions between transcription factors, coregulators and the basal transcription complex, and by regulating transcription factor DNA binding (Whitmarsh and Davis 2000).

2 The p38 MAPK signaling pathway

2.1 MAPK pathways

The family of mitogen-activated protein kinases (MAPKs) controls an enormous number of processes, ranging from gene expression, metabolism, cell proliferation, division and differentiation, to apoptosis, embryogenesis or synaptic plasticity among others (Pearson et al. 2001; Hazzalin and Mahadevan 2002; Johnson and Lapadat 2002; Edmunds and Mahadevan 2004; Thomas and Huganir 2004). MAP kinases are regulated by phosphorylation by upstream kinases, which together encompass a three-component linear kinase cascade (Pearson et al. 2001; Johnson and Lapadat 2002). Each of the MAP kinases has at least two cognate MAP kinase kinases (MKKs or MAP2Ks) and several MAP kinase kinase kinases (MKKKs or MAP3Ks). Four different MAP kinase pathways have been described: extracellular signal-regulated kinases (ERKs), stress-activated protein kinases (SAPKs) comprising the c-Jun N-terminal (JNK) and p38 MAP kinases, respectively, and ERK5/big MAP kinase 1 (BMK1).

2.2 p38 MAPKs

The p38 MAPK pathway was initially described as being preferentially activated by different types of stress and cytokines. Since then, numerous studies have implicated this pathway in the regulation of a wide spectrum of cellular processes, including cell cycle arrest, apoptosis, senescence, regulation of RNA splicing, tumorigenesis, as well as growth and differentiation of particular cell types such as adipocytes, cardiomyocytes and myoblasts (reviewed in Han and Molkentin 2000; Nebreda and Porras 2000; Bulavin and Fornace; Olson and Hallahan 2004; Schieven 2005; Zarubin and Han 2005; Keren et al. 2006). In mammals, there are four p38 MAP kinases, p38α, p38β, p38γ (SAPK3, ERK6), and p38δ (SAPK4). p38α is ubiquitously expressed whereas p38β, p38γ, and p38δ have a more restricted pattern of expression (Schieven 2005). The identification of physiological substrates for p38 MAP kinases has been facilitated by the availability of specific pyridinyl imidazole inhibitors such as SB203580 and SB202190 (which specifically inhibit p38α and p38β isoforms), and the recently reported inhibitor of the four p38 isoforms (BIRB0796) (Kuma et al. 2005) (Fig.1) (See Table 1 for knockout mice phenotypes).

2.3 Upstream kinases

There are two main MKKs known to activate p38s, MKK3, and MKK6. MKK6 activates all four p38s and MKK3 activates p38α, p38γ, and p38δ. MKK4, an upstream kinase of JNK, can also, in some cases, activate p38α and p38δ (Nebreda and Porras 2000; Zarubin and Han 2005). In addition, two MKK-independent

Table 1. Phenotypes of knockout mice of p38 MAP kinase pathway components

Genes	Phenotype	References
MKK3	Viable and fertile. Defects in Th1 $CD4^+$ immune response	(Lu et al. 1999; Wysk et al. 1999)
MKK6	Viable and fertile. Defects in thymocyte development	(Tanaka et al. 2002; Suzuki et al. 2003)
MKK3/6	Die at midgestation. Defects in angiogenesis, placental insufficiency and anemia	(Brancho et al. 2003)
p38α	Die at midgestation. Defects in angiogenesis, placental insufficiency and anemia	(Adams et al. 2000; Allen et al. 2000; Mudgett et al. 2000; Tamura et al. 2000)
p38β	Viable and fertile. No obvious phenotype	(Beardmore et al. 2005)
p38γ	Viable and fertile. No obvious phenotype	(Sabio et al. 2005)
p38δ	Viable and fertile. No obvious phenotype	(Sabio et al. 2005)
p38γ/δ	Viable and fertile. No obvious phenotype	(Sabio et al. 2005)

mechanisms of p38 MAPK activation have been described: (i) p38α autophosphorylation mediated by TAB1 (transforming growth factor-β-activated protein kinase 1 (TAK1)-binding protein) (Ge et al. 2002) and (ii) T-cell receptor (TCR)-induced atypical phosphorylation of p38α/p38β by ZAP70 which then induces its autophosphorylation in T cells (Salvador et al. 2005). Several MKKKs participate in the activation of the p38 cascade, TAK1, ASK1/MAPKKK5, DLK/MUK/ZPK, MEKK4, MLK3, and ZAK (Nebreda and Porras 2000; Gallo and Johnson 2002; Zarubin and Han 2005). Also contributing to p38 activation upstream of MKKs are the low molecular weight GTP-binding proteins of the Rho family such as Rac1 and Cdc42, which can then bind to MEKK1, MLK1, or p21-activated kinases (PAKs), resulting in activation of p38 via MKKKs (Kyriakis and Avruch 2001).

2.4 Downstream substrates

The p38 MAPKs phosphorylate Ser/Thr residues of their substrates, which include protein kinases, transcription factors, cell cycle regulators and cytoskeletal proteins among others.

2.4.1 Protein kinases

Multiple protein kinases are known to be phosphorylated by p38α/β, such as MAPK-activated protein kinase-2 (MK2), MK3 and MK5/PRAK, MAPK-integrating kinase (MNK) 1 and 2, mitogen and stress-activated protein kinase

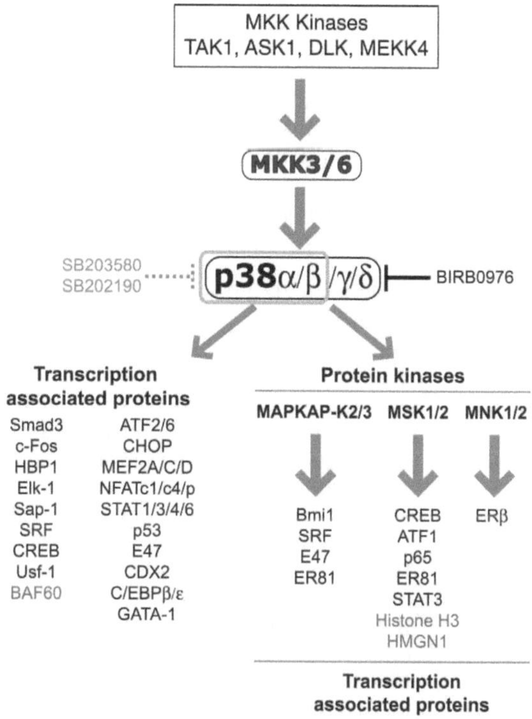

Fig. 1. Transcription-associated protein targets of p38 MAPKs in mammalian cells. In mammals, there are four p38 MAP kinases, p38α, p38β, p38γ, and p38δ, which are all phosphorylated and activated by the MAPK kinase MKK6. Another p38 MAPK kinase is MKK3, which activates p38α, p38γ and p38δ. Once activated, p38 MAPKs phosphorylate Ser/Thr residues of their substrates, which include transcription-associated proteins (mainly transcription factors –in black-, and the SWI/SNF chromatin remodeling complex component BAF60 –in grey-), as well as protein kinases, which in turn phosphorylate additional transcription-associated proteins (transcription factors –in black-, and histone H3 and HMGN1 –in grey-). SB203580 and SB202190 are specific inhibitors of the p38α and p38β isoforms while BIRB0796 is an inhibitor of the four p38 isoforms.

(MSK) 1 and 2, and casein kinase (CK2) 2 (Kyriakis and Avruch 2001; Zarubin and Han 2005). p38δ phosphorylates eEF2 kinase. After activation, MK2/MK3, MSK1/2 and MNK1/2 can phosphorylate numerous proteins implicated in transcriptional regulation (Fig. 1) (Roux and Blenis 2004).

2.4.2 Transcription factors

Numerous transcription factors can be directly phosphorylated and activated by p38 MAP kinases (See Table 2 for details).

Table 2. Transcription-associated proteins directly regulated by p38 MAP kinases

Factor	Regulation	Pathway	References
MEF2A/C/D	Activation	Skeletal and cardiac muscle differentiation Cardiac hypertrophy	(Ornatsky et al. 1999; Yang et al. 1999; Zhao et al. 1999; Han and Molkentin 2000; Penn et al. 2004)
E47	Activation, heterodimerization	Skeletal muscle differentiation	(Lluis et al. 2005)
MRF4	Repression	Skeletal muscle differentiation	(Suelves et al. 2004)
BAF60	Activation	Skeletal muscle differentiation	(Simone et al. 2004)
C/EBPβ, CHOP	Activation	Adipocyte differentiation, cellular stress	(Wang and Ron 1996; Engelman et al. 1998)
NFATc4	Inactivation	Downregulation of adipocyte differentiation	(Yang et al. 2002)
NFATc1, MITF	Activation	RANKL[a] signaling in osteoclast differentiation	(Mansky et al. 2002; Matsumoto et al. 2004)
MafA/B, c-Maf	Activation	Lens differentiation	(Sii-Felice et al. 2005)
GATA-1	Activation	IL-9 expression in mast cells	(Stassen et al. 2006)
C/EBPε	Activation	Neutrophil differentiation	(Williamson et al. 2005)
PGC-1α	Activation	Metabolism, mitochondrial biogenesis	(Puigserver et al. 2001; Fan et al. 2004; Akimoto et al. 2005)
Smad3	Activation	TGF-β signaling	(Furukawa et al. 2003; Wang et al. 2006)
p53	Activation	UV-induced cellular stress, G2/M checkpoint	(Bulavin et al. 1999; Huang et al. 1999; Pedraza-Alva et al. 2006)
HBP1	Stabilization	G1 arrest	(Yee et al. 2004)
STAT1/3/4/6	Activation	IFN-α, IFN-γ, IL-2, IL-4, IL-12, and IL-13 responses	(Goh et al. 1999; Gollob et al. 1999; Pesu et al. 2002; Ramsauer et al. 2002; Xu et al. 2003)
c-Fos	Activation	UV response, cellular stress	(Tanos et al. 2005)
NFATp	Nuclear export	Downregulation of calcium signaling	(Gomez del Arco et al. 2000)
ATF2	Activation	Cellular stress, insulin resistance	(Cuenda and Cohen 1999; Ouwens et al. 2002; Shen et al. 2006)
ATF6	Activation	ER[b] stress	(Luo and Lee 2002)
Elk1, Sap-1, SRF, CREB	Activation	Cellular stress	(Whitmarsh et al. 1997; Yang et al. 1998; Bebien et al. 2003)
Usf-1	Activation	UV response, cellular stress	(Galibert et al. 2001; Corre et al. 2004)

Note: [a] Receptor activator of NF-κB ligand; [b] Endoplasmic reticulum

2.4.3 Other substrates

p38α has been reported to regulate several cell cycle related proteins, like Cdc25B and Cdc25C (Bulavin et al. 2001), cyclin D1 (Casanovas et al. 2000), and cyclin D3 (Casanovas et al. 2004), p21/Cip1 (Kim et al. 2002), and p27/Kip1 (Faust et al. 2005), being then implicated in G1 and G2/M arrests (Wang et al. 2000; Bulavin et al. 2001; Yee et al. 2004) and in contact inhibition (Faust et al. 2005). Other substrates include cytosolic phospholipase A2 (cPLA2) (Kramer et al. 1996), peroxisome proliferator-activated receptor (PPARα) (Barger et al. 2001), phosphorylated heat- and acid-stable protein 1 (PHAS-1) (Jiang et al. 1997), Na^+/H^+ exchanger isoform-1 (NHE-1) (Khaled et al. 2001), p47phox (El Benna et al. 1996), TAK1 binding protein (TAB-1) (Cheung et al. 2003), Stathmin/Op18 (Mizumura et al. 2006), and keratin 8 (Ku et al. 2002). p38α, p38γ, and p38δ phosphorylate the microtubule-associated protein Tau (Buee-Scherrer and Goedert 2002) while p38β phosphorylates glycogen synthase (Kuma et al. 2004). p38γ is able to bind and phosphorylate PDZ domain-containing proteins like α1-syntrophin (Hasegawa et al. 1999), synapse-associated protein 90 (SAP90/PSD95) (Sabio et al. 2004) and synapse-associated protein 97 (SAP97/hDlg) (Sabio et al. 2005). On the other hand, p38δ specifically phosphorylates fibroblast growth factor homologous factor 2 (FHF2) (Schoorlemmer and Goldfarb 2002) and Stathmin (Parker et al. 1998).

2.5 Downregulators

Dephosphorylation has been described as the major system to downregulate MAPK activities. Among the dual-specificity phosphatases specific to the MAPKs (MAP kinase phosphatases, MKP) (Keyse 2000; Tamura et al. 2002), several members can efficiently dephosphorylate p38α and p38β; however, p38γ and p38δ are resistant to dephosphorylation by all known MKP family members (Tanoue et al. 2001). In addition, other phosphatases, such as the members of the PP2C family, PP2Cα and PP2Cδ/Wip1, have been shown to dephosphorylate p38α (Tamura et al. 2002; Yamaguchi et al. 2005).

3 Consequences of p38 MAPK activation on growth and differentiation of mammalian cells

The genes regulated by p38 MAPK in mammalian cells code for a wide range of cytokines, extracellular matrix proteins, cell surface receptors, and transcription factors. Notably, most of the p38 direct phosphorylation substrates are also transcription factors (see Table 2), suggesting the implication of this signaling pathway in the regulation of mammalian gene transcription. p38-mediated posttranscriptional regulation is also an important step contributing to the overall expression of certain genes, especially of inflammatory cytokines. Beyond its classical function in inflammatory and stress responses, p38 regulates the growth and differentiation-specific gene expression programs of particular cell types. In this section, we will discuss the functional consequences of ectopic p38 activation and inhibition in skeletal myogenesis, adipogenesis, and cardiac hypertrophy, as well as the underlying transcriptional regulatory mechanisms, with a special emphasis on the emerging discrepancies on the positive and negative function of p38 signaling in these three cellular processes.

3.1 Skeletal muscle proliferation and differentiation

Skeletal myogenesis is controlled by a family of muscle-specific regulatory factors (MRFs), including MyoD, Myf5, myogenin, and MRF4, which in concert with members of the ubiquitous E2A and myocyte enhancer factor 2 (MEF2) families, activates the differentiation program by inducing transcription of regulatory and structural muscle-specific genes (Sartorelli and Caretti 2005). Numerous studies have shown that the p38 MAPK signaling pathway is a critical regulator of skeletal muscle differentiation. Treatment with the p38 inhibitor SB203580 prevented the fusion of myoblasts into myotubes, as well as the induction of muscle-specific genes (Cuenda and Cohen 1999; Zetser et al. 1999; Li et al. 2000; Wu et al. 2000). Moreover, myoblasts obtained from p38α deficient mice showed delayed cell cycle exit and continued proliferation, as well as impaired differentiation and fusion *in vitro*, correlating with extended activation of the JNK pathway, upregulation of cyclin D1 and hyperphosphorylation of pRb (Perdiguero et al. 2007).

3.1.1 Regulation of myogenic transcription factor activity by p38 MAPK

The first potential explanation for the positive effect of p38 in myogenesis arose from the finding that activation of p38 induced the transcriptional activity of MEF2 proteins (Han et al. 1997) (Fig. 2). The interactions between MEF2 and MRFs during muscle differentiation (Black and Olson 1998) raised the possibility that p38-mediated phosphorylation of MEF2 family members might contribute to the transcriptional synergy between MyoD and MEF2 (Forcales and Puri 2005; Keren et al. 2006). In addition, phosphorylation of the obligate MyoD partner E47 by p38 has been shown to have important consequences on muscle gene transcription (Lluis et al. 2005). In particular, p38-mediated phosphorylation of E47 at Ser140 induced MyoD/E47 heterodimer formation, subsequent binding to the cognate E-box on muscle promoters and activation of muscle-specific gene transcription. Studies using forced dimers of MyoD and/or E47 fusion proteins have demonstrated that heterodimers may provide an scaffolding that is conformationally preferable than the homodimers for establishing a coactivator bridge between the proximal E-boxes and the minimal promoter to activate transcription, through p300 HAT (Dilworth et al. 2004). In this scenario, p38 might be favoring both the formation of the functional MyoD/E47 transcription factor as well as the subsequent interaction with DNA and chromatin-associated proteins.

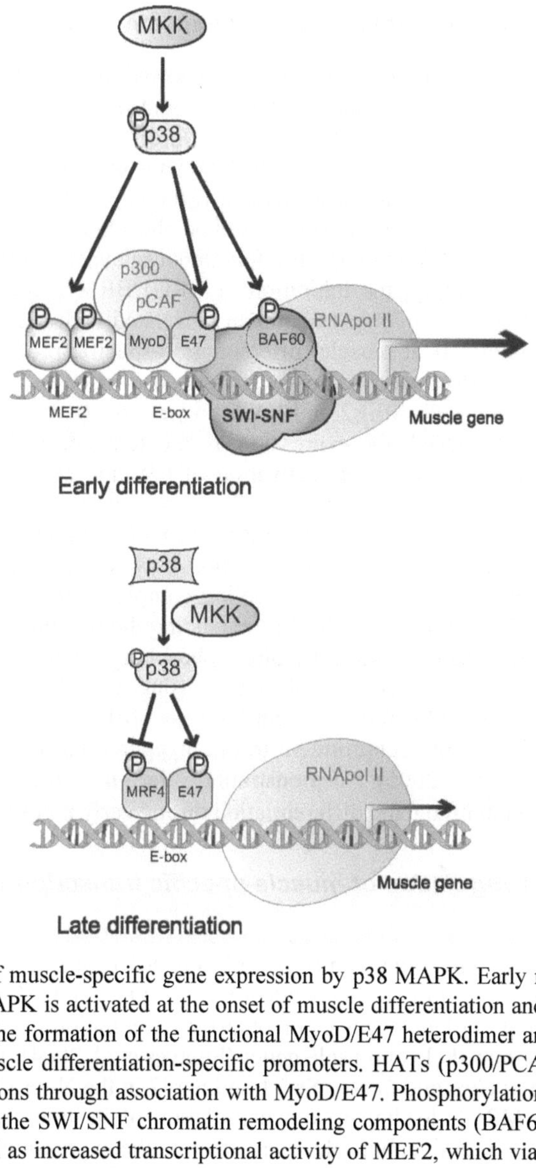

Fig. 2. Control of muscle-specific gene expression by p38 MAPK. Early myogenic differentiation: p38 MAPK is activated at the onset of muscle differentiation and phosphorylates E47, promoting the formation of the functional MyoD/E47 heterodimer and subsequent to the E-box on muscle differentiation-specific promoters. HATs (p300/PCAF) are recruited to regulatory regions through association with MyoD/E47. Phosphorylation by p38 also allows targeting of the SWI/SNF chromatin remodeling components (BAF60) to the muscle promoters as well as increased transcriptional activity of MEF2, which via interaction with MyoD contributes to the overall induction of muscle-specific gene transcription. Finally, p38 facilitates the progression of RNA polymerase II, in a MyoD-mediated feed-forward circuit. Late myogenic differentiation: Since E47 can heterodimerize with MRF4 (whose expression is induced at late stages of differentiation), it is proposed that p38 phosphorylation of E47 will promote the formation of the MRF4/E47 heterodimers (as it does with MyoD/E47), and the subsequent binding to the E-boxes on target genes. MRF4 phosphorylation by p38 in its N-terminal transactivating domain leads to an overall reduction of MRF4-mediated muscle transcription.

3.1.2 Regulation of chromatin-associated activities by p38 MAPK

An important study by Simone and colleagues showed that SB203580 treatment of myoblasts prevented the association of MyoD and the ATPase subunits of the SWI/SNF complex, BRG1 and BRM, to muscle differentiation-specific loci, although neither the acetylation status of histones nor the recruitment of p300 and PCAF coactivators/acetylaces were affected (Simone et al. 2004). Interestingly, the SWI/SNF structural subunit BAF60 could be phosphorylated by p38 MAPK *in vitro* (Simone et al. 2004), although the functional relevance of this event has yet to be established. Notably, the inhibition of BRG1/BRM recruitment to muscle promoters by SB203580 myoblast treatment correlated with reduced levels of active RNA polymerase II at muscle differentiation-specific promoters (Simone et al. 2004) (Fig. 2). Moreover, a muscle gene transcription circuitry initiated by MyoD in association with MEF2 and RNA polymerase II was facilitated by p38 (Penn et al. 2004). Indeed, the expression of late-activated differentiation genes could be advanced by precocious activation of p38 and expression of MEF2D, demonstrating a mechanism for temporally patterning muscle gene expression through a MyoD-mediated feed-forward circuit involving p38 (Penn et al. 2004). Taken together, these findings establish a link between differentiation-activated p38 and recruitment of chromatin-remodeling complexes to transcriptionally active loci during skeletal myogenesis, thus extending the function of p38 beyond its ability to activate gene expression by direct phosphorylation of transcription factors. Importantly, de Angelis and colleagues, using a MEF2 transgenic reporter mouse, showed that SB203580 treatment blocked MEF2 activity and differentiation in the somites, while commitment to the myogenic lineage was not affected (de Angelis et al. 2005), demonstrating the *in vivo* functionality of p38α/β activation in myogenic differentiation during early mouse development.

3.1.3 Negative regulation of muscle-specific transcription by p38 MAPK

An unexpected inhibitory function of p38 MAPK at late stages of myogenesis has been recently reported in two independent studies. One study showed that inhibition of p38 activity at late stages of myoblast differentiation resulted in increased expression of certain skeletal muscle genes (Suelves et al. 2004). In particular, p38 phosphorylated *in vitro* and *in vivo* Ser31 and Ser42 in the N-terminal transactivating domain of MRF4, leading to downregulation of its transcriptional activity, which induced a repressive, but selective, effect on the expression of muscle genes during terminal differentiation (Suelves et al. 2004) (Fig. 2). Using a different approach, Weston and colleagues showed that treatment of primary limb mesenchymal cultures, which should differentiate to cartilage, with p38α/p38β inhibitors, enhanced muscle formation rather than promoting chondrogenesis (Weston et al. 2003). Likewise, the transcriptional activity of MEF2-GAL4 fusion proteins expressed in primary cultures was also enhanced by treatment with the p38α/p38β inhibitors (Weston et al. 2003).

3.2 Adipocyte differentiation

The peroxisome proliferator-activated receptors (PPARs) and the CCAAT/enhancer-binding protein (C/EBP) families of transcription factors are master regulators of adipocyte differentiation (Rosen and MacDougald 2006). The C/EBPβ and C/EBPδ isoforms are rapidly induced upon adipocyte differentiation and are responsible for activating the late adipogenic regulators C/EBPα and PPARγ, which together activate the majority of genes expressed in differentiating adipocytes (Wu et al. 1995; Darlington et al. 1998; Engelman et al. 1998). Independent studies have demonstrated a positive role for the p38 MAPK signaling pathway in the induction of adipogenic differentiation, although recent evidence has began to challenge this conclusion. p38 activity was shown to increase at the onset of adipocyte differentiation, induced by adequate hormonal stimulation of 3T3-L1 fibroblasts, decreasing at later differentiation stages. In accordance with it, SB203580 treatment of 3T3-L1 fibroblasts during only the early stages of differentiation was sufficient to block adipogenesis, as evidenced by the lack of adipocyte-specific genes and a fibroblast-like morphological appearance (Engelman et al. 1998). Conversely, overexpression of constitutively active MKK6 in fibroblasts induced extensive p38-dependent adipocyte conversion in the absence of hormonal stimulation (Engelman et al. 1999). However, persistent activation of p38 in adipocytes led to cell death. Thus, p38 hyperactivation promoted two distinct responses in the 3T3-L1 model system: adipocyte conversion in fibroblast and cell death in fully differentiated adipocytes.

3.2.1 Regulation of adipogenic differentiation by posttranslational modification of transcription factors by p38 MAPK

Subsequent studies tried to provide mechanistic explanations to the early proadipogenic role of p38. C/EBPβ, the key initial adipogenic factor, was phosphorylated by p38 *in vitro* and *in vivo*, in a SB203580-dependent fashion (Engelman et al. 1998). Moreover, the transcriptional induction of PPARγ, a gene whose expression is induced by C/EBPβ, was impaired in the presence of p38 inhibitors in 3T3-L1 fibroblasts, and this impairment was attributed to a decreased phosphorylation and posttranslational activation of C/EBPβ.

At variance with the proadipogenic effect of p38, alternative studies showed that p38 activity was capable of inhibiting adipocyte differentiation through an indirect mechanism, involving the dominant-negative regulator of C/EBP, CHOP. CHOP could be phosphorylated by p38 on Ser78 and Ser81, serving as a link between p38 and cellular growth and differentiation processes (Wang and Ron 1996). In the context of adipogenesis, while overexpression of CHOP was capable of blocking adipogenic differentiation of 3T3-L1 cells (Batchvarova et al. 1995), a p38-nonphosphorylatable form of CHOP was a poor inhibitor of adipogenesis. Taken together, this suggested that p38 activity could prevent adipocyte differentiation via CHOP posttranslational modification. Interestingly, a similar mechanism has been proposed for the function of the NFATc4 protein in adipocyte differentiation (Ho et al. 1998; Yang et al. 2002). Dephosphorylation of NFATc4 was shown to promote its nuclear localization, thereby increasing PPARγ expression, via two distinct NFAT binding elements in the PPARγ2 gene promoter, which resulted in induction of adipocyte differentiation (Yang et al. 2002). Conversely, phosphorylation of NFATc4 at multiple residues, including Ser168 and Ser170, by p38 kinase promoted NFAT nuclear export and inhibited adipose cell differentiation of fibroblasts (Yang et al. 2002). More recently, novel evidence for a negative role of p38 in adipogenesis has been reported using various experimental models, including embryonic to adult stage cells (Aouadi et al. 2006b). In this study, pharmacological inhibition of p38 or knockout of the p38α gene increased adipogenesis in primary embryonic fibroblasts and preadipocytes, as well as in embryonic stem cell lines, via enhancement of the expression and transcriptional activity of C/EBPβ and PPARγ (Aouadi et al. 2006b); importantly, these effects were blunted by rescue of p38 in p38α-deficient cells, indicating that p38 activity negatively regulates adipogenesis via inhibition of both adipogenic transcription factors (Aouadi et al. 2006b). Taken together, the different studies evidence discrepancies on the role of p38 in adipogenesis: p38 inhibition decreased PPARγ activity in one cellular model, 3T3-L1 cells, and had the opposite effect in primary cells. Since the referred primary cultures were obtained from mouse embryos to adults, they likely represent more physiological experimental models than the established 3T3-L1 cell line (Aouadi et al. 2006a). These observations strongly suggest that the regulation of PPARγ by the p38 pathway involves molecular partners that are different in 3T3-L1 cells from the other primary cell models. It is tempting to propose that p38 might be acting on PPARγ coactivators involved in adipogenesis, such as steroid receptor coactivator, CREB-binding protein, or p300, although no clear evidence exists of their phosphorylation by p38. PPARγ corepressors might alternatively be involved in p38 regulation, especially because they regulate the PPARγ transcriptional activity in adipocytes (Guan et al. 2005).

3.2.2 Regulation of mitochondrial gene expression in brown adipose tissue by p38 MAPK

PPAR-interacting protein 1α (PGC-1α) was isolated as a coativator of PPARγ and is considered a master regulator of mitochondrial gene expression, thereby activating mitochondrial biogenesis and oxidative metabolism (Puigserver et al. 1998; Wu et al. 1999; Akimoto et al. 2005). Key cellular signals that control thermogenesis, energy and nutrient homeostasis, such as cAMP and cytokine pathways, were shown to strongly activate PGC-1α in brown adipose tissue (BAT). (Puigserver and Spiegelman 2003). Activated PGC-1α was also shown to stimulate fiber-type switching in skeletal muscle (see below), and multiple aspects of the fasting response in liver. Transcription of the uncoupling protein-1 (UCP1) gene is almost exclusively restricted to BAT where its major function in diet-induced and in non-shivering thermogenesis depends on the capacity to uncouple oxidative phosphorylation from ATP generation (Dulloo et al. 2004). Several physiological stimuli including cold exposure and sympathetic signaling via norepinephrine and cAMP as well as thyroid hormone regulate UCP1 gene expression in rodents. The cAMP response of the UPC1 gene is achieved predominantly through an enhancer region (Kliewer and Willson 1998; Puigserver et al. 2001), which contains two key elements: a peroxisome proliferator response element (PPRE) and a cAMP response element (CRE). Importantly, the cAMP-dependent transcription of the UPC1 gene was shown to be regulated through these two elements by p38 MAPK in a coordinated manner (Cao et al. 2004). In a first step, p38 phosphorylates PGC-1α, thereby enhancing its activity as a nuclear coactivator of gene transcription in coordination with PPARγ, which in turn binds to the PPRE element on the UPC1 promoter. In a second step, p38 directly stimulates expression of the UPC1 gene through phosphorylation of the transcription factor ATF2, which binds to a CRE site (Cao et al. 2004). Finally, the PGC-1α gene itself also possesses a CRE, serving as a target for positive autoregulation in a p38-dependent manner (Puigserver and Spiegelman 2003) (Fig. 3). Taken together, a model for the mechanistic implication of p38 in BAT thermogenesis emerges: by increasing the overall amount of PGC-1 (via induced transcriptional synthesis and protein stability, see below), p38 may prime the cell for a persistent increase in UCP1 expression. Despite the better understanding of the role of p38 in the regulation of the UPC1 and PGC-1α genes in BAT, the detailed sequence of signaling events downstream of cAMP/PKA which result in p38 activation remains to be known.

3.2.3 Regulation of PGC-1α coactivator activity by p38 MAPK

How is PGC-1a activity regulated by p38 MAPK? Fundamental studies by Spiegelman's group demonstrated that p38 was able to phosphorylate PGC-1α in three residues (Thr262, Ser265, and Thr298) leading to increased protein stability (Puigserver et al. 2001). Interestingly, these phosphorylations occurred in a region (aa 170-350) mediating a repressive effect in transcription, which also served for docking of different transcription factors (Puigserver et al. 1999). Transcription factor docking on PGC-1α was shown to cause a conformational change that accelerated the binding of other transcriptional effector proteins into this complex, including CBP/p300 and SRC-1 (Puigserver and Spiegelman 2003). Further studies identified the 160 myb binding protein ($p160^{MBP}$) as a repressor of PGC-1α (Fan et al. 2004). Importantly, the binding and repression of PGC-1α by $p160^{MBP}$ could be disrupted by p38 phosphorylation of PGC-1α (Fig. 3). Of note, this repression did not require removal of PGC-1α from chromatin (Fan et al. 2004), suggesting that $p160^{MBP}$ may be, or engage, a direct transcriptional suppressor. Overall, $p160^{MBP}$ constitutes a potent negative regulator of PGC-1α function, providing a molecular mechanism for the activation of PGC-1α by p38 MAPK.

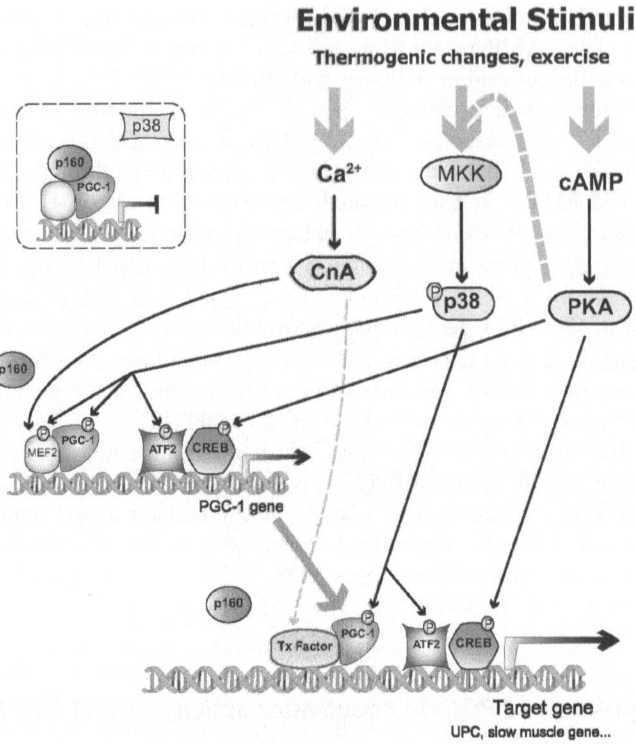

Fig. 3. p38 MAPK induces a positive autoregulatory loop for the transcriptional coactivator PGC-1α. p38 activation induced either by exercise (in slow-twitch muscle) or by environmental stimuli such as cold or diet (in brown adipose tissue) can promote PGC-1α-mediated transcription. There are three crucial targets of p38 in the induction of PCG-1α-mediated transcription. One target is ATF2, which is phosphorylated by p38 and coordinates the transcriptional induction of the PGC-1α gene itself and on the mitochrondrial genes through their CREs. The second target is the MEF2 transcription factor in muscle, which once phosphorylated and activated by p38, can directly interact with PGC-1α on the PGC-1α promoter, resulting in a positive autofeedback regulation of PGC-1α transcription. The third p38 target is the PGC-1α protein itself. Stimulation of p38 directly phosphorylates the PGC-1α protein, inhibiting the binding of $p160^{MBP}$, which derepresses PGC-1α function, resulting in its activation and stabilization (see dashed box). Activation of PGC-1α then drives transcription of target genes, including 1) the PGC-1α gene itself; 2) the UCP1 gene in BAT, through its association with PPARγ bound to the UCP1 promoter; 2) oxidative slow fiber-type genes in muscle, though its association with MEF2. In brown adipose tissue, β-Adrenergic stimulation triggers activation of a cAMP and PKA signaling cascade, in addition to p38. PKA phosphorylates CREB transcription factor, which is also involved in the induction of PGC-1α gene expression. In slow-twitch muscle, besides activating p38, exercise-induced neuromuscular activity elevates intracellular Ca^{+2}, which in turn results in activation of the calcineurin pathway, which in parallel, can stimulate MEF2 activity and promote PGC-1α transcription.

3.2.4 Regulation of oxidative metabolism in muscle by p38 MAPK

Similarly to BAT, PGC-1α has the ability to promote mitochondrial biogenesis and respiration in skeletal muscle. Slow-twitch muscle is composed primarily of type I fibers that are rich in mitochondria and rely on oxidative metabolism, while fast-twitch muscle is mainly composed by type 2b highly glycolytic fibers. Interestingly, PGC-1α was found expressed at much higher levels in slow-twitch muscle, suggesting a role for this protein in fiber-type regulation (reviewed in Bassel-Duby and Olson 2006). This interpretation was supported by muscle-specific overexpression of PGC1α in transgenic mice, which resulted in increased mitochondrial biogenesis and slow-twitch fibers (Lin et al. 2002). Interestingly, endurance training exercise has been shown to induce PGC-1α mRNA and protein expression in rats and human, at least in part, because of increased transcription (Pilegaard et al. 2003). Transcriptional analysis revealed that PGC-1α regulates its own promoter activity in a positive autoregulatory loop, through MEF2 binding, possibly via a MEF2-PGC-1α direct interaction (Fig. 3) (Michael et al. 2001; Handschin et al. 2003). Given that both MEF2 and PGC-1α are phosphorylation substrates of p38, and since p38 activity was found to be induced by endurance training (Yu et al. 2001), a p38/MEF2/PGC-1α pathway could be proposed underlying exercise-stimulated slow-twitch-fiber oxidative metabolism. Recent studies have further shown that exercise-induced activation of p38 promoted PGC-1α gene transcription in slow-twitch muscle by phosphorylating and activating ATF2 and MEF2 directly, and by inhibiting repression of PGC-1α function by $p160^{MBP}$, which in turn exerted a positive autofeedback regulation of PGC-1α transcription, possibly through interaction with and activation of MEF2 (Akimoto et al. 2005) (Fig. 3). Taken together, these studies provide evidence linking activation of the p38 MAPK pathway to brown adipose tissue and muscle adaptation through its regulatory control of PGC-1α expression and posttranslational activity, thereby influencing the subsequent mitochondrial and oxidative gene transcription response.

3.3 Cardiomyocyte hypertrophy

3.3.1 Positive regulation of hypertrophic growth in neonatal cardiomyocytes by p38 MAPK

The heart is capable of significant cellular adaptation and molecular reprogramming following both physiologic and pathologic stimulation. Postnatal cardiomyocytes are terminally differentiated cells incapable of cell division, which respond to external stimuli by adaptative growth (hypertrophy) (Chien et al. 1991). Usually, these adaptations are initiated by stress-responsive signaling pathways, including that of p38 MAPK. Despite the important research effort in analyzing the role of p38 signaling in regulating cardiac hypertrophy, conflicting conclusions have been obtained as to whether p38 exerts a positive or negative effect on this process. The first studies implicating p38 signaling as an effector of the hypertrophic cardiac response were mainly performed in neonatal cardiomyocytes. Synthetic p38 inhibitors or catalytically inactive forms of p38α/p38β were reported to attenuate agonist-induced cardiomyocyte hypertrophy *in vitro* (Zechner et al. 1997; Nemoto et al. 1998; Wang et al. 1998; Liang and Gardner 1999; Liang et al. 2000). Conversely, overexpression of activated MKK3 or MKK6 induced hypertrophy and atrial natriuretic factor (ANF) expression in cultured neonatal cardiomyocytes, reinforcing the implication of p38 as a positive transducer of the cardiomyocyte growth response (Zechner et al. 1997; Nemoto et al. 1998; Wang et al. 1998). Since the MEF2 transcription factor regulates the expression of most cardiac expressed genes and its transcriptional activity is positively regulated by direct p38 phosphorylation, MEF2 constituted a likely effector of p38 in the cardiac hypertrophic gene expression response (Han and Molkentin 2000). In agreement with this, MEF2A and MEF2C, the predominant forms upregulated during cardiac hypertrophy, were shown to be required for normal postnatal growth of the myocardium (Kolodziejczyk et al. 1999). Taken together, these data strongly suggested that the p38 pathway transduced signals leading to the hypertrophic response in cultured neonatal cardiomyocytes.

3.3.2 Negative regulation of hypertrophic growth in mouse cardiac tissue by p38 MAPK

At variance with the above conclusions, overexpression of either activated MKK3 or MKK6 by transgenesis in the mouse heart did not induce hypertrophic growth, suggesting that p38 activation is not causal in the cardiac growth process *in vivo* (Liao et al. 2001). In addition, mice expressing catalytically inactive forms of p38α or p38β specifically in the heart presented a similar cardiac hypertrophic response compared to wild type controls (Zhang et al. 2003). Furthermore, transgenic mice overexpressing catalytically inactive forms of MKK3, MKK6 and p38α they all exhibited an importantly enhanced cardiac hypertrophy both at baseline and following different hypertrophic stimuli (Braz et al. 2003; Liang and Molkentin 2003). Which is then the mechanism responsible for the enhanced hypertrophy upon inhibition of p38 signaling in the heart? Numerous studies point towards the transcription factor NFAT as the p38 phosphorylation target and effector of this response. It is well known that N-terminal phosphorylation of NFAT transcription factors by p38 hampers their dephosphorylation by the phosphatase calcineurin and their subsequent nuclear translocation (Gomez del Arco et al. 2000; Porter et al. 2000; Yang et al. 2002). Indeed, mice with cardiac-specific deficiency in NFATc3 presented a reduced ability to hypertrophy in response to activated calcineurin and pressure-overload stimulation (Wilkins et al. 2002). In agreement with the proposed p38-NFAT pathway, transgenic mice overexpressing catalytically inactive p38α exhibited enhanced NFAT transcriptional activity in the heart, suggesting that p38 prevents the prohypertrophic-stimulating effect of NFAT transcription factors in cardiac tissue. This conclusion was further reinforced by the reversal of the p38α inhibition-mediated hypertrophic response by genetic disruption of the calcineurin Aβ gene (Braz et al. 2003). Taken together,

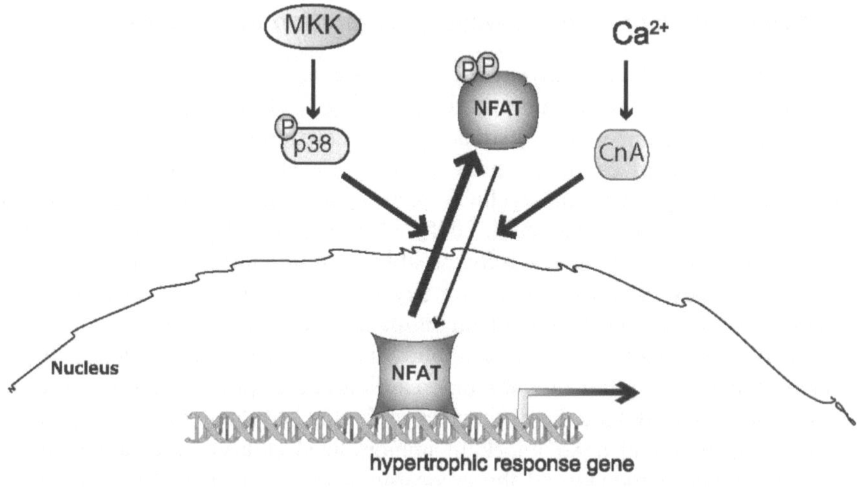

Fig. 4. p38 prevents cardiomyocyte growth by antagonizing calcineurin-mediated NFAT nuclear translocation. p38 phosphorylates NFAT factors, promoting their cytoplasmic sequestration and/or nuclear export, thus antagonizing the effect of calcineurin-mediated dephosphorylation, which is activated by calcium signals, for the induction of hypertrophic response genes in cardiomyocytes.

these results strongly suggest that NFAT proteins are central mediators of the cardiac hypertrophic response and that alterations in the activity p38 MAPK can modify the growth response through a mechanism involving enhanced calcineurin-NFAT signaling (Fig. 4) (Wilkins and Molkentin 2004). Interestingly, several studies have demonstrated that MEF2 transcription factors are also downstream targets of calcineurin signaling, although the underlying mechanism has not been elucidated. Consequently, MEF2 proteins could potentiate the activity of NFATs on the promoters of genes involved in the cardiomyocyte hypertrophic response (McKinsey et al. 2002a; Czubryt and Olson 2004).

Nevertheless, the involvement of p38 in cardiac hypertrophic growth is still far from clear. In a recent study using cardiac-specific p38α-deficient mice, pressure overload induced normal hypertrophic responses in these mice but also cardiac dysfunction and heart dilatation, indicating that p38 plays a critical role in the cardiomyocyte survival pathway in response to pressure overload, while hypertrophic growth is unaffected despite the absence of p38α (Nishida et al. 2004). While part of the controversy may be ascribed to the inherent differences between the *in vitro* and *in vivo* models used for cardiomyocyte hypertrophy (Liang and Molkentin 2003), differences in hypertrophic stimuli, in the degree of p38 activation or inhibition resulting from the different pharmacological and transgenic strategies, as well as in the genetic background of the mice used in these studies, could all account for the referred discrepancies.

3.3.3 Regulation of cardiomyocyte proliferation by p38 MAPK

In contrast to adult cardiomyocytes, which are terminally differentiated cells, mammalian cardiomyocytes do proliferate during fetal development. An elegant recent study using mice with a cardiac-specific deletion of p38α showed induced proliferation of fetal cardiomyocytes *in vitro*, correlating with upregulation of Cdc2, Cdc25B, cyclin D, and cyclin B expression, which are required factors for cell cycle progression (Engel et al. 2005). More importantly, post-mitotic, differentiated adult cardiomyocytes were capable of proliferation in the absence of p38α. This study established p38 as a key negative regulator of cardiomyocyte proliferation, most likely by modulating important cell cycle factors.

A very recent study by Ambrosino and colleagues has demonstrated a housekeeping function of p38α in embryonic cardiomyocyte proliferation (Ambrosino et al. 2006). By analyzing the differences in the transcriptome of normally proliferating wild type and p38α knockout immortalized embryonic cardiomyocytes, many potential components of the myocardium extracellular matrix were found upregulated in the absence of p38α, such as collagen 1A1 (COL1A1). TEF-1, a known regulator of heart-specific gene expression, and C/EBPβ, were identified as the two transcription factors whose binding sites were most enriched in the promoters of p38α-regulated genes. In particular, the upregulation of COL1A1 gene expression in p38α-deficient cardiomyocytes could be related to the binding of the transcriptional activator TEF-1 and the simultaneous release of the cardiac transcriptional repressor C/EBPβ at the COL1A1 loci, respectively (Ambrosino et al. 2006). Thus, the basal level of p38 activity in non-stimulated cardiomyocytes may be implicated in the inhibition of profibrotic and prohypertrophic pathways, which are activated upon heart injury and play a key role in the hypertrophic and repair response (McKinsey and Olson 1999; Ambrosino et al. 2006). These studies further indicate that C/EBPβ and TEF-1 transcription factors are important mediators of p38 MAPK function in cardiomyocytes.

4 Conclusions and perspectives

The p38 MAP kinase signaling pathway has been implicated in a broad variety of cellular responses, including inflammation, cell cycle, cell death, senescence, tumorigenesis as well as development and cell growth and differentiation. Accordingly, its activity controls the expression of a large number of genes, acting at different steps along the gene expression pathway (i.e. transcription, mRNA stability and translation). Although new evidence linking p38 to senescence and cancer as well as cytokine regulation at the posttranscriptional level has shed more light on p38 function and regulation, its implication as a regulator of transcription factor activity on the differentiation and growth of specific cell types is still obscure. Indeed, its predicted positive influence in tissue-specific processes such as cardiac hypertrophy and adipogenesis has been challenged in the last years. The postulated p38/MEF2 axis acting positively on cardiomyocyte hypertrophy in cell culture has not been validated *in vivo*; indeed, transgenic mice expressing mutant version of the p38 pathway components rather support a negative role of p38 in this process via phosphorylation of the transcription factor NFAT. Similarly, the reported positive influence of p38 on the C/EBP/PPAR adioponenic regulators leading to adipocyte differentiation needs to be reexamined based on divergent conclusions emerging from different studies. Most of the discrepancies could well arise from the variable levels of p38 inhibition or activation resulting from the different experimental strategies utilized and, even more, from the use of *in vitro* versus *in vivo* models. Furthermore, the nature of the cell culture model (established cell line versus primary cell) might also influence the final output and subsequent conclusions. Despite these cautious arguments, both cell culture and transgenic mouse approaches should be valuable tools to get a better understanding of p38-mediated cell differentiation processes if uniformly and coordinately used.

At variance with the nebulous picture on the role of p38 in the transcriptional control of cardiac- and adipogenic-specific genes, findings from last years have unambiguously demonstrated that p38 plays a fundamental role in skeletal muscle differentiation, and have shed more light on its underlying mechanisms of action. Indeed, p38 emerges as a pivotal molecule orchestrating sequential events in the myogenic transcription pathway, by regulating key myogenic transcription factors (including MEF2, E47/MyoD and MRF4) as well as chromatin remodeling activities associated to the muscle differentiation gene program.

When analyzing processes as complex as tissue-specific differentiation and growth responses, it should be emphasized that the p38 pathway does not operate in isolation. p38 can be activated by numerous extracellular upstream inputs. Moreover, the activating intracellular upstream signals can be general (MKK4), specific (MMK3/6) or MAPKK-independent (TAB1). In addition, an extra level of divergence occurs downstream of the p38 pathway, and each of the multiple p38 protein targets can interact with other cellular components, to coordinate diverse cellular processes. Moreover, p38 interacts with other MAPK kinases and a plethora of other signaling pathways. In view of the controversy concerning the positive or negative influence of p38 on certain cellular growth and differentiation transcriptional processes, future work should focus on its interaction with the different signaling pathways and downstream transcriptional effectors. In addition, the relative contribution of the four p38 MAPK family members to cell-specific differentiation processes remains to be elucidated. Most of the work that demonstrates the involvement of p38 in distinct cell responses is based on the use of synthetic compounds such as SB203580, which only inhibit the activity of p38α and p38β, or on the use of constitutively active or catalytically inactive MKK3/MKK6 forms. Whether different p38 isoforms regulate the expression of particular subsets of genes, at different stages of differentiation, and whether they possess inducing or repressing activities, remain to be determined. The identification of new substrates for the different p38 family members as well as upstream mechanisms and crosstalk with other signaling pathways will undoubtedly increase our understanding of how p38 regulates cell type-specific growth and differentiation.

Acknowledgements

We sincerely thank Dr. Antonio Serrano for helpful discussions. Work in the authors' laboratory was supported by MDA, SAF2004-06983, AFM, Fundación MM and R. Pascual. EP is recipient of a postdoctoral contract from CIBERNED.

References

Adams RH, Porras A, Alonso G, Jones M, Vintersten K, Panelli S, Valladares A, Perez L, Klein R, Nebreda AR (2000) Essential role of p38alpha MAP kinase in placental but not embryonic cardiovascular development. Mol Cell 6:109-116

Akimoto T, Pohnert SC, Li P, Zhang M, Gumbs C, Rosenberg PB, Williams RS, Yan Z (2005) Exercise stimulates Pgc-1alpha transcription in skeletal muscle through activation of the p38 MAPK pathway. J Biol Chem 280:19587-19593

Allen M, Svensson L, Roach M, Hambor J, McNeish J, Gabel CA (2000) Deficiency of the stress kinase p38alpha results in embryonic lethality: characterization of the kinase dependence of stress responses of enzyme-deficient embryonic stem cells. J Exp Med 191:859-870

Ambrosino C, Iwata T, Scafoglio C, Mallardo M, Klein R, Nebreda AR (2006) TEF-1 and C/EBPbeta are major p38alpha MAPK-regulated transcription factors in proliferating cardiomyocytes. Biochem J 396:163-172

Aouadi M, Binetruy B, Caron L, Le Marchand-Brustel Y, Bost F (2006a) Role of MAPKs in development and differentiation: lessons from knockout mice. Biochimie 88:1091-1098

Aouadi M, Laurent K, Prot M, Le Marchand-Brustel Y, Binetruy B, Bost F (2006b) Inhibition of p38MAPK increases adipogenesis from embryonic to adult stages. Diabetes 55:281-289

Barger PM, Browning AC, Garner AN, Kelly DP (2001) p38 mitogen-activated protein kinase activates peroxisome proliferator-activated receptor alpha: a potential role in the cardiac metabolic stress response. J Biol Chem 276:44495-44501

Bassel-Duby R, Olson EN (2006) Signaling pathways in skeletal muscle remodeling. Annu Rev Biochem 75:19-37

Batchvarova N, Wang XZ, Ron D (1995) Inhibition of adipogenesis by the stress-induced protein CHOP (Gadd153). EMBO J 14:4654-4661

Beardmore VA, Hinton HJ, Eftychi C, Apostolaki M, Armaka M, Darragh J, McIlrath J, Carr JM, Armit LJ, Clacher C, Malone L, Kollias G, Arthur JS (2005) Generation and characterization of p38beta (MAPK11) gene-targeted mice. Mol Cell Biol 25:10454-10464

Bebien M, Salinas S, Becamel C, Richard V, Linares L, Hipskind RA (2003) Immediate-early gene induction by the stresses anisomycin and arsenite in human osteosarcoma cells involves MAPK cascade signaling to Elk-1, CREB and SRF. Oncogene 22:1836-1847

Black BL, Olson EN (1998) Transcriptional control of muscle development by myocyte enhancer factor-2 (MEF2) proteins. Annu Rev Cell Dev Biol 14:167-196

Brancho D, Tanaka N, Jaeschke A, Ventura JJ, Kelkar N, Tanaka Y, Kyuuma M, Takeshita T, Flavell RA, Davis RJ (2003) Mechanism of p38 MAP kinase activation *in vivo*. Genes Dev 17:1969-1978

Braz JC, Bueno OF, Liang Q, Wilkins BJ, Dai YS, Parsons S, Braunwart J, Glascock BJ, Klevitsky R, Kimball TF, Hewett TE, Molkentin JD (2003) Targeted inhibition of p38 MAPK promotes hypertrophic cardiomyopathy through upregulation of calcineurin-NFAT signaling. J Clin Invest 111:1475-1486

Buee-Scherrer V, Goedert M (2002) Phosphorylation of microtubule-associated protein tau by stress-activated protein kinases in intact cells. FEBS Lett 515:151-154

Bulavin DV, Fornace AJ Jr (2004) p38 MAP kinase's emerging role as a tumor suppressor. Adv Cancer Res 92:95-118

Bulavin DV, Higashimoto Y, Popoff IJ, Gaarde WA, Basrur V, Potapova O, Appella E, Fornace AJ Jr (2001) Initiation of a G2/M checkpoint after ultraviolet radiation requires p38 kinase. Nature 411:102-107

Bulavin DV, Saito S, Hollander MC, Sakaguchi K, Anderson CW, Appella E, Fornace AJ Jr (1999) Phosphorylation of human p53 by p38 kinase coordinates N-terminal phosphorylation and apoptosis in response to UV radiation. EMBO J 18:6845-6854

Cao W, Daniel KW, Robidoux J, Puigserver P, Medvedev AV, Bai X, Floering LM, Spiegelman BM, Collins S (2004) p38 mitogen-activated protein kinase is the central regulator of cyclic AMP-dependent transcription of the brown fat uncoupling protein 1 gene. Mol Cell Biol 24:3057-3067

Casanovas O, Jaumot M, Paules AB, Agell N, Bachs O (2004) P38SAPK2 phosphorylates cyclin D3 at Thr-283 and targets it for proteasomal degradation. Oncogene 23:7537-7544

Casanovas O, Miro F, Estanyol JM, Itarte E, Agell N, Bachs O (2000) Osmotic stress regulates the stability of cyclin D1 in a p38SAPK2-dependent manner. J Biol Chem 275:35091-35097

Corre S, Primot A, Sviderskaya E, Bennett DC, Vaulont S, Goding CR, Galibert MD (2004) UV-induced expression of key component of the tanning process, the POMC and MC1R genes, is dependent on the p-38-activated upstream stimulating factor-1 (USF-1). J Biol Chem 279:51226-51233

Cuenda A, Cohen P (1999) Stress-activated protein kinase-2/p38 and a rapamycin-sensitive pathway are required for C2C12 myogenesis. J Biol Chem 274:4341-4346

Czubryt MP, Olson EN (2004) Balancing contractility and energy production: the role of myocyte enhancer factor 2 (MEF2) in cardiac hypertrophy. Recent Prog Horm Res 59:105-124

Cheung PC, Campbell DG, Nebreda AR, Cohen P (2003) Feedback control of the protein kinase TAK1 by SAPK2a/p38alpha. EMBO J 22:5793-5805

Chien KR, Knowlton KU, Zhu H, Chien S (1991) Regulation of cardiac gene expression during myocardial growth and hypertrophy: molecular studies of an adaptive physiologic response. FASEB J 5:3037-3046

Darlington GJ, Ross SE, MacDougald OA (1998) The role of C/EBP genes in adipocyte differentiation. J Biol Chem 273:30057-30060

de Angelis L, Zhao J, Andreucci JJ, Olson EN, Cossu G, McDermott JC (2005) Regulation of vertebrate myotome development by the p38 MAP kinase-MEF2 signaling pathway. Dev Biol 283:171-179

Dilworth FJ, Seaver KJ, Fishburn AL, Htet SL, Tapscott SJ (2004) *In vitro* transcription system delineates the distinct roles of the coactivators pCAF and p300 during MyoD/E47-dependent transactivation. Proc Natl Acad Sci USA 101:11593-11598

Dulloo AG, Seydoux J, Jacquet J (2004) Adaptive thermogenesis and uncoupling proteins: a reappraisal of their roles in fat metabolism and energy balance. Physiol Behav 83:587-602

Edmunds JW, Mahadevan LC (2004) MAP kinases as structural adaptors and enzymatic activators in transcription complexes. J Cell Sci 117:3715-3723

El Benna J, Han J, Park JW, Schmid E, Ulevitch RJ, Babior BM (1996) Activation of p38 in stimulated human neutrophils: phosphorylation of the oxidase component p47phox by p38 and ERK but not by JNK. Arch Biochem Biophys 334:395-400

Engel FB, Schebesta M, Duong MT, Lu G, Ren S, Madwed JB, Jiang H, Wang Y, Keating MT (2005) p38 MAP kinase inhibition enables proliferation of adult mammalian cardiomyocytes. Genes Dev 19:1175-1187

Engelman JA, Berg AH, Lewis RY, Lin A, Lisanti MP, Scherer PE (1999) Constitutively active mitogen-activated protein kinase kinase 6 (MKK6) or salicylate induces spontaneous 3T3-L1 adipogenesis. J Biol Chem 274:35630-35638

Engelman JA, Lisanti MP, Scherer PE (1998) Specific inhibitors of p38 mitogen-activated protein kinase block 3T3-L1 adipogenesis. J Biol Chem 273:32111-32120

Fan M, Rhee J, St-Pierre J, Handschin C, Puigserver P, Lin J, Jaeger S, Erdjument-Bromage H, Tempst P, Spiegelman BM (2004) Suppression of mitochondrial respiration through recruitment of p160 myb binding protein to PGC-1alpha: modulation by p38 MAPK. Genes Dev 18:278-289

Faust D, Dolado I, Cuadrado A, Oesch F, Weiss C, Nebreda AR, Dietrich C (2005) p38alpha MAPK is required for contact inhibition. Oncogene 24:7941-7945

Forcales SV, Puri PL (2005) Signaling to the chromatin during skeletal myogenesis: novel targets for pharmacological modulation of gene expression. Semin Cell Dev Biol 16:596-611

Furukawa F, Matsuzaki K, Mori S, Tahashi Y, Yoshida K, Sugano Y, Yamagata H, Matsushita M, Seki T, Inagaki Y, Nishizawa M, Fujisawa J, Inoue K (2003) p38 MAPK mediates fibrogenic signal through Smad3 phosphorylation in rat myofibroblasts. Hepatology 38:879-889

Galibert MD, Carreira S, Goding CR (2001) The Usf-1 transcription factor is a novel target for the stress-responsive p38 kinase and mediates UV-induced Tyrosinase expression. EMBO J 20:5022-5031

Gallo KA, Johnson GL (2002) Mixed-lineage kinase control of JNK and p38 MAPK pathways. Nat Rev Mol Cell Biol 3:663-672

Ge B, Gram H, Di Padova F, Huang B, New L, Ulevitch RJ, Luo Y, Han J (2002) MAPKK-independent activation of p38alpha mediated by TAB1-dependent autophosphorylation of p38alpha. Science 295:1291-1294

Goh KC, Haque SJ, Williams BR (1999) p38 MAP kinase is required for STAT1 serine phosphorylation and transcriptional activation induced by interferons. EMBO J 18:5601-5608

Gollob JA, Schnipper CP, Murphy EA, Ritz J, Frank DA (1999) The functional synergy between IL-12 and IL-2 involves p38 mitogen-activated protein kinase and is associated with the augmentation of STAT serine phosphorylation. J Immunol 162:4472-4481

Gomez del Arco P, Martinez-Martinez S, Maldonado JL, Ortega-Perez I, Redondo JM (2000) A role for the p38 MAP kinase pathway in the nuclear shuttling of NFATp. J Biol Chem 275:13872-13878

Guan HP, Ishizuka T, Chui PC, Lehrke M, Lazar MA (2005) Corepressors selectively control the transcriptional activity of PPARgamma in adipocytes. Genes Dev 19:453-461

Han J, Jiang Y, Li Z, Kravchenko VV, Ulevitch RJ (1997) Activation of the transcription factor MEF2C by the MAP kinase p38 in inflammation. Nature 386:296-299

Han J, Molkentin JD (2000) Regulation of MEF2 by p38 MAPK and its implication in cardiomyocyte biology. Trends Cardiovasc Med 10:19-22

Handschin C, Rhee J, Lin J, Tarr PT, Spiegelman BM (2003) An autoregulatory loop controls peroxisome proliferator-activated receptor gamma coactivator 1alpha expression in muscle. Proc Natl Acad Sci USA 100:7111-7116

Hasegawa M, Cuenda A, Spillantini MG, Thomas GM, Buee-Scherrer V, Cohen P, Goedert M (1999) Stress-activated protein kinase-3 interacts with the PDZ domain of alpha1-syntrophin. A mechanism for specific substrate recognition. J Biol Chem 274:12626-12631

Hazzalin CA, Mahadevan LC (2002) MAPK-regulated transcription: a continuously variable gene switch? Nat Rev Mol Cell Biol 3:30-40

Ho IC, Kim JH, Rooney JW, Spiegelman BM, Glimcher LH (1998) A potential role for the nuclear factor of activated T cells family of transcriptional regulatory proteins in adipogenesis. Proc Natl Acad Sci USA 95:15537-15541

Huang C, Ma WY, Maxiner A, Sun Y, Dong Z (1999) p38 kinase mediates UV-induced phosphorylation of p53 protein at serine 389. J Biol Chem 274:12229-12235

Jiang Y, Li Z, Schwarz EM, Lin A, Guan K, Ulevitch RJ, Han J (1997) Structure-function studies of p38 mitogen-activated protein kinase. Loop 12 influences substrate specificity and autophosphorylation, but not upstream kinase selection. J Biol Chem 272:11096-11102

Johnson GL, Lapadat R (2002) Mitogen-activated protein kinase pathways mediated by ERK, JNK, and p38 protein kinases. Science 298:1911-1912

Kadam S, Emerson BM (2003) Transcriptional specificity of human SWI/SNF BRG1 and BRM chromatin remodeling complexes. Mol Cell 11:377-389

Keren A, Tamir Y, Bengal E (2006) The p38 MAPK signaling pathway: A major regulator of skeletal muscle development. Mol Cell Endocrinol 252:224-230

Keyse SM (2000) Protein phosphatases and the regulation of mitogen-activated protein kinase signalling. Curr Opin Cell Biol 12:186-192

Khaled AR, Moor AN, Li A, Kim K, Ferris DK, Muegge K, Fisher RJ, Fliegel L, Durum SK (2001) Trophic factor withdrawal: p38 mitogen-activated protein kinase activates NHE1, which induces intracellular alkalinization. Mol Cell Biol 21:7545-7557

Kim GY, Mercer SE, Ewton DZ, Yan Z, Jin K, Friedman E (2002) The stress-activated protein kinases p38 alpha and JNK1 stabilize p21(Cip1) by phosphorylation. J Biol Chem 277:29792-29802

Kliewer SA, Willson TM (1998) The nuclear receptor PPARgamma - bigger than fat. Curr Opin Genet Dev 8:576-581

Kolodziejczyk SM, Wang L, Balazsi K, DeRepentigny Y, Kothary R, Megeney LA (1999) MEF2 is upregulated during cardiac hypertrophy and is required for normal post-natal growth of the myocardium. Curr Biol 9:1203-1206

Kramer RM, Roberts EF, Um SL, Borsch-Haubold AG, Watson SP, Fisher MJ, Jakubowski JA (1996) p38 mitogen-activated protein kinase phosphorylates cytosolic phospholipase A2 (cPLA2) in thrombin-stimulated platelets. Evidence that proline-directed phosphorylation is not required for mobilization of arachidonic acid by cPLA2. J Biol Chem 271:27723-27729

Ku NO, Azhar S, Omary MB (2002) Keratin 8 phosphorylation by p38 kinase regulates cellular keratin filament reorganization: modulation by a keratin 1-like disease causing mutation. J Biol Chem 277:10775-10782

Kuma Y, Campbell DG, Cuenda A (2004) Identification of glycogen synthase as a new substrate for stress-activated protein kinase 2b/p38beta. Biochem J 379:133-139

Kuma Y, Sabio G, Bain J, Shpiro N, Marquez R, Cuenda A (2005) BIRB796 inhibits all p38 MAPK isoforms *in vitro* and *in vivo*. J Biol Chem 280:19472-19479

Kyriakis JM, Avruch J (2001) Mammalian mitogen-activated protein kinase signal transduction pathways activated by stress and inflammation. Physiol Rev 81:807-869

Levine M, Tjian R (2003) Transcription regulation and animal diversity. Nature 424:147-151

Li Y, Jiang B, Ensign WY, Vogt PK, Han J (2000) Myogenic differentiation requires signalling through both phosphatidylinositol 3-kinase and p38 MAP kinase. Cell Signal 12:751-757

Liang F, Gardner DG (1999) Mechanical strain activates BNP gene transcription through a p38/NF-kappaB-dependent mechanism. J Clin Invest 104:1603-1612

Liang F, Lu S, Gardner DG (2000) Endothelin-dependent and -independent components of strain-activated brain natriuretic peptide gene transcription require extracellular signal regulated kinase and p38 mitogen-activated protein kinase. Hypertension 35:188-192

Liang Q, Molkentin JD (2003) Redefining the roles of p38 and JNK signaling in cardiac hypertrophy: dichotomy between cultured myocytes and animal models. J Mol Cell Cardiol 35:1385-1394

Liao P, Georgakopoulos D, Kovacs A, Zheng M, Lerner D, Pu H, Saffitz J, Chien K, Xiao RP, Kass DA, Wang Y (2001) The *in vivo* role of p38 MAP kinases in cardiac remodeling and restrictive cardiomyopathy. Proc Natl Acad Sci USA 98:12283-12288

Lin J, Wu H, Tarr PT, Zhang CY, Wu Z, Boss O, Michael LF, Puigserver P, Isotani E, Olson EN, Lowell BB, Bassel-Duby R, Spiegelman BM (2002) Transcriptional co-activator PGC-1 alpha drives the formation of slow-twitch muscle fibres. Nature 418:797-801

Lu HT, Yang DD, Wysk M, Gatti E, Mellman I, Davis RJ, Flavell RA (1999) Defective IL-12 production in mitogen-activated protein (MAP) kinase kinase 3 (Mkk3)-deficient mice. EMBO J 18:1845-1857

Luo S, Lee AS (2002) Requirement of the p38 mitogen-activated protein kinase signalling pathway for the induction of the 78 kDa glucose-regulated protein/immunoglobulin heavy-chain binding protein by azetidine stress: activating transcription factor 6 as a target for stress-induced phosphorylation. Biochem J 366:787-795

Lluis F, Ballestar E, Suelves M, Esteller M, Munoz-Canoves P (2005) E47 phosphorylation by p38 MAPK promotes MyoD/E47 association and muscle-specific gene transcription. EMBO J 24:974-984

Mansky KC, Sankar U, Han J, Ostrowski MC (2002) Microphthalmia transcription factor is a target of the p38 MAPK pathway in response to receptor activator of NF-kappa B ligand signaling. J Biol Chem 277:11077-11083

Matsumoto M, Kogawa M, Wada S, Takayanagi H, Tsujimoto M, Katayama S, Hisatake K, Nogi Y (2004) Essential role of p38 mitogen-activated protein kinase in cathepsin K

gene expression during osteoclastogenesis through association of NFATc1 and PU.1. J Biol Chem 279:45969-45979
McKinsey TA, Olson EN (1999) Cardiac hypertrophy: sorting out the circuitry. Curr Opin Genet Dev 9:267-274
McKinsey TA, Zhang CL, Olson EN (2002a) MEF2: a calcium-dependent regulator of cell division, differentiation and death. Trends Biochem Sci 27:40-47
McKinsey TA, Zhang CL, Olson EN (2002b) Signaling chromatin to make muscle. Curr Opin Cell Biol 14:763-772
Michael LF, Wu Z, Cheatham RB, Puigserver P, Adelmant G, Lehman JJ, Kelly DP, Spiegelman BM (2001) Restoration of insulin-sensitive glucose transporter (GLUT4) gene expression in muscle cells by the transcriptional coactivator PGC-1. Proc Natl Acad Sci USA 98:3820-3825
Mizumura K, Takeda K, Hashimoto S, Horie T, Ichijo H (2006) Identification of Op18/stathmin as a potential target of ASK1-p38 MAP kinase cascade. J Cell Physiol 206:363-370
Mudgett JS, Ding J, Guh-Siesel L, Chartrain NA, Yang L, Gopal S, Shen MM (2000) Essential role for p38alpha mitogen-activated protein kinase in placental angiogenesis. Proc Natl Acad Sci USA 97:10454-10459
Narlikar GJ, Fan HY, Kingston RE (2002) Cooperation between complexes that regulate chromatin structure and transcription. Cell 108:475-487
Nebreda AR, Porras A (2000) p38 MAP kinases: beyond the stress response. Trends Biochem Sci 25:257-260
Nemoto S, DiDonato JA, Lin A (1998) Coordinate regulation of IkappaB kinases by mitogen-activated protein kinase kinase kinase 1 and NF-kappaB-inducing kinase. Mol Cell Biol 18:7336-7343
Nishida K, Yamaguchi O, Hirotani S, Hikoso S, Higuchi Y, Watanabe T, Takeda T, Osuka S, Morita T, Kondoh G, Uno Y, Kashiwase K, Taniike M, Nakai A, Matsumura Y, Miyazaki J, Sudo T, Hongo K, Kusakari Y, Kurihara S, Chien KR, Takeda J, Hori M, Otsu K (2004) p38alpha mitogen-activated protein kinase plays a critical role in cardiomyocyte survival but not in cardiac hypertrophic growth in response to pressure overload. Mol Cell Biol 24:10611-10620
Olson JM, Hallahan AR (2004) p38 MAP kinase: a convergence point in cancer therapy. Trends Mol Med 10:125-129
Ornatsky OI, Cox DM, Tangirala P, Andreucci JJ, Quinn ZA, Wrana JL, Prywes R, Yu YT, McDermott JC (1999) Post-translational control of the MEF2A transcriptional regulatory protein. Nucleic Acids Res 27:2646-2654
Orphanides G, Reinberg D (2002) A unified theory of gene expression. Cell 108:439-451
Ouwens DM, de Ruiter ND, van der Zon GC, Carter AP, Schouten J, van der Burgt C, Kooistra K, Bos JL, Maassen JA, van Dam H (2002) Growth factors can activate ATF2 via a two-step mechanism: phosphorylation of Thr71 through the Ras-MEK-ERK pathway and of Thr69 through RalGDS-Src-p38. EMBO J 21:3782-3793
Parker CG, Hunt J, Diener K, McGinley M, Soriano B, Keesler GA, Bray J, Yao Z, Wang XS, Kohno T, Lichenstein HS (1998) Identification of stathmin as a novel substrate for p38 delta. Biochem Biophys Res Commun 249:791-796
Pearson G, Robinson F, Beers Gibson T, Xu BE, Karandikar M, Berman K, Cobb MH (2001) Mitogen-activated protein (MAP) kinase pathways: regulation and physiological functions. Endocr Rev 22:153-183

Pedraza-Alva G, Koulnis M, Charland C, Thornton T, Clements JL, Schlissel MS, Rincon M (2006) Activation of p38 MAP kinase by DNA double-strand breaks in V(D)J recombination induces a G2/M cell cycle checkpoint. EMBO J 25:763-773

Penn BH, Bergstrom DA, Dilworth FJ, Bengal E, Tapscott SJ (2004) A MyoD-generated feed-forward circuit temporally patterns gene expression during skeletal muscle differentiation. Genes Dev 18:2348-2353

Perdiguero E, Ruiz-Bonilla V, Gresh L, Hui L, Ballestar E, Sousa-Victor P, Baeza-Raja B, Jardi M, Bosch-Comas A, Esteller M, Caelles C, Serrano AL, Wagner EF, Munoz-Canoves P (2007) Genetic analysis of p38 MAP kinases in myogenesis: fundamental role of p38alpha in abrogating myoblast proliferation. EMBO J 7:1245-1256

Pesu M, Aittomaki S, Takaluoma K, Lagerstedt A, Silvennoinen O (2002) p38 Mitogen-activated protein kinase regulates interleukin-4-induced gene expression by stimulating STAT6-mediated transcription. J Biol Chem 277:38254-38261

Pilegaard H, Saltin B, Neufer PD (2003) Exercise induces transient transcriptional activation of the PGC-1alpha gene in human skeletal muscle. J Physiol 546:851-858

Porter CM, Havens MA, Clipstone NA (2000) Identification of amino acid residues and protein kinases involved in the regulation of NFATc subcellular localization. J Biol Chem 275:3543-3551

Puigserver P, Adelmant G, Wu Z, Fan M, Xu J, O'Malley B, Spiegelman BM (1999) Activation of PPARgamma coactivator-1 through transcription factor docking. Science 286:1368-1371

Puigserver P, Rhee J, Lin J, Wu Z, Yoon JC, Zhang CY, Krauss S, Mootha VK, Lowell BB, Spiegelman BM (2001) Cytokine stimulation of energy expenditure through p38 MAP kinase activation of PPARgamma coactivator-1. Mol Cell 8:971-982

Puigserver P, Spiegelman BM (2003) Peroxisome proliferator-activated receptor-gamma coactivator 1 alpha (PGC-1 alpha): transcriptional coactivator and metabolic regulator. Endocr Rev 24:78-90

Puigserver P, Wu Z, Park CW, Graves R, Wright M, Spiegelman BM (1998) A cold-inducible coactivator of nuclear receptors linked to adaptive thermogenesis. Cell 92:829-839

Ramsauer K, Sadzak I, Porras A, Pilz A, Nebreda AR, Decker T, Kovarik P (2002) p38 MAPK enhances STAT1-dependent transcription independently of Ser-727 phosphorylation. Proc Natl Acad Sci USA 99:12859-12864

Rosen ED, MacDougald OA (2006) Adipocyte differentiation from the inside out. Nat Rev Mol Cell Biol 7:885-896

Roux PP, Blenis J (2004) ERK and p38 MAPK-activated protein kinases: a family of protein kinases with diverse biological functions. Microbiol Mol Biol Rev 68:320-344

Sabio G, Arthur JS, Kuma Y, Peggie M, Carr J, Murray-Tait V, Centeno F, Goedert M, Morrice NA, Cuenda A (2005) p38gamma regulates the localisation of SAP97 in the cytoskeleton by modulating its interaction with GKAP. EMBO J 24:1134-1145

Sabio G, Reuver S, Feijoo C, Hasegawa M, Thomas GM, Centeno F, Kuhlendahl S, Leal-Ortiz S, Goedert M, Garner C, Cuenda A (2004) Stress- and mitogen-induced phosphorylation of the synapse-associated protein SAP90/PSD-95 by activation of SAPK3/p38gamma and ERK1/ERK2. Biochem J 380:19-30

Salvador JM, Mittelstadt PR, Guszczynski T, Copeland TD, Yamaguchi H, Appella E, Fornace AJ Jr, Ashwell JD (2005) Alternative p38 activation pathway mediated by T cell receptor-proximal tyrosine kinases. Nat Immunol 6:390-395

Sartorelli V, Caretti G (2005) Mechanisms underlying the transcriptional regulation of skeletal myogenesis. Curr Opin Genet Dev 15:528-535

Schieven GL (2005) The biology of p38 kinase: a central role in inflammation. Curr Top Med Chem 5:921-928

Schoorlemmer J, Goldfarb M (2002) Fibroblast growth factor homologous factors and the islet brain-2 scaffold protein regulate activation of a stress-activated protein kinase. J Biol Chem 277:49111-49119

Shen YH, Zhang L, Gan Y, Wang X, Wang J, LeMaire SA, Coselli JS, Wang XL (2006) Up-regulation of PTEN (phosphatase and tensin homolog deleted on chromosome ten) mediates p38 MAPK stress signal-induced inhibition of insulin signaling. A cross-talk between stress signaling and insulin signaling in resistin-treated human endothelial cells. J Biol Chem 281:7727-7736

Sii-Felice K, Pouponnot C, Gillet S, Lecoin L, Girault JA, Eychene A, Felder-Schmittbuhl MP (2005) MafA transcription factor is phosphorylated by p38 MAP kinase. FEBS Lett 579:3547-3554

Simone C, Forcales SV, Hill DA, Imbalzano AN, Latella L, Puri PL (2004) p38 pathway targets SWI-SNF chromatin-remodeling complex to muscle-specific loci. Nat Genet 36:738-743

Stassen M, Klein M, Becker M, Bopp T, Neudorfl C, Richter C, Heib V, Klein-Hessling S, Serfling E, Schild H, Schmitt E (2006) p38 MAP kinase drives the expression of mast cell-derived IL-9 via activation of the transcription factor GATA-1. Mol Immunol PAGES

Suelves M, Lluis F, Ruiz V, Nebreda AR, Munoz-Canoves P (2004) Phosphorylation of MRF4 transactivation domain by p38 mediates repression of specific myogenic genes. EMBO J 23:365-375

Suzuki H, Wu J, Hossain K, Ohhata T, Du J, Akhand AA, Hayakawa A, Kimura H, Hagiwara M, Nakashima I (2003) Involvement of MKK6 in TCRalphabeta(int)CD69lo: a target population for apoptotic cell death in thymocytes. FASEB J 17:1538-1540

Tamura K, Sudo T, Senftleben U, Dadak AM, Johnson R, Karin M (2000) Requirement for p38alpha in erythropoietin expression: a role for stress kinases in erythropoiesis. Cell 102:221-231

Tamura S, Hanada M, Ohnishi M, Katsura K, Sasaki M, Kobayashi T (2002) Regulation of stress-activated protein kinase signaling pathways by protein phosphatases. Eur J Biochem 269:1060-1066

Tanaka N, Kamanaka M, Enslen H, Dong C, Wysk M, Davis RJ, Flavell RA (2002) Differential involvement of p38 mitogen-activated protein kinase kinases MKK3 and MKK6 in T-cell apoptosis. EMBO Rep 3:785-791

Tanos T, Marinissen MJ, Leskow FC, Hochbaum D, Martinetto H, Gutkind JS, Coso OA (2005) Phosphorylation of c-Fos by members of the p38 MAPK family. Role in the AP-1 response to UV light. J Biol Chem 280:18842-18852

Tanoue T, Yamamoto T, Maeda R, Nishida E (2001) A Novel MAPK phosphatase MKP-7 acts preferentially on JNK/SAPK and p38 alpha and beta MAPKs. J Biol Chem 276:26629-26639

Thomas GM, Huganir RL (2004) MAPK cascade signalling and synaptic plasticity. Nat Rev Neurosci 5:173-183

Wang FM, Hu T, Tan H, Zhou XD (2006) p38 Mitogen-activated protein kinase affects transforming growth factor-beta/Smad signaling in human dental pulp cells. Mol Cell Biochem 291:49-54

Wang X, McGowan CH, Zhao M, He L, Downey JS, Fearns C, Wang Y, Huang S, Han J (2000) Involvement of the MKK6-p38gamma cascade in gamma-radiation-induced cell cycle arrest. Mol Cell Biol 20:4543-4552

Wang XZ, Ron D (1996) Stress-induced phosphorylation and activation of the transcription factor CHOP (GADD153) by p38 MAP kinase. Science 272:1347-1348

Wang Y, Huang S, Sah VP, Ross J Jr, Brown JH, Han J, Chien KR (1998) Cardiac muscle cell hypertrophy and apoptosis induced by distinct members of the p38 mitogen-activated protein kinase family. J Biol Chem 273:2161-2168

Weston AD, Sampaio AV, Ridgeway AG, Underhill TM (2003) Inhibition of p38 MAPK signaling promotes late stages of myogenesis. J Cell Sci 116:2885-2893

Whitmarsh AJ, Davis RJ (2000) Regulation of transcription factor function by phosphorylation. Cell Mol Life Sci 57:1172-1183

Whitmarsh AJ, Yang SH, Su MS, Sharrocks AD, Davis RJ (1997) Role of p38 and JNK mitogen-activated protein kinases in the activation of ternary complex factors. Mol Cell Biol 17:2360-2371

Wilkins BJ, De Windt LJ, Bueno OF, Braz JC, Glascock BJ, Kimball TF, Molkentin JD (2002) Targeted disruption of NFATc3, but not NFATc4, reveals an intrinsic defect in calcineurin-mediated cardiac hypertrophic growth. Mol Cell Biol 22:7603-7613

Wilkins BJ, Molkentin JD (2004) Calcium-calcineurin signaling in the regulation of cardiac hypertrophy. Biochem Biophys Res Commun 322:1178-1191

Williamson EA, Williamson IK, Chumakov AM, Friedman AD, Koeffler HP (2005) CCAAT/enhancer binding protein epsilon: changes in function upon phosphorylation by p38 MAP kinase. Blood 105:3841-3847

Wu Z, Puigserver P, Andersson U, Zhang C, Adelmant G, Mootha V, Troy A, Cinti S, Lowell B, Scarpulla RC, Spiegelman BM (1999) Mechanisms controlling mitochondrial biogenesis and respiration through the thermogenic coactivator PGC-1. Cell 98:115-124

Wu Z, Woodring P, Bhakta K, Tamura K, Wen F, Feramisco JR, Karin M, Wang JY, Puri PL (2000) p38 and extracellular signal-regulated kinases regulate the myogenic program at multiple steps. Mol Cell Biol 20:3951-3964

Wu Z, Xie Y, Bucher NL, Farmer SR (1995) Conditional ectopic expression of C/EBP beta in NIH-3T3 cells induces PPAR gamma and stimulates adipogenesis. Genes Dev 9:2350-2363

Wysk M, Yang DD, Lu HT, Flavell RA, Davis RJ (1999) Requirement of mitogen-activated protein kinase kinase 3 (MKK3) for tumor necrosis factor-induced cytokine expression. Proc Natl Acad Sci USA 96:3763-3768

Xu B, Bhattacharjee A, Roy B, Xu HM, Anthony D, Frank DA, Feldman GM, Cathcart MK (2003) Interleukin-13 induction of 15-lipoxygenase gene expression requires p38 mitogen-activated protein kinase-mediated serine 727 phosphorylation of Stat1 and Stat3. Mol Cell Biol 23:3918-3928

Yamaguchi H, Minopoli G, Demidov ON, Chatterjee DK, Anderson CW, Durell SR, Appella E (2005) Substrate specificity of the human protein phosphatase 2Cdelta, Wip1. Biochemistry 44:5285-5294

Yang SH, Galanis A, Sharrocks AD (1999) Targeting of p38 mitogen-activated protein kinases to MEF2 transcription factors. Mol Cell Biol 19:4028-4038

Yang SH, Whitmarsh AJ, Davis RJ, Sharrocks AD (1998) Differential targeting of MAP kinases to the ETS-domain transcription factor Elk-1. EMBO J 17:1740-1749

Yang TT, Xiong Q, Enslen H, Davis RJ, Chow CW (2002) Phosphorylation of NFATc4 by p38 mitogen-activated protein kinases. Mol Cell Biol 22:3892-3904

Yee AS, Paulson EK, McDevitt MA, Rieger-Christ K, Summerhayes I, Berasi SP, Kim J, Huang CY, Zhang X (2004) The HBP1 transcriptional repressor and the p38 MAP kinase: unlikely partners in G1 regulation and tumor suppression. Gene 336:1-13

Yu M, Blomstrand E, Chibalin AV, Wallberg-Henriksson H, Zierath JR, Krook A (2001) Exercise-associated differences in an array of proteins involved in signal transduction and glucose transport. J Appl Physiol 90:29-34

Zarubin T, Han J (2005) Activation and signaling of the p38 MAP kinase pathway. Cell Res 15:11-18

Zechner D, Thuerauf DJ, Hanford DS, McDonough PM, Glembotski CC (1997) A role for the p38 mitogen-activated protein kinase pathway in myocardial cell growth, sarcomeric organization, and cardiac-specific gene expression. J Cell Biol 139:115-127

Zetser A, Gredinger E, Bengal E (1999) p38 mitogen-activated protein kinase pathway promotes skeletal muscle differentiation. Participation of the Mef2c transcription factor. J Biol Chem 274:5193-5200

Zhang S, Weinheimer C, Courtois M, Kovacs A, Zhang CE, Cheng AM, Wang Y, Muslin AJ (2003) The role of the Grb2-p38 MAPK signaling pathway in cardiac hypertrophy and fibrosis. J Clin Invest 111:833-841

Zhao M, New L, Kravchenko VV, Kato Y, Gram H, Di Padova F, Olson EN, Ulevitch RJ, Han J (1999) Regulation of the MEF2 family of transcription factors by p38. Mol Cell Biol 19:21-30

Perdiguero, Eusebio
Program on Differentiation and Cancer, Center for Genomic Regulation (CRG), UPF/PRBB, Barcelona and CIBER de Enfermedades Neurodegenerativas (CIBERNED), Spain

Muñoz-Cánoves, Pura
Program on Differentiation and Cancer, Center for Genomic Regulation (CRG), UPF/PRBB, Barcelona and CIBER de Enfermedades Neurodegenerativas (CIBERNED), Spain
pura.munoz@crg.es

Regulation of gene expression in response to osmostress by the yeast stress-activated protein kinase Hog1

Eulàlia de Nadal and Francesc Posas

Abstract

Exposure of cells to an increase in extracellular osmolarity results in rapid activation of a highly conserved stress-activated protein kinase (SAPK) pathway, known in the budding yeast *Saccharomyces cerevisiae* as the HOG pathway. Activation of the HOG pathway is essential for the induction of adaptive responses required for cell survival upon osmostress. Regulation of gene expression is a major adaptive response required for cell survival in response to osmostress. The HOG pathway controls gene expression through several mechanisms, such as direct regulation of transcription factor activities, the recruitment of the RNA Polymerase II at osmoresponsive promoters and chromatin modification via recruitment of the Rpd3 histone deacetylase complex. Moreover, recent studies have shown that SAPKs can be recruited to the coding regions of osmostress genes acting as a transcriptional elongation factor. This review summarizes the current knowledge on the regulation of gene expression by the SAPK Hog1 and discusses new insights from yeast that could be relevant to mammals.

1 Introduction

MAP kinase (MAPK) cascades are common signaling modules found in both higher and lower eukaryotic organisms and are composed of three consecutively activated tiers of kinases: MAPKKK, MAPKK, and MAPK. Eukaryotic organisms contain multiple MAPK families organized in discrete cascades. One of these is the stress-activated protein kinase (SAPK) cascade that plays an essential role for proper cell adaptation to extracellular stimuli (Kyriakis and Avruch 2001). A prototype of the SAPK family is the yeast *Saccharomyces cerevisiae* Hog1 MAPK, the homolog of the mammalian p38, which specifically responds to increases in extracellular osmolarity and is required for cell survival under these conditions. Several lines of evidence clearly demonstrated a functional preservation of MAP Kinases and adaptative responses from yeast to mammals (reviewed in Sheikh-Hamad and Gustin 2004). Thus, new insights from yeast could be relevant to understand the osmostress response in mammals.

Topics in Current Genetics, Vol. 20
F. Posas, A. R. Nebreda (Eds.): Stress-Activated Protein Kinases
DOI 10.1007/4735_2007_0244 / Published online: 1 August 2007

Adaptation to osmostress requires the accumulation of intracellular osmolytes such as glycerol to increase the total intracellular solute concentration, thereby providing osmotic stabilization. The HOG (High Osmolarity Glycerol) pathway is the signal transduction network necessary to achieve this end. However, osmostress not only induces osmolyte accumulation, but also has a great impact on cellular physiology, such as cytoskeleton reorganization, changes in cell-wall dynamics, alteration of ion homeostasis, metabolic adjustments and cell-cycle arrest, as well as a very notable effect on gene expression (reviewed in Hohmann 2002). In this review, we focus on the role of Hog1 SAPK in the regulation of gene expression upon osmostress.

2 Regulation of HOG signaling

There are two branches of the HOG pathway that can lead to the phosphorylation and activation of the MAPKK Pbs2 and its cognate MAPK Hog1 (Fig. 1) (Dard and Peter 2006; de Nadal et al. 2002; Hohmann 2002; Westfall et al. 2004). The first branch leads to the activation of the Ssk2 and Ssk22 MAPKKKs that sequentially activate Pbs2. This branch involves a "two-component" osmosensor, composed of the Sln1-Ypd-Ssk1 proteins (Maeda et al. 1994; Posas et al. 1996; Posas and Saito 1998). The Sln1 transmembrane protein has intrinsic histidine kinase activity and, using a phospho-relay mechanism involving the Ypd1 and Ssk1 proteins, it controls the activity of Ssk1, which in turn interacts with and regulates the Ssk2 and Ssk22 MAPKKKs. The second branch involves the transmembrane protein Sho1 and possibly the Msb2 protein (a mucin-like protein with a single transmembrane segment) (Cullen et al. 2004; Maeda et al. 1995; O'Rourke and Herskowitz 2002), although additional components still need to be identified. Sho1-dependent signaling requires the G-protein Cdc42, the Ste50 protein and the Ste20 and Cla4 kinases, members of the PAK (p21-activate protein kinase) family (Dan et al. 2001; Reiser et al. 2000; Posas et al. 1998; Raitt et al. 2000; Truckses et al. 2006; Winters and Pryciak 2005). When stimulated by osmostress, the Sho1 branch activates the Ste11 MAPKKK and subsequently Pbs2 (Posas and Saito 1997). Although the basic mechanism of Pbs2 activation by the Sho1 module remains unclear, it has recently been described that Cdc42, Ste50, and Sho1 act as adaptor proteins that control the flow of the osmostress signal from Ste20/Cla4 to Ste11, and then on to Pbs2 (Tatebayashi et al. 2006; Zarrinpar et al. 2004).

Hog1 is phosphorylated by Pbs2 within seconds of exposing cells to osmostress. The magnitude and duration of signaling through SAPKs are critical determinants for Hog1-mediated biological effects and depend on the severity of the stress. In response to a mild osmotic shock (e.g. 0.4 M NaCl), Hog1 phosphorylation is rapid and very transient. In response to a more severe stress (e.g. 1 M NaCl), signaling is stronger and clearly more sustained (Hohmann 2002; Klipp et al. 2005). In addition, the HOG pathway is controlled by specific feedback mechanisms that downregulate signaling, both under basal conditions and during adaptation. Two major families of phosphatases interact with and inactivate Hog1:

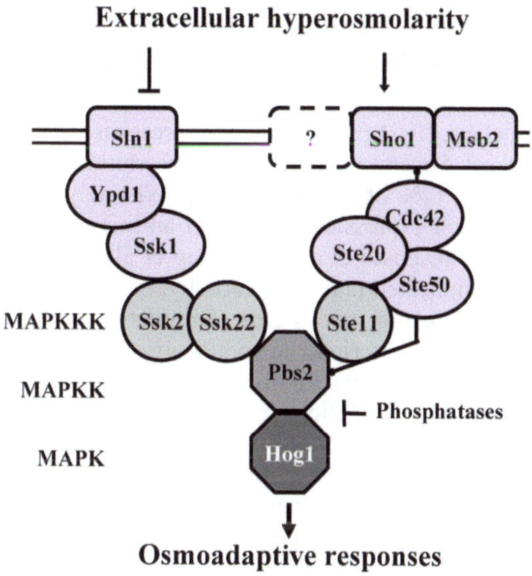

Fig. 1. Schematic diagram of the yeast HOG pathway. Pbs2 integrates signals from two major independent upstream osmosensing mechanisms, which leads to the activation of specific MAPKKKs. Under osmostress, activated Pbs2 activates the Hog1 MAPK, which induces a set of osmoadaptive response.

the Ser/Thr protein phosphatase type 2C (mainly Ptc1) and the protein tyrosine phosphatases (Ptp2 and Ptp3) (Jacoby et al. 1997; Warmka et al. 2001; Wurgler-Murphy et al. 1997). Recent data suggested that the main role of the phosphatases is to constantly counteract pathway activation to reduce noise instead of providing a pathway-intrinsic feedback loop. Thus, additional systems apart from phosphatases must exist to account for the downregulation of signaling (Klipp et al. 2005).

Studies using a Hog1-green fluorescent protein (GFP) fusion revealed that under basal conditions the MAPK Hog1 cycles between cytoplasmic and nuclear compartments. When activated, Hog1 concentrates in the nucleus from which it is reexported after adaptation. Hog1 translocation occurs very rapidly, is transient and correlates with its phosphorylation. The duration of Hog1 nuclear residence is modulated by the presence of nuclear targets (Ferrigno et al. 1998; Reiser et al. 1999). Moreover, the protein tyrosine phosphatases Ptp2 and Ptp3 regulate the MAPK localization by binding Hog1. It has been reported that Ptp2 is a nuclear tether for Hog1 while Ptp3 is a cytoplasmic anchor (Mattison and Ota 2000).

3 The Hog1 MAPK as a central component of transcription activation upon osmostress

Once inside the nucleus, Hog1 regulates gene expression allowing cells to adapt to osmostress conditions. Several genome-wide transcription studies revealed that a large number of genes (7% of the genome approximately) show significant but transient changes in their expression levels after an osmotic shock (Causton et al. 2001; Gasch et al. 2000; Posas et al. 2000; Rep et al. 2000). These osmostress-regulated genes are implicated mainly in carbohydrate metabolism, general stress protection, protein production and signal transduction (reviewed in Hohmann 2002). Significantly, the transcriptional induction of the majority of genes that strongly respond to osmostress (approximately 75%) is highly or fully dependent on the presence of the MAPK, indicating that the HOG pathway plays a key role in global gene regulation under osmostress conditions.

It has been described that initial exposure of cells to osmostress causes nonspecific dissociation of many proteins from chromatin. Indeed, as an immediate response to osmostress, Hog1 reconstitutes the transcriptional capacity of the cell by modulating the ionic balance inside the cell via the control of both a Na^{+}-H^{+} antiporter (Nha1) and a potassium channel (Tok1). Thus, it is proposed that Hog1-stimulated Na^{+} efflux allows DNA binding proteins to reassociate with DNA as a short-term response (Proft and Struhl 2004).

There is no unique mechanism by which Hog1 modulates gene expression. Once in the nucleus, Hog1 directly phosphorylates and modifies the transcriptional properties of several DNA binding proteins. However, the finding that Hog1 itself associates with chromatin has added a new and exciting dimension to gene regulation by signaling kinases. Next, we will discuss in detail each of the distinct proposed mechanisms by which the MAPK Hog1 modulates gene expression during osmostress signaling.

4 Transcriptional regulators downstream of the HOG pathway

At least five transcription factors have been reported to be controlled by the Hog1 MAPK: the bZIP protein Sko1/Acr1, the MADS box protein Smp1, the Hot1 protein and the redundant zinc finger proteins Msn2 and Msn4. These factors are unrelated, and the mechanisms by which Hog1 regulates their function may differ from one to another. Global gene expression analyses carried out to dissect the specific roles of these transcription factors have revealed that each one of them can account for only a small subset of the osmostress inducible genes. Actually, additional transcription factors probably remain to be described for the regulation of gene expression upon stress by Hog1.

4.1 The bZIP protein Sko1

Sko1, also known as Acr1, is a protein belonging to the ATF/CREB family of AP1-related transcription factors that contains a leucine zipper for dimerization and an adjacent basic transcription activation domain (Nehlin et al. 1992; Vincent and Struhl 1992). The analysis of the *ENA1* promoter, which encodes a plasma membrane Na-export pump, showed that regulation by the HOG pathway was mediated through a cyclic AMP response element (CRE)-like sequence that was bound by the bZIP transcriptional factor Sko1 protein (Proft and Serrano 1999). The significant role of Sko1 in mediating Hog1-dependent response was confirmed by other studies, which identified upstream CRE sequences in the *GRE2* and *HAL1* genes (Marquez et al. 1998; Rep et al. 2001). Actually, by combining chromatin immunoprecipitation (ChIP) and microarrays analyses, it has been estimated that yeast cells contain approximately forty Sko1 target promoters (Proft et al. 2005).

Genetic data indicated that Sko1 represses stress gene expression from CRE sites by recruiting the general corepressor complex Ssn6-Tup1 to target promoters. Moreover, Sko1 physically interacts with Tup1 (Pascual-Ahuir et al. 2001). Induction of Sko1 dependent osmo-regulated genes involves the release of repression, and this process is completely dependent on the HOG pathway (Garcia-Gimeno and Struhl 2000; Proft and Serrano 1999). *In vivo* coprecipitation and phosphorylation studies showed that Sko1 and Hog1 interact and that Sko1 is phosphorylated upon osmostress by the MAPK (Proft et al. 2001). Hence, the role of Sko1 in regulation of gene expression upon stress goes further than recruiting Ssn6-Tup1 to repressed promoters. Indeed, Hog1 dependent phosphorylation switches Sko1 from a repressor into an activator. During osmostress, Tup1 remains bound to target promoters together with Sko1. Phosphorylation of Sko1 by Hog1 modifies its association with Tup1-Ssn6 (Proft et al. 2001) and allows for the recruitment of the remodeling complexes SAGA and SWI/SNF to osmostress inducible promoters to activate gene expression (Proft and Struhl 2002).

4.2 The MADS box protein Smp1

Smp1 is a member of the MEF2C family of transcription factors (Yu et al. 1992) reported to act downstream of Hog1. It was identified by a genetic screen that isolated genes whose overexpression was able to induce *STL1* expression (see below). *In vivo* coprecipitation and phosphorylation studies showed that Smp1 and Hog1 interacted and that Smp1 was directly phosphorylated upon osmostress in a Hog1-dependent manner. Phosphorylation of Smp1 by the MAPK is important for its function, since a mutant allele unable to be phosphorylated by the MAPK displayed impaired stress gene expression. Furthermore, the same study reported that Smp1 might play an important role not only in osmostress responses, but also in a new function for the Hog1 MAPK required for cell survival in stationary phase (de Nadal et al. 2003). In mammals, regulation of MEF2A and MEF2C factors has been shown to be under the control of the p38 MAPK (Kyriakis and Avruch 2001;

McKinsey et al. 2002). There is evidence indicating a crucial role for p38 MAPK signaling via the MEF2 transcriptional regulators during early mammalian somite development and myotome formation (de Angelis et al. 2005; Berkes and Tapscott 2005).

4.3 The zinc finger proteins Msn2 and Msn4

Several stressful conditions, including osmostress, induce the expression of genes controlled through the so-called Stress Response Element (STRE) via the transcriptional regulators Msn2 and Msn4 (Martinez-Pastor et al. 1996; Schmitt and McEntee 1996). Under normal conditions, these transcription factors reside in the cytosol. However, upon environmental stress or when glucose is suddenly withdrawn, they are rapidly concentrated to the nucleus. Localization and activity of Msn2 is controlled by the protein kinase A, the protein kinase Snfl and the protein phosphatase 1 by modulating the phosphorylation state of several sites in its nuclear localization signal (De, V et al. 2005; Gorner et al. 2002; Gorner et al. 1998). In addition, it has been demonstrated that Msn2 activity is also regulated by degradation (Durchschlag et al. 2004). A more detailed description on the function and control of these transcription factors can be found in a chapter of this volume (G. Ammerer).

Global expression analyses have uncovered a remarkable correlation between *MSN2*/*MSN4* and *HOG1*-dependent gene expression (Rep et al. 2000). For instance, expression of genes such as *CTT1*, which encodes cytosolic catalase, or *HSP12*, a small heat shock protein, is strongly induced in response to osmotic stress in a Hogl and Msn2/Msn4 dependent manner. Actually, transcription regulation of these genes is thought to be achieved through controlled recruitment of Msn2 to respective target promoters (see below).

4.4 The Hot1 transcription factor

Hot1 (High-Osmolarity-induced Transcription) is a nuclear transcription factor that was identified in a two-hybrid screening for proteins interacting with Hogl (Rep et al. 1999). It was reported that Hot1 affects expression of a small subset of Hogl dependent genes, including *GPD1* and *GPP2*, involved in glycerol biosynthesis, and *STL1*, which encodes a glycerol proton symporter (Ferreira et al. 2005; Rep et al. 2000).

As revealed by ChIP analyses, Hot1 binds to the *STL1* promoter only under osmostress and needs the MAPK Hogl to associate with the chromatin. On the other hand, binding of Hot1 to the *GPD1* promoter occurs under all conditions, although the level of the transcription factor at the promoter increases under stress. It is worth noting that Hot1 is necessary for the recruitment of Hogl to Hot1-responsive genes (Alepuz et al. 2001). However, although the Hot1 transcription factor is phosphorylated by the MAPK, its phosphorylation does not seem to be a crucial step in transcriptional activation (Alepuz et al. 2003). Thus, in addition to the role of Hog1 in the direct modification of transcription factor activities, there must exist other mechanisms by which the MAPK modulates gene expression upon osmostress (see below).

5 Hog1 is part of the transcription complexes at the promoters of osmostress genes

One of the most remarkable discoveries about how the MAPK Hog1 modulates gene expression was the fact that during stress induction the Hog1 kinase itself cross-linked with several target promoters such as *GPD1*, *STL1*, *HSP12*, and *CTT1*. Thus, while in a more traditional scenario, a MAP kinase would control the activity of a transcriptional regulator, the appearance of Hog1 at target promoters indicated that Hog1 itself might take part in the activation process (Alepuz et al. 2001; Chellappan 2001). The association of Hog1 to chromatin depends on its activity but not on its nuclear accumulation because the artificial increase of the amount of nuclear Hog1 by adding a nuclear localization signal on the MAPK did not result in enhanced chromatin association (Alepuz et al. 2001). Actually, recruitment of the MAPK to target promoters is mediated through physical interaction with specific transcription factors that function as anchors to chromatin. For instance, recruitment of Hog1 to the *STL1* promoter depends on the activator Hot1, whereas recruitment of the kinase to the *CTT1* promoter depends on the transcriptions factors Msn2 and Msn4 (Alepuz et al. 2001). Recent evidence has shown that binding of Hog1 to stress promoters is general but restricted to osmoresponsive genes (Pascual-Ahuir et al. 2006; Pokholok et al. 2006; Proft et al. 2006) and that other MAP kinases such as Fus3 and Kss1 also are recruited to chromatin (Pokholok et al. 2006).

Recruitment of the active Hog1 MAPK by the Hot1 activator is critical for gene expression. However, phosphorylation of Hot1 or Msn2 and Msn4 by the MAPK seems not to be essential for gene expression. The observation that the kinase activity of Hog1 is needed for transcriptional activation but that the phosphorylation of transcription factors is not an absolute requirement indicated that Hog1 must induce activation of gene expression by a mechanism other than phosphorylation of the activator (Alepuz et al. 2003). In fact, it has been reported that a critical step to induce gene expression upon stress is the recruitment of the RNA Polymerase II complex by Hog1 to the promoters. Similar to the MAPK and the transcriptional regulators, the RNA Polymerase II machinery is recruited to osmoresponsive genes in response to stress and this association depends on both active Hog1 and the presence of the specific transcription factors. Furthermore, Hog1 interacts with the RNA Polymerase II holoenzyme upon osmostress. The functional relevance of the interaction of Hog1 with RNA Polymerase II has been further exemplified by the fact that artificial recruitment of Hog1 to DNA is able to induce gene expression upon stress (Alepuz et al. 2003).

The p38 MAPK interacted with the core of the RNA Polymerase II in human cells (Alepuz et al. 2003). These data suggested that a novel conserved mechanism for regulation of gene transcription mediated by stress-activated MAPKs could exist among eukaryotic cells. Indeed, during skeletal myogenesis, p38 kinases were recruited to the chromatin and selectively targeted the SWI-SNF chromatin-remodeling complex to muscle-regulatory elements (Simone et al. 2004).

6 Regulation of chromatin remodeling by the Hog1 MAPK

The specific chromatin association of Hogl to stress-responsive promoters suggested that the MAPK could be playing a role in chromatin modification (Alepuz et al. 2001; Chellappan 2001). A genetic screen designed to identify mutations that render cells osmosensitive at high osmolarity showed that the Rpd3 histone deacetylase complex plays an important role in osmostress gene expression (de Nadal et al. 2004). Rpd3 is a member of a family of five related histone deacetylases in yeast that also comprises of Hda1, Hos1, Hos2, and Hos3 that has been demonstrated to regulate the expression of a large number of genes (Bernstein et al. 2000; Sabet et al. 2004). There are two known Rpd3 complexes that share a core of three subunits including Rpd3, Sin3, and Ume1. Whereas the large Rpd3L complex is recruited to the promoters and functions in transcription initiation, the small Rpd3S functions together with the histone methyltransferase Set2 in the elongation process (Carrozza et al. 2005; Keogh et al. 2005). It has been described that cells lacking genes that encode the Rpd3 histone deacetylase, and other components of the large Rpd3L complex, rendered cells osmosensitive and showed compromised expression of osmostress genes. Actually, microarrays analyses showed that more than 90% of genes induced in response to stress that are Hogl dependent have a significant reduction in expression in a *RPD3* mutant strain. Moreover, Hogl interacts physically with Rpd3 and, on stress, targets the deacetylase to specific osmostress-responsive gene promoters. Binding of the Rpd3/Sin3 complex to stress-specific promoters is dependent on the presence of Hogl. Therefore, upon osmostress, Hogl is recruited to osmoresponsive promoters by specific transcription factors. Hogl binding then facilitates direct recruitment of the Rpd3 deacetylase complex to these promoters, leading to histone deacetylation, entry of RNA Polymerase II and induction of gene expression (de Nadal et al. 2004).

Although histone deacetylation has been traditionally associated with the repression of gene expression (Robyr et al. 2002), the histone deacetylase Hos2, that belongs to the same family as Rpd3, was shown to induce transcription of the *INO* and *GAL* genes (Wang et al. 2002). In response to osmostress, as well as to heat stress, it seems that the Rpd3 complex functions positively to induce transcription. Moreover, there are many examples of particular genes in which a decrease in histone acetylation is associated with transcription induction (Bernstein et al. 2000) (Deckert and Struhl 2001). On the other hand, the role of Rpd3 in osmostress promoters might not be restricted to altering chromatin structure but might also provide a unique binding surface or recognition motifs for the recruitment of activators, as it has been proposed for acetylation and deacetylation in gene expression (Millar and Grunstein 2006).

In addition to Rpd3 complex, two major complexes have been identified as essential for cell viability in high osmolarity conditions: the SAGA and Mediator complexes. Several observations indicate that whereas Mediator is crucial for proper gene induction upon both mild and high osmostress conditions, the role of SAGA at the promoters seems to be dependent on the extent of the osmostress. This leads to different promoter regulation as a function of the extent of the stimuli perceived by the cell (Zapater, Posas, and de Nadal; unpublished data). Therefore, recruitment of chromatin-remodeling/modifying activities such Rpd3 (de Nadal et al. 2004), SWI/SNF (Proft and Struhl 2002), SAGA, or Mediator (Zapater, Posas and de Nadal; unpublished data) by Hog1 may lead to alteration of local chromatin structure and/or recruitment of other coactivators, facilitating the additional role of Hog1 in recruitment and activation of the RNA Polymerase II holoenzyme.

Recent evidence has shown that p38 also modulates gene expression through remodeling the structure of chromatin (reviewed in (Clayton and Mahadevan, 2003; de Nadal et al. 2002; Edmunds and Mahadevan 2004)). For instance, p38 has been shown to recruit the SWI/SNF complex to transcriptionally activate several loci during skeletal myogenesis (Simone et al. 2004). Likewise, p38, through the downstream mitogen and stress-activated kinase 1/2 (MSK1/2), elicits phosphorylation of the high-mobility-group protein HMG-14 at Ser6 and of distinct pools of histone H3 at Ser10 or Ser28 concomitant with gene induction (Dyson et al. 2005; Soloaga et al. 2003).

7 Hog1 MAPK and transcription elongation

In addition to its various functions during transcription initiation, Hog1 also behaves as a transcriptional elongation factor that is selective for genes induced in osmostress conditions. Elongation factors are defined by their ability to associate with or affect the activity of the transcriptional elongation complex and they can affect the elongation rate or processivity of RNA Polymerase II or mRNA maturation (Mason and Struhl 2005); (Sims III et al. 2004). Recently, it has been described that Hog1 interacted with elongating RNA Polymerase II (phosphorylated at serine 2 and 5 of the C-terminal domain) as well as with general components of the transcription elongation complex upon osmostress (Proft et al. 2006). Additionally, genome-wide location analysis have revealed binding of stress-activated Hog1 to promoters but also to ORF regions of osmoresponsive genes (Pascual-Ahuir et al. 2006; Pokholok et al. 2006; Proft et al. 2006). Thus, Hog1 is both recruited to promoters and coding regions of genes whose expression is induced upon osmotic shock and selectively travels with elongating RNA Polymerase II through stress-responsive genes. Binding of the MAPK to coding regions is independent on the promoter bound-specific transcription factors but depends on the 3'UTR region of osmostress genes. However, the mechanism by which the MAPK is recruited to the 3'regions of osmoresponsive gene remains unknown (Proft et al. 2006).

By uncoupling Hog1-dependent transcription initiation from transcription elongation, it has been demonstrated that Hog1 at coding regions is essential for proper accumulation of stress-responsive mRNAs. Furthermore, Hog1 is important for increased association of RNA Polymerase II and elongation factors at coding regions, strongly suggesting that it directly affects the process of RNA Polymerase II elongation (Proft et al. 2006). Several lines of evidence have suggested that the requirement of the MAPK Hog1 in elongation could be important for the function of the transcriptional elongation complex at the initial stages of elongation. Similarly to transcription initiation, one candidate for such a function could be the modification of the nucleosomes occupying the stress-responsive ORF regions during the initial rounds of transcription. Alternatively, Hog1, perhaps by phosphorylation of some unknown substrate, may mediate the correct assembly of processing activities required for efficient mRNA production in response to stress (Proft et al. 2006; Edmunds and Mahadevan 2006).

Different studies have highlighted the role for the MAPK in the posttranscriptional control of gene expression. For instance, inhibition of the HOG pathway leads to destabilization on the ARE-bearing transcript (AU-rich element, present in 3'UTR transcripts) (Vasudevan and Peltz 2001). Likewise, it has been demonstrated that Hog1 regulates 3'-UTR-mediated translation by modulating recruitment of Pab1 and Pub1, which can interact with the translation machinery (Vasudevan et al. 2005). Nevertheless, more detailed studies have to be carried out to shed some light on the regulation of mRNA stability and the translational process upon stress conditions.

8 Conclusions and perspectives

It has become clear that the MAPK Hog1 regulates gene expression upon osmostress by different mechanisms (Fig. 2). This MAPK not only modifies directly stress-dependent transcriptional regulators by phosphorylation, but also associates specifically to stress responsive promoters to recruit the Rpd3 histone deacetylase complex and the preinitiation transcriptional machinery. Moreover, the role of Hog1 in the regulation of the transcription cycle is not limited to transcription initiation but rather extents to the process of transcription elongation acting as a selective elongation factor for genes induced by osmostress.

Some key questions regarding the control of gene expression by the MAPK Hog1 still remain to be elucidated. A major issue is the identification of MAPK targets that are specifically modified to regulate transcription initiation as well as during elongation or termination. Although the binding proteins required for recruitment of Hog1 at the promoters seem to be identified, at least in part, the identity of the proteins/mechanisms that define its selectivity during the process of elongation remain uncharacterized. A second issue that remains open is the identification of new transcriptional regulators or chromatin remodeling/modifying

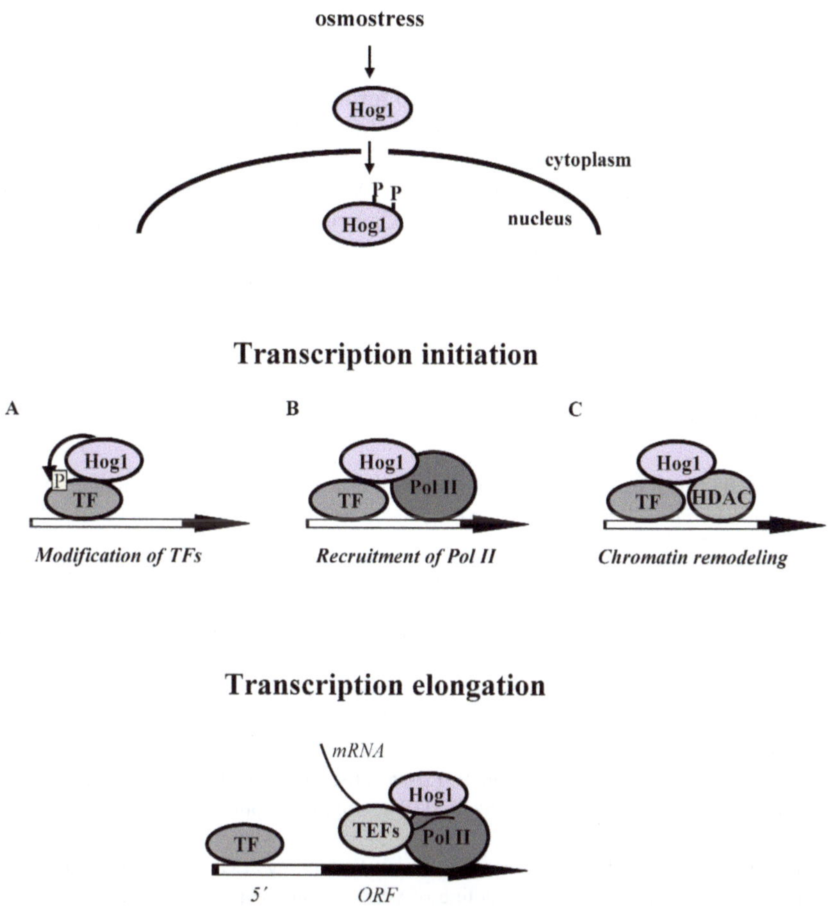

Fig. 2. Activation of the Hog1 MAPK has an essential role in transcription regulation of osmostress genes. In osmostress conditions, Hog1 is activated and concentrates into the nucleus, where it regulates different aspects of the transcription cycle. Hog1 modulates transcription initiation by direct regulation of transcription factor activities (A), stimulation of the recruitment of the RNA Polymerase II at the promoters (B), recruitment of the Rpd3 histone deacetylase complex, and modification of chromatin (C). Furthermore, the MAPK is recruited to the coding regions of osmostress genes behaving as selective a transcriptional elongation factor.

factors necessary for a proper transcriptional response upon osmostress. Furthermore, it is also important to characterize the processes regulated by the MAPK Hog1 after the pre-mRNA molecule is already produced; for instance mRNA maturation, mRNA export or mRNA stability. Last but not least, it is of great interest to study the role of Hog1 in the control of mRNA translation in response to stress.

Acknowledgement

We special thank the members of our group for their contribution and helpful support from G. Ammerer, D. Shore, K. Struhl, M. Proft, S. Hohmann, M. Peter, A. Aguilera, and S. Chavez groups. We apologize to those who might feel that their work was not adequately taken into account. Work in our laboratory is supported by grants from Ministerio de Educación y Ciencia to Francesc Posas and Eulàlia de Nadal, from DURSI (Generalitat de Catalunya) and from a EURYI scheme award (www.esf.org/euryi) to Francesc Posas. Eulàlia de Nadal is supported by the "Programa Ramón y Cajal".

References

Alepuz PM, de Nadal E, Zapater M, Ammerer G, Posas F (2003) Osmostress-induced transcription by Hot1 depends on a Hog1-mediated recruitment of the RNA Pol II. EMBO J 22:2433-2442

Alepuz PM, Jovanovic A, Reiser V, Ammerer G (2001) Stress-induced MAP kinase Hog1 is part of transcription activation complexes. Mol Cell 7:767-777

Berkes CA, Tapscott SJ (2005) MyoD and the transcriptional control of myogenesis. Semin Cell Dev Biol 16:585-595

Bernstein BE, Tong JK, Schreiber SL (2000) Genomewide studies of histone deacetylase function in yeast. Proc Natl Acad Sci USA 97:13708-13713

Carrozza MJ, Li B, Florens L, Suganuma T, Swanson SK, Lee KK, Shia WJ, Anderson S, Yates J, Washburn MP, Workman JL (2005) Histone H3 methylation by Set2 directs deacetylation of coding regions by Rpd3S to suppress spurious intragenic transcription. Cell 123:581-592

Causton HC, Ren B, Koh SS, Harbison CT, Kanin E, Jennings EG, Lee TI, True HL, Lander ES, Young RA (2001) Remodeling of yeast genome expression in response to environmental changes. Mol Biol Cell 12:323-337

Chellappan SP (2001) *HOG* on the promoter: regulation of the osmotic stress response. Sci STKE 31:PE1

Clayton AL, Mahadevan LC (2003) MAP kinase-mediated phosphoacetylation of histone H3 and inducible gene regulation. FEBS Lett 546:51-58

Cullen PJ, Sabbagh W Jr, Graham E, Erick MM, van Olden EK, Neal C, Delray J, Bardwell L, Sprague GF Jr (2004) A signaling mucin at the head of the Cdc42- and MAPK-dependent filamentous growth pathway in yeast. Genes Dev 18:1695-1708

Dan I, Watanabe NM, Kurume A (2001) The Ste20 group kinases as regulators of MAP kinase cascades. Trends Cell Biol 11:220-230

Dard N, Peter M (2006) Scaffold proteins in MAP kinase signaling: more than simple passive activating platforms. Bioassays 28:146-156

de Angelis L, Zhao J, Androecia JJ, Olson EN, Cosset G, McDermott JC (2005) Regulation of vertebrate myotome development by the p38 MAP kinase-MEF2 signaling pathway. Dev Biol 283:171-179

de Nadal E, Alepuz PM, Posas F (2002) Dealing with osmostress through MAP kinase activation. EMBO Rep 3:735-740 de Nadal E, Casadome L, Posas F (2003) Targeting the MEF2-like transcription factor Smp1 by the stress-activated Hog1 mitogen-activated protein kinase. Mol Cell Biol 23:229-237

de Nadal E, Zapater M, Alepuz PM, Sumoy L, Mas G, Posas F (2004) The MAPK Hog1 recruits Rpd3 histone deacetylase to activate osmoresponsive genes. Nature 427:370-374

De W, Reiter W, Ballerina A, Ammerer G, Brocade C (2005) A dual role for PP1 in shaping the Msn2-dependent transcriptional response to glucose starvation. EMBO J 24:4115-4123

Deckert J, Struhl K (2001) Histone acetylation at promoters is differentially affected by specific activators and repressors. Mol Cell Biol 21:2726-2735

Durchschlag E, Reiter W, Ammerer G, Schuller C (2004) Nuclear localization destabilizes the stress-regulated transcription factor Msn2. J Biol Chem 279:55425-55432

Dyson MH, Thomson S, Inagaki M, Goto H, Arthur SJ, Nightingale K, Iborra FJ, Mahadevan LC (2005) MAP kinase-mediated phosphorylation of distinct pools of histone H3 at S10 or S28 via mitogen- and stress-activated kinase 1/2. J Cell Sci 118:2247-2259

Edmunds JW, Mahadevan LC (2004) MAP kinases as structural adaptors and enzymatic activators in transcription complexes. J Cell Sci 117:3715-3723

Edmunds JW, Mahadevan LC (2006) Cell signaling. Protein kinases seek close encounters with active genes. Science 313:449-451

Ferreira ME, Hermann S, Prochasson P, Workman JL, Berndt KD, Wright AP (2005) Mechanism of transcription factor recruitment by acidic activators. J Biol Chem 280:21779-21784

Ferrigno P, Posas F, Koepp D, Saito H, Silver PA (1998) Regulated nucleo/cytoplasmic exchange of HOG1 MAPK requires the importin beta homologs *NMD5* and *XPO1*. EMBO J 17:5606-5614

Garcia-Gimeno MA, Struhl K (2000) Aca1 and Aca2, ATF/CREB activators in *Saccharomyces cerevisiae*, are important for carbon source utilization but not the response to stress. Mol Cell Biol 20:4340-4349

Gasch AP, Spellman PT, Kao CM, Carmel-Harel O, Eisen MB, Storz G, Botstein D, Brown PO (2000) Genomic expression programs in the response of yeast cells to environmental changes. Mol Biol Cell 11:4241-4257

Gorner W, Durchschlag E, Martinez-Pastor MT, Estruch F, Ammerer G, Hamilton B, Ruis H, Schuller C (1998) Nuclear localization of the C2H2 zinc finger protein Msn2p is regulated by stress and protein kinase A activity. Genes Dev 12:586-597

Gorner W, Durchschlag E, Wolf J, Brown EL, Ammerer G, Ruis H, Schuller C (2002) Acute glucose starvation activates the nuclear localization signal of a stress-specific yeast transcription factor. EMBO J 21:135-144

Hohmann S (2002) Osmotic stress signaling and osmoadaptation in yeasts. Microbiol Mol Biol Rev 66:300-372

Jacoby T, Flanagan H, Faykin A, Seto AG, Mattison C, Ota I (1997) Two protein-tyrosine phosphatases inactivate the osmotic stress response pathway in yeast by targeting the mitogen-activated protein kinase, Hog1. J Biol Chem 272:17749-17755

Keogh MC, Kurdistani SK, Morris SA, Ahn SH, Podolny V, Collins SR, Schuldiner M, Chin K, Punna T, Thompson NJ, Boone C, Emili A, Weissman JS, Hughes TR, Strahl BD, Grunstein M, Greenblatt JF, Buratowski S, Krogan NJ (2005) Cotranscriptional

Set2 methylation of histone H3 lysine 36 recruits a repressive Rpd3 complex. Cell 123:593-605

Klipp E, Nordlander B, Kruger R, Gennemark P, Hohmann S (2005) Integrative model of the response of yeast to osmotic shock. Nat Biotechnol 23:975-982

Kyriakis JM, Avruch J (2001) Mammalian mitogen-activated protein kinase signal transduction pathways activated by stress and inflammation. Physiol Rev 81:807-869

Maeda T, Takekawa M, Saito H (1995) Activation of yeast PBS2 MAPKK by MAPKKKs or by binding of an SH3-containing osmosensor. Science 269:554-558

Maeda T, Wurgler-Murphy SM, Saito H (1994) A two-component system that regulates an osmosensing MAP kinase cascade in yeast. Nature 369:242-245

Marquez JA, Pascual-Ahuir A, Proft M, Serrano R (1998) The Ssn6-Tup1 repressor complex of *Saccharomyces cerevisiae* is involved in the osmotic induction of HOG-dependent and -independent genes. EMBO J 17:2543-2553

Martinez-Pastor MT, Marchler G, Schuller C, Marchler-Bauer A, Ruis H, and Estruch F (1996) The *Saccharomyces cerevisiae* zinc finger proteins Msn2p and Msn4p are required for transcriptional induction through the stress response element (STRE). EMBO J 15:2227-2235

Mason PB, Struhl K (2005) Distinction and relationship between elongation rate and processivity of RNA polymerase II *in vivo*. Mol Cell 17:831-840

Mattison CP, Ota IM (2000) Two protein tyrosine phosphatases, Ptp2 and Ptp3, modulate the subcellular localization of the Hog1 MAP kinase in yeast. Genes Dev 14:1229-1235

McKinsey TA, Zhang CL, Olson EN (2002) MEF2: a calcium-dependent regulator of cell division, differentiation and death. Trends Biochem Sci 27:40-47

Millar CB, Grunstein M (2006) Genome-wide patterns of histone modifications in yeast. Nat Rev Mol Cell Biol 7:657-666

Nehlin JO, Carlberg M, Ronne H (1992) Yeast *SKO1* gene encodes a bZIP protein that binds to the CRE motif and acts as a repressor of transcription. Nucleic Acids Res 20:5271-5278

O'Rourke SM, Herskowitz I (2002) A third osmosensing branch in *Saccharomyces cerevisiae* requires the Msb2 protein and functions in parallel with the Sho1 branch. Mol Cell Biol 22:4739-4749

Pascual-Ahuir A, Posas F, Serrano R, Proft M (2001) Multiple levels of control regulate the yeast cAMP-response element-binding protein repressor Sko1p in response to stress. J Biol Chem 276:37373-37378

Pascual-Ahuir A, Struhl K, Proft M (2006) Genome-wide location analysis of the stress-activated MAP kinase Hog1 in yeast. Methods 40 272-278

Pokholok DK, Zeitlinger J, Hannett NM, Reynolds DB, Young RA (2006) Activated signal transduction kinases frequently occupy target genes. Science 313:533-536

Posas F, Chambers JR, Heyman JA, Hoeffler JP, de Nadal E, Arino J (2000) The transcriptional response of yeast to saline stress. J Biol Chem 275:17249-17255

Posas F, Saito H (1997) Osmotic activation of the HOG MAPK pathway via Ste11p MAPKKK: scaffold role of Pbs2p MAPKK. Science 276:1702-1705

Posas F, Saito H (1998) Activation of the yeast SSK2 MAP kinase kinase kinase by the SSK1 two-component response regulator. EMBO J 17:1385-1394

Posas F, Witten EA, Saito H (1998) Requirement of STE50 for osmostress-induced activation of the STE11 mitogen-activated protein kinase kinase kinase in the high-osmolarity glycerol response pathway. Mol Cell Biol 18:5788-5796

Posas F, Wurgler-Murphy SM, Maeda T, Witten EA, Thai TC, Saito H (1996) Yeast HOG1 MAP kinase cascade is regulated by a multistep phosphorelay mechanism in the SLN1-YPD1-SSK1 "two-component" osmosensor. Cell 86:865-875

Proft M, Gibbons FD, Copeland M, Roth FP, Struhl K (2005) Genomewide identification of Sko1 target promoters reveals a regulatory network that operates in response to osmotic stress in *Saccharomyces cerevisiae*. Eukaryot Cell 4:1343-1352

Proft M, Mas G, de Nadal E, Vendrell A, Noriega N, Struhl K, Posas F (2006) The stress-activated Hog1 kinase is a selective transcriptional elongation factor for genes responding to osmotic stress. Mol Cell 23:241-250

Proft M, Pascual-Ahuir A, de Nadal E, Arino J, Serrano R, Posas F (2001) Regulation of the Sko1 transcriptional repressor by the Hog1 MAP kinase in response to osmotic stress. EMBO J 20:1123-1133

Proft M, Serrano R (1999) Repressors and upstream repressing sequences of the stress-regulated *ENA1* gene in *Saccharomyces cerevisiae*: bZIP protein Sko1p confers HOG-dependent osmotic regulation. Mol Cell Biol 19:537-546

Proft M, Struhl K (2002) Hog1 kinase converts the Sko1-Cyc8-Tup1 repressor complex into an activator that recruits SAGA and SWI/SNF in response to osmotic stress. Mol Cell 9:1307-1317

Proft M, Struhl K (2004) MAP kinase-mediated stress relief that precedes and regulates the timing of transcriptional induction. Cell 118:351-361

Raitt DC, Posas F, Saito H (2000) Yeast Cdc42 GTPase and Ste20 PAK-like kinase regulate Sho1-dependent activation of the Hog1 MAPK pathway. EMBO J 19:4623-4631

Reiser V, Ruis H, Ammerer G (1999) Kinase activity-dependent nuclear export opposes stress-induced nuclear accumulation and retention of Hog1 mitogen-activated protein kinase in the budding yeast *Saccharomyces cerevisiae*. Mol Biol Cell 10:1147-1161

Reiser V, Salah SM, Ammerer G (2000) Polarized localization of yeast Pbs2 depends on osmostress, the membrane protein Sho1 and Cdc42. Nat Cell Biol 2:620-627

Rep M, Krantz M, Thevelein JM, Hohmann S (2000) The transcriptional response of *Saccharomyces cerevisiae* to osmotic shock. Hot1p and Msn2p/Msn4p are required for the induction of subsets of high osmolarity glycerol pathway-dependent genes. J Biol Chem 275:8290-8300

Rep M, Proft M, Remize F, Tamas M, Serrano R, Thevelein JM, Hohmann S (2001) The *Saccharomyces cerevisiae* Sko1p transcription factor mediates HOG pathway-dependent osmotic regulation of a set of genes encoding enzymes implicated in protection from oxidative damage. Mol Microbiol 40:1067-1083

Rep M, Reiser V, Gartner U, Thevelein JM, Hohmann S, Ammerer G, Ruis H (1999) Osmotic stress-induced gene expression in *Saccharomyces cerevisiae* requires Msn1p and the novel nuclear factor Hot1p. Mol Cell Biol 19:5474-5485

Robyr D, Suka Y, Xenarios I, Kurdistani SK, Wang A, Suka N, Grunstein M (2002) Microarray deacetylation maps determine genome-wide functions for yeast histone deacetylases Cell 109:437-446

Sabet N, Volo S, Yu C, Madigan JP, Morse RH (2004) Genome-wide analysis of the relationship between transcriptional regulation by Rpd3p and the histone H3 and H4 amino termini in budding yeast. Mol Cell Biol 24:8823-8833

Schmitt AP, McEntee K (1996) Msn2p, a zinc finger DNA-binding protein, is the transcriptional activator of the multistress response in *Saccharomyces cerevisiae*. Proc Natl Acad Sci USA 93:5777-5782

Sheikh-Hamad D, Gustin MC (2004) MAP kinases and the adaptive response to hypertonicity: functional preservation from yeast to mammals. Am J Physiol Renal Physiol 287:F1102-F1110

Simone C, Forcales SV, Hill DA, Imbalzano AN, Latella L, Puri PL (2004) p38 pathway targets SWI-SNF chromatin-remodeling complex to muscle-specific loci. Nat Genet 36:738-743

Sims RJ III, Belotserkovskaya R, Reinberg D (2004) Elongation by RNA polymerase II: the short and long of it. Genes Dev 18:2437-2468

Soloaga A, Thomson S, Wiggin GR, Rampersaud N, Dyson MH, Hazzalin CA, Mahadevan LC, Arthur JS (2003) MSK2 and MSK1 mediate the mitogen- and stress-induced phosphorylation of histone H3 and HMG-14. EMBO J 22:2788-2797

Tatebayashi K, Yamamoto K, Tanaka K, Tomida T, Maruoka T, Kasukawa E, Saito H (2006) Adaptor functions of Cdc42, Ste50, and Sho1 in the yeast osmoregulatory HOG MAPK pathway. EMBO J 25:3033-3044

Truckses DM, Bloomekatz JE, Thorner J (2006) The RA domain of Ste50 adaptor protein is required for delivery of Ste11 to the plasma membrane in the filamentous growth signaling pathway of the yeast *Saccharomyces cerevisiae*. Mol Cell Biol 26:912-928

Vasudevan S, Garneau N, Tu KD, Peltz SW (2005) p38 mitogen-activated protein kinase/Hog1p regulates translation of the AU-rich-element-bearing *MFA2* transcript. Mol Cell Biol 25:9753-9763

Vasudevan S, Peltz SW (2001) Regulated ARE-mediated mRNA decay in *Saccharomyces cerevisiae*. Mol Cell 7:1191-1200

Vincent AC, Struhl K (1992) *ACR1*, a yeast ATF/CREB repressor. Mol Cell Biol 12:5394-5405

Wang A, Kurdistani SK, Grunstein M (2002) Requirement of Hos2 histone deacetylase for gene activity in yeast. Science 298:1412-1414

Warmka J, Hanneman J, Lee J, Amin D, Ota I (2001) Ptc1, a type 2C Ser/Thr phosphatase, inactivates the HOG pathway by dephosphorylating the mitogen-activated protein kinase Hog1. Mol Cell Biol 21:51-60

Westfall PJ, Ballon DR, Thorner J (2004) When the stress of your environment makes you go HOG wild. Science 306:1511-1512

Winters MJ, Pryciak PM (2005) Interaction with the SH3 domain protein Bem1 regulates signaling by the *Saccharomyces cerevisiae* p21-activated kinase Ste20. Mol Cell Biol 25:2177-2190

Wurgler-Murphy SM, Maeda T, Witten EA, Saito H (1997). Regulation of the *Saccharomyces cerevisiae* HOG1 mitogen-activated protein kinase by the PTP2 and PTP3 protein tyrosine phosphatases. Mol Cell Biol 17:1289-1297

Yu YT, Breitbart RE, Smoot LB, Lee Y, Mahdavi V, Nadal-Ginard B (1992) Human myocyte-specific enhancer factor 2 comprises a group of tissue-restricted MADS box transcription factors. Genes Dev 6:1783-1798

Zarrinpar A, Bhattacharyya RP, Nittler MP, Lim WA (2004) Sho1 and Pbs2 act as coscaffolds linking components in the yeast high osmolarity MAP kinase pathway. Mol Cell 14:825-832

de Nadal, Eulàlia
Cell Signaling Unit, Departament de Ciències Experimentals i de la Salut, Universitat Pompeu Fabra (UPF), E-08003 Barcelona, Spain.
eulalia.nadal@upf.edu

Posas, Francesc
Cell Signaling Unit, Departament de Ciències Experimentals i de la Salut, Universitat Pompeu Fabra (UPF), E-08003 Barcelona, Spain.
francesc.posas@upf.edu

Regulation of tumorigenesis by p38α MAP kinase

Ignacio Dolado and Angel R. Nebreda

Abstract

p38α is a stress-activated protein kinase that can suppress tumor formation by negatively regulating cell cycle progression or by inducing apoptosis. More recently, the ability of p38α to induce cell differentiation has also been connected to tumor suppression. Accordingly, several proteins that can potentially downregulate the activity of p38α have been found overexpressed in human tumors and cancer cell lines. However, p38α can impinge on cancer progression by modulating other cellular responses, in addition to proliferation and differentiation, such as cell migration as well as the processes of invasion and inflammation. This could explain why, in some cancer types, p38α activation has been correlated with malignancy and poor prognosis rather than with tumor suppression. Here, we will review the evidence connecting p38α to distinct cancer traits and will discuss the mechanisms that may account for the oncogenic and tumor suppressor roles of p38α.

1 Introduction

p38α mitogen-activated protein kinase (MAPK), also known as RK, CSBP, and SAPK2a, was originally identified as a 38-kDa protein that mediated the inflammatory effect of several cytokines, as well as the target of anti-inflammatory drugs and an important regulator of the cellular responses to stress (Freshney et al. 1994; Han et al. 1994; Lee et al. 1994; Rouse et al. 1994). Three other p38 MAPK family members were subsequently identified (p38β, p38γ, and p38δ), with p38β showing the highest similarity to p38α in terms of substrate specificity and sensitivity to inhibitors (Nebreda and Porras 2000; Ono and Han 2000).

In addition to the p38 MAPK family, three other major MAPK pathways have been characterized in mammals: the mitogenic extracellular-signal regulated kinases (ERK1 and ERK2), the stress-activated c-jun-N-terminal kinases (JNK1, JNK2, and JNK3), and the ERK5/BMK1 cascade. These four MAPK pathways are structurally related and share some regulatory mechanisms, but have been shown to play different roles in tumorigenesis. For instance, constitutive activation of the ERK1/ERK2 pathway is necessary and sufficient for cell transformation (Cowley et al. 1994; Mansour et al. 1994) and has been detected in many human tumors (Gollob et al. 2006). Consequently, inhibitors of the ERK1/ERK2 pathway have been developed for cancer therapeutics (Sebolt-Leopold and Herrera 2004).

Topics in Current Genetics, Vol. 20
F. Posas, A. R. Nebreda (Eds.): Stress-Activated Protein Kinases
DOI 10.1007/4735_2007_0245 / Published online: 1 August 2007

The role of the JNK pathway in cancer is less straight forward, as reviewed elsewhere (Engelberg 2004; Manning and Davis 2003; Rennefahrt et al. 2005). Essentially, JNKs can have both oncogenic effects, based on their ability to activate the proto-oncogene c-Jun, and tumor suppressor roles, due to their pro-apoptotic activity (Fan and Chambers 2001; Kennedy et al. 2003; Tront et al. 2006). It is therefore likely that the overall effect of JNKs in tumor development will depend on the balance between oncogenic and tumor-suppressive functions, which in turn may be affected by factors such as signal intensity and JNK isoform cross-talk. For example, apoptosis induction has been associated only with sustained JNK activation (Brozovic et al. 2004; Ventura et al. 2006). Furthermore, whereas JNK2 has been proposed as an oncogene, JNK1 and JNK3 are tumor suppressor candidates (Rennefahrt et al. 2005).

Evidence accumulated over the last years support the function of p38α as a tumor suppressor (Bulavin and Fornace 2004), which seems to be mostly based on the ability of p38α to negative regulate cell cycle progression as well as to mediate apoptosis induction. For instance, mouse embryo fibroblasts (MEFs) deficient in either p38α (Bulavin et al. 2002) or the p38 MAPK activators MKK3 and MKK6 (Brancho et al. 2003) are able to produce higher oncogene-induced tumor burden in nude mice than their wild type counterparts. In addition, genetic inactivation in mice of the PPM1D/Wip1 phosphatase, which can de-phosphorylate p38 MAPKs, results in reduced mammary tumorigenesis that correlates with increased p38 MAPK activity levels (Bulavin et al. 2004). p38α can also suppress tumor formation by inducing terminal differentiation (Puri et al. 2000). In agreement with these observations, several proteins that can potentially downregulate p38α signaling have been found overexpressed in human tumors and cancer cell lines, including the phosphatases PPM1D/Wip1 and DUSP26/Mkp8 (Bulavin et al. 2002; Li et al. 2002; Yu et al. 2006) and the ASK1 inhibitors Gstm1 and Gstm2 (Dolado et al. 2007). Furthermore, some human tumors, such as hepatocellular carcinomas (Iyoda et al. 2003), have lower p38 MAPK activity levels than the corresponding non-tumorigenic tissues. While these results are all consistent with a tumor suppressor role of p38α, the picture is more complex if one takes into account the function of p38α in cell migration as well as in key processes for cancer progression such as invasion and inflammation. In this review we will discuss the evidence linking p38α to distinct cancer features.

2 Cell cycle regulation

Cell cycle deregulation is considered one of the hallmarks of cancer and is normally associated with uncontrolled cell proliferation and checkpoint failure. Indeed, more than half of human cancer cells are thought to have impaired G1 checkpoints (Kawabe 2004). Interestingly, p38α can negatively regulate cell cycle progression both at the G1/S and the G2/M transitions, and several mechanisms have been proposed to be involved in each case.

2.1 Inhibition of the G1/S transition

p38α can induce a G1/S delay in response to stress either at the transcriptional level, by downregulating cyclin D1 (Lavoie et al. 1996) or by upregulating the cyclin-dependent kinase (CDK) inhibitor $p16^{INK4a}$ (Bulavin et al. 2004; Wang et al. 2002), or by phosphorylation-mediated mechanisms that trigger the degradation of D-type cyclins (Casanovas et al. 2000, 2004) and Cdc25A (Goloudina et al. 2003) or the stabilization of the CDK inhibitor $p21^{Cip1}$ (Kim et al. 2002a) and the cyclin D transcriptional repressor HBP1 (Yee et al. 2004). In addition, p38α can modulate the expression and activity of the tumor suppressor p53 both at the transcriptional and posttranslational levels (Bulavin and Fornace 2004), which also contributes to the p38α-mediated G1 cell cycle arrest through the induction of $p21^{Cip1}$. Of note, the upregulation of $p21^{Cip1}$ expression mediated by p38α, but in a p53-independent manner, is also essential for $H\text{-}Ras^{G12V}$-induced cell cycle arrest (Nicke et al. 2005). Furthermore, the induction of G1 cell cycle arrest under conditions of high cellular density, a process called cell-cell contact inhibition, is also regulated by p38α and probably involves the CDK inhibitor $p27^{Kip1}$ (Faust et al. 2005). Noteworthy, the ability of p38α to regulate progression through the G1 phase of the cell cycle has been recently shown to be relevant for both myogenesis and hematopoiesis *in vivo*. Thus, in response to the accumulation of reactive oxygen species (ROS), p38α induces the upregulation of the G1 cell cycle inhibitors $p16^{INK4a}$ and $p19^{ARF}$, which in turn limit the lifespan of hematopoietic stem cells (Ito et al. 2006). Conversely, p38α-deficient myoblasts are impaired in cell cycle exit and continuously proliferate in differentiation-promoting conditions, which may be accounted for by the enhanced activation of JNK/c-Jun in the absence of p38α (Perdiguero et al. 2007).

2.2 Inhibition of the G2/M transition

p38α has also been associated with G2/M cell cycle arrest or delay induced by a variety of stresses, including ultraviolet (UV) light, methylating agents, osmotic shock, and inhibitors of topoisomerase II or histone deacetylases (Mikhailov et al. 2005). At the mechanistic level, MAPK-activated protein kinase 2 (MAPKAP-K2, lately referred to as MK-2) has been proposed as a key mediator of the p38α-induced G2/M arrest (Manke et al. 2005; Reinhardt et al. 2007). MK-2 is activated by p38α phosphorylation and in turn can phosphorylate Cdc25B and Cdc25C, which induces their binding to 14-3-3 proteins. This prevents Cdc25 from activating the Cdc2/Cyclin B complex and inhibits mitosis entry. Interestingly, Cdc25B can be also directly phosphorylated by p38α, independently of MK-2 (Lemaire et al. 2006), but the contribution of this phosphorylation to the p38α-induced G2/M arrest needs further investigation. In addition, downregulation of cyclins A and B expression may also contribute to the G2 cell cycle arrest induced by p38α *in vitro* (Garner et al. 2002) as well as *in vivo* (Engel et al. 2005). Moreover, recent studies in mouse thymocytes support an *in vivo* role for p38α in the p53-mediated G2/M arrest in response to DNA breaks (Pedraza-Alva et al. 2006).

Table 1. p38α as a mediator of cell cycle arrest

Cell type[a]	Cell line	Malignancy[b]	Inducer	Phenotype (mechanism)[c]	Reference
FI	BJ	-	Oncogenic H-Ras, Active MKK3 and MKK6	Irreversible cell cycle arrest (↑ $p16^{INK4a}$)	(Wang et al. 2002)
	MEFs	-	Ppm1d/Wip1 deletion	Irreversible cell cycle arrest (↑ $p16^{INK4a}$ and $p19^{ARF}$)	(Bulavin et al. 2004)
	CCL-39	-	Active MKK3	G1 delay (↓ cyclin D1)	(Lavoie et al. 1996)
	CCL-39	-	Active MEKK3	G1/S arrest (↑ $p21^{Cip1}$)	(Todd et al. 2004)
	CCL-39	-	Active MEKK3	G2/M arrest (↓ cyclins A and B1)	(Garner et al. 2002)
	NIH 3T3	-	MKK3, MKK6, Cdc42	G1/S arrest	(Molnar et al. 1997)
	NIH 3T3	-	Sodium arsenite	Inhibition of proliferation (↑ $p21^{Cip1}$)	(Kim et al. 2002b)
	NIH 3T3	-	Active MKK6	Inhibition of proliferation	(Chen et al. 2000)
	DFs	-	UV irradiation	G2/M arrest (Cdc25B phosphorylation)	(Bulavin et al. 2001)
EN	Cardio myocytes	-	Active MKK3	Inhibition of proliferation (↓ cyclins A2 and B)	(Engel et al. 2005)
	Myoblasts	-	Serum withdrawal	Cell cycle exit (↓JNK/c-Jun pathway)	(Perdiguero et al. 2007)
HE	T cells	-	Active MKK6	Mitotic arrest (↑ p53)	(Pedraza-Alva et al. 2006)
	Granta 519	+	Osmotic shock	G1 delay (↓ cyclin D1)	(Casanovas et al. 2000)
	Daudi	+	Etoposide	G2/M arrest (Cdc25 phosphorylation)	(Kurosu et al. 2005)
EP	Ptk1	-	Topoisomerase II and HDACs inhibitors	G2/M arrest	(Mikhailov et al. 2004)
	mIMCD3	-	Osmotic shock	G2/M arrest	(Dmitrieva et al. 2002)
	HOSE	-	Oncogenic H-Ras	Inhibition of proliferation (↑ $p21^{Cip1}$)	(Nicke et al. 2005)
	RIE-1	+	Active MKK6	Inhibition of proliferation (↓ cyclin D1)	(Pruitt et al. 2002)
	T24	+	Sodium arsenite Active MKK6	Inhibition of proliferation	(Chen et al. 2000)
	U2OS	+	UV irradiation	G2/M arrest (Cdc25 phosphorylation)	(Manke et al. 2005)

Cell type[a]	Cell line	Malig-nancy[b]	Inducer	Phenotype (mecha-nism)[c]	Reference
	HeLa	+	UV irradiation	G2/M arrest (Cdc25B phosphorylation)	(Bulavin et al. 2001)
	HeLa	+	Osmotic shock	S phase arrest (Cdc25A degradation)	(Goloudina et al. 2003)
	U87MG	+	Temozolomide	G2/M arrest (Cdc25C phosphorylation)	(Hirose et al. 2003)

[a] FI, fibroblast; EN, endothelial; HE, hematopoietic; EP, epithelial
[b] -, non-transformed; +, transformed/tumorigenic
[c] ↑, upregulation; ↓, downregulation

2.3 Stimulation of cell cycle progression

As mentioned above, the ability of p38α to phosphorylate specific proteins is important for the induction of cell cycle arrest. Intriguingly, it has been recently proposed that p38α positively regulates mitotic progression in HeLa cells in a kinase-independent fashion (Fan et al. 2005). The relevance of this observation and the putative mechanism involved remain to be elucidated. However, it should be noted that the *S. cerevisiae* p38 MAPK homolog Hog1 might regulate transcription by a mechanism that does not require its kinase activity (De Nadal et al. 2004; Proft et al. 2006).

It is conceivable that the function of p38α in cell cycle progression could depend on both the type and the malignant stage of the cell. Accordingly, p38α stimulates proliferation in transformed chondrosarcoma cells, but not in primary non-tumorigenic chondrocytes (Halawani et al. 2004; Yosimichi et al. 2001). On the other hand, p38α activation has no effect in the proliferation of hepatocarcinoma cells (Aguirre-Ghiso et al. 2004) or E1A/H-Ras^{G12V}-transformed MEFs, but inhibits cell cycle progression in primary MEFs (Bulavin et al. 2004).

An extensive review of the literature (Table 1) argues that p38α activation in fibroblasts, hematopoietic and epithelial cells, either by UV light, osmotic shock, oncogenic H-Ras or active MKK3 and MKK6, leads in all cases to cell cycle arrest or delay, which is in agreement with the above mentioned p38α anti-proliferative roles. Noteworthy, this effect appears to be independent of the transformed phenotype of the cells (Table 1). In contrast, p38α has a positive role in the proliferation of cytokine-stimulated hematopoietic cells and human breast cancer cells (Table 2). Thus, p38α implication in both the production of inflammatory cytokines and the signal rely from inflammatory cytokine receptors (see Section 5.1) may determine its pro-oncogenic activity in cytokine-sensitive neoplasias, such as hematological malignancies (Platanias 2003), prostate cancer (Ricote et al. 2006), and melanoma (Recio and Merlino 2002). Perhaps, low levels of active p38α (i.e. after cytokine stimulation) can stimulate proliferation, in contrast to the strong, stress-induced p38α activation that mediates cell cycle arrest. On the other hand, the pro-tumorigenic effect of p38α in human breast cancer (Lee et al. 1999; Neve et al. 2002) might be accounted for by the participation of p38α-activated transcription factors, such as the cAMP-responsive element binding protein (CREB)

Table 2. p38α as an inducer of cell proliferation

Cell type[a]	Cell line	Malignancy[b]	Inducer	Mechanism[c]	Reference
HE	CT6	-	IL-2, IL-7	N.D.	(Crawley et al. 1997)
	BaF3	-	GCSF	N.D.	(Rausch and Marshall 1999)
	OCI-AML5	+	GCSF	N.D.	(Srinivasa and Doshi 2002)
	Mo7e	+	TNF-α	N.D.	(Liu et al. 2000)
	MM.1 S	+	-	↑ paracrine IL-6 expression (by bone marrow stromal cells)	(Hideshima et al. 2003)
EP	22Rv1	+	Oncostatin-M (OSM)	N.D.	(Godoy-Tundidor et al. 2005)
	T47D	+	Heregulinβ1 (HRG)	↑ cyclins D1 and D2	(Neve et al. 2002)
	MCF-7	+	Estradiol Spermine	↑ cyclin D1	(Lewis et al. 2005)
	MCF-7	+	pp60$^{v\text{-}src}$	↑ cyclin D1	(Lee et al. 1999)

[a] HE, hematopoietic; EP, epithelial
[b] -, non-transformed; +, transformed/tumorigenic
[c] N.D., no determined; ↑, upregulation

(Sabbah et al. 1999), in the proliferative response triggered by the oestrogen receptor, which is usually overexpressed in breast cancer cells. Consequently, p38α has been found over-activated and associated with poor outcome in 20% of primary human breast carcinomas (Esteva et al. 2004) and has been suggested as an early marker for mammoplasty screening (Gauthier et al. 2005).

3 Regulation of cell survival and apoptosis

Alterations in cell survival programs are thought to play important roles in cancer. In fact, whereas the acquisition of invasiveness is regarded as the deadliest characteristic of cancer, the development of drug resistance may account for the failure of treatment, and ultimately death, of more than 90% of metastatic cancer patients (Dean et al. 2005). It is therefore not surprising that numerous proteins involved in the regulation of apoptosis have been found silenced or deregulated in animal cancer models and human tumors (Johnstone et al. 2002); hence, providing a molecular basis for the common clinical occurrence of cancer drug resistance and recurrence. In this context, the modulation of p38α activity might be a strategy worth exploring for sensitizing cancer cells to apoptotic death, which might prove useful for cancer therapy.

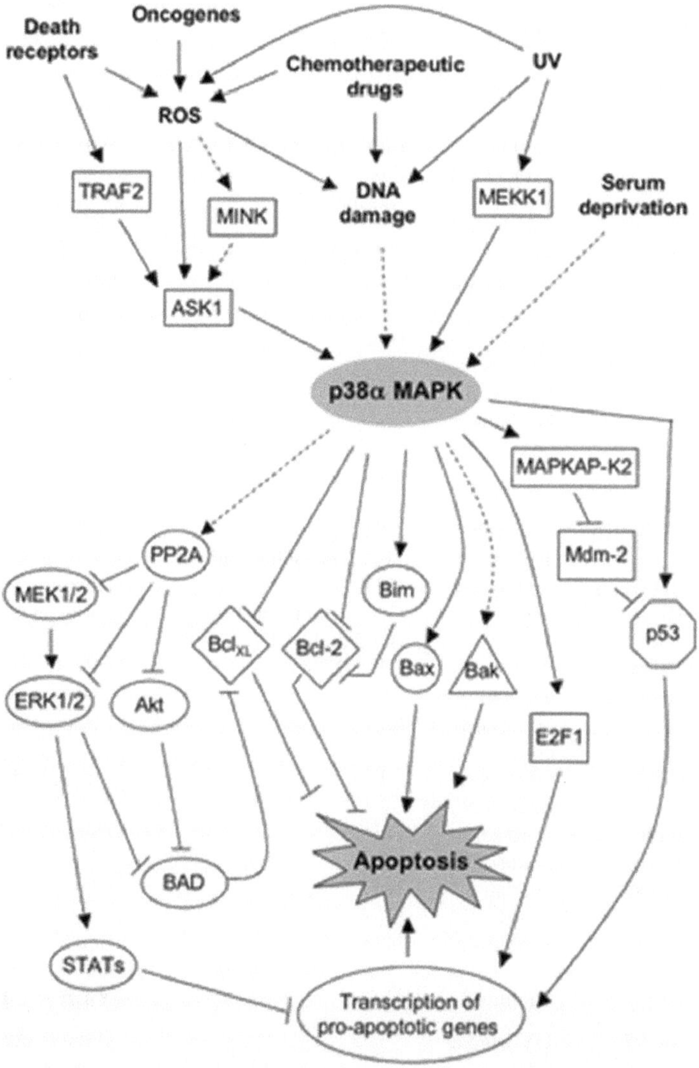

Fig. 1. Pathways potentially involved in the pro-apoptotic effect of p38α. A variety of apoptotic stimuli can activate p38α and some of the upstream kinases involved in particular pathways have been identified, such as MEKK1 for UV irradiation (Zhuang et al. 2006) or MINK (Nicke et al. 2005) and ASK1 (Matsukawa et al. 2004) for oxidative stress. In turn, p38α activation can induce apoptosis by several mechanisms; see text for details. The dashed lines indicate mechanisms that are not well characterized.

Table 3. p38α as a mediator of the apoptotic response

Cell type[a]	Cell line	Inducer	Mechanism[b]	Reference
FI	NIH3T3	Cisplatin	↑ p53	(Sanchez-Prieto et al. 2000)
	MEFs	Serum withdrawal	ERK inhibition, ↑ Fas and Bax	(Porras et al. 2004)
	MEFs	Serum withdrawal	Akt inhibition	(Zuluaga et al. 2007)
	MEFs	Serum withdrawal	Bcl-2 phosphorylation	(De Chiara et al. 2006)
EN	Cardio-myocytes	Serum withdrawal, UV irradiation, Anti-Fas antibody	ERK inhibition, ↑ Fas and Bax	(Porras et al. 2004)
	Cardio-myocytes	H_2O_2	ERK inhibition	(Liu and Hofmann 2004)
	EA.hy926	Doxorubicin TNF-α	↓ BCL_{XL} ↓BAD phosphorylation	(Grethe et al. 2004; Grethe et al. 2006)
	HDMECs	γ-irradiation	N.D.	(Kumar et al. 2004)
HE	Thymo-cytes	FasL	Cytosol translocation of Bcl-2 and BCL_{XL}	(Farley et al. 2006)
	Thymo-cytes	Active MKK6	↓ Bcl-2	(Merritt et al. 2000)
	S49.1	Glucocorti-coids	N.D.	(Miller et al. 2005)
	CEM	Glucocorti-coids	↑ Bim	(Lu et al. 2006)
	Jurkat	Cannabi-noids (THC)	N.D.	(Herrera et al. 2005)
	Jurkat	Oxidative stress	N.D.	(Chen et al. 2006)
	ML-1	Vinblastine	N.D.	(Stadheim et al. 2001)
	U937	Paclitaxel	N.D.	(Yu et al. 2001)
EP	MCF7	UV irradia-tion	↑ p53	(Bulavin et al. 1999)
	HepG2	Methyl-cholan-threne Stauro-sporine, H_2O_2	↑ p53 Bax phosphorylation	(Kwon et al. 2002) (Kim et al. 2006)
	PC12	Arsenite	Bim phosphorylation	(Cai et al. 2006)

Cell type[a]	Cell line	Inducer	Mechanism[b]	Reference
	H1299, H460	γ-irradiation	Bak and Bax conformational activation	(Choi et al. 2006)
	HeLa	Taxol, Nocodazole	Bak and Bax conformational activation	(Deacon et al. 2003)
	A2780S	Cisplatin	Bak and Bax conformational activation	(Yuan et al. 2003)
	2008	Cisplatin	↑ FasL	(Mansouri et al. 2003)
	SW480	Cisplatin	↓EGFR pro-survival signaling	(Zwang and Yarden 2006)
	WM35, MeWo	mda-7/IL-24	↑ GADD proteins	(Sarkar et al. 2002)
	CGNs	Anti-Fas antibody	Rb inhibition	(Hou et al. 2002)
	CRC cells	Serum withdrawal, substrate detachment	N.D.	(Fassetta et al. 2006)
	HN4	Deferoxamine	N.D.	(Lee et al. 2006)
	Eca109, HaCaT, HeLa, TOV21G, OV-90, SK-OV-3	Cisplatin	N.D.	(Brozovic et al. 2004; Coltella et al. 2006; Losa et al. 2003; Zhang et al. 2005)
	MCF7, TOV21G, OV-90, SK-OV-3	Paclitaxel	N.D.	(Bacus et al. 2001; Coltella et al. 2006)

[a] FI, fibroblast; EN, endothelial; HE, hematopoietic; EP, epithelial
[b] N.D., not determined;↑, upregulation; ↓, downregulation

3.1 Apoptosis induction

p38α mediates apoptosis induction in normal and tumorigenic cell lines following various stimuli (Table 3), including chemotherapeutic drugs (Olson and Hallahan 2004), death receptor signals (Farley et al. 2006; Grethe and Porn-Ares 2006; Hou et al. 2002; Porras et al. 2004), UV irradiation (Bulavin et al. 1999), and conditions that mimic the tumor environment such as serum withdrawal and substrate detachment (Fassetta et al. 2006; Porras et al. 2004). In some cases, apoptotic stimuli trigger p38α activation via secondary effects such as ROS production or the induction of DNA damage (Fig. 1). In particular, the ability of p38α to sense oncogene-induced ROS and to induce apoptosis is likely to play a key role in the suppression of tumor initiation (Dolado et al. 2007).

Several mechanisms have been proposed to underlie the induction of apoptosis by p38α (Fig. 1), including the phosphorylation-dependent inactivation of the pro-survival proteins Bcl-2 and Bcl_{xL}, as well as the activation of the pro-apoptotic proteins BAD, Bim, Bax, and Bak. BAD activation may be triggered by the PP2A-mediated inhibition of ERK1/ERK2 and Akt, three kinases that can phosphorylate and directly inhibit BAD function (Table 3). In addition, *de novo* transcription of pro-apoptotic genes such as Fas, Bax, and Apaf-1 by the transcription factors p53, E2F1, or STAT3 can mediate the regulation of both the extrinsic and intrinsic apoptotic pathways by p38α (Bulavin et al. 1999; Hou et al. 2002; Porras et al. 2004; Sanchez-Prieto et al. 2000). Furthermore, a recent report has found that p38α may sensitize cells to apoptosis by phosphorylating the epidermal growth factor receptor (EGFR) and inducing its internalization (Zwang and Yarden 2006). Studies in mice also support a pro-apoptotic role for p38α *in vivo* (Wada and Penninger 2004; Yang et al. 2006).

Whether these different mechanisms operate simultaneously in p38α-induced apoptosis or their individual contribution is regulated in a context-dependent manner are still open questions. It is noteworthy, however, that whereas p38α has been shown to promote apoptosis in fibroblasts and endothelial cells through the p38α-induced and PP2A-dependent inactivation of the ERK1/ERK2 and Akt survival pathways (Grethe and Porn-Ares 2006; Porras et al. 2004; Zuluaga et al. 2007), this mechanism might not be operative in tumorigenic cell lines (Li et al. 2003). Indeed, substrate attachment has been shown to be necessary for the PP2A-mediated dephosphorylation of Akt in non-transformed cells (Zuluaga et al. 2007), suggesting that alterations in cell adhesion proteins usually found in cancer cells (i.e. downregulation of E-cadherin or miss-expression of integrins) may contribute to impair, at least in part, p38α-mediated pro-apoptotic signaling in tumor cells.

In addition to the malignant state of the cell, the stimulus type may also determine how p38α induces apoptosis. For instance, the p38α-mediated inhibition of the ERK1/ERK2 pathway seems to be consistently used for membrane receptor-induced apoptosis (Grethe and Porn-Ares 2006; Porras et al. 2004), whereas p38α and p53 have been linked in several cases of DNA damage-induced apoptosis (Bulavin et al. 2001; Kwon et al. 2002; Sanchez-Prieto et al. 2000). As mentioned above, p38α can contribute to p53 stabilization and activation by several mechanisms, including direct phosphorylation by p38α (Bulavin and Fornace 2004) as well as MK-2 mediated inhibition of Hdm-C (Weber et al. 2005). Furthermore, we have recently identified a new p38α substrate named ZnF-HIT1 or $p18^{Hamlet}$ that contributes to DNA damage-induced apoptosis by stimulating the expression of p53-regulated pro-apoptotic genes (Cuadrado et al. 2007).

3.2 Anti-apoptotic roles

In contrast to the well-established pro-apoptotic effects of p38α, several studies have also described previously unexpected pro-survival roles (Table 4). These anti-apoptotic effects seem to be mainly related to the inflammatory response (Kumar et al. 2003; Saklatvala 2004) (see Section 5), as p38α mediates cell

Table 4. p38α as an inhibitor of apoptosis

Cell line	Inducer	Mechanism[a]	Reference
Myocytes	TNF-α	↑ NF-κB/IL-6	(Craig et al. 2000)
Jurkat T cells	-	N.D.	(Nemoto et al. 1998)
Leukocytes	-	Caspases-3/8 inactivation by phosphorylation	(Alvarado-Kristensson et al. 2004)
CLL B cells	-	↑ MMP-9	(Ringshausen et al. 2004)
CLL B cells	mda-7/IL-24	↑ IL-2	(Sainz-Perez et al. 2006)
Thymocytes, Jurkat	Photodynamic therapy	N.D.	(Cappellini et al. 2005)
Macrophages	Lipopolysaccharide	↑ NF-κB-regulated genes	(Park et al. 2002b)
HeLa, T24	Photodynamic therapy	↑ COX-2	(Hendrickx et al. 2003)
Daudi	Etoposide	Induction of G2/M arrest	(Kurosu et al. 2005)
U1810 NSCLC	γ-irradiation	↑ Ku86, ↑ Ku-DNA-binding (increased DNA repair)	(Cosaceanu et al. 2006)
PC-12	TNF-α	Induction of differentiation	(Park et al. 2002a)
MCF-7	TNF-α	↑ NF-κB	(Weldon et al. 2004)

[a] N.D., not determined; ↑, upregulation

survival in several cases by inducing anti-apoptotic inflammatory signals, such as the cytokine interleukin (IL)-6 or the transcriptional regulator nuclear factor *k*B (NF-*k*B) (Table 4). However, independently of its role in inflammation, p38α can also protect differentiated PC12 tumor cells from apoptosis induced by tumor necrosisfactor (TNF)-α (Park et al. 2002a). Since differentiated, non-proliferative cells are usually refractory to apoptosis, it is plausible that p38α may sometimes mediate cell survival by inducing differentiation, for example in neuronal cells (Mao et al. 1999), and cell cycle arrest. Accordingly, p38α can induce a quiescent cancer state, known as cancer dormancy, which has been proposed to be important for the acquisition of drug resistance in cancer (Ranganathan et al. 2006).

p38α has also been involved in the establishment of the G2/M checkpoint in response to various cytotoxic agents (see Section 2.2). This cellular response is aimed at repairing damaged DNA and it is therefore necessary for normal cell homeostasis, but it may also potentially lead to apoptosis resistance in cancer cells by antagonizing chemotherapy-induced DNA damage (Kawabe 2004; Reinhardt et al. 2007). Indeed, one of the hallmarks of cancer cell resistance to apoptosis is enhanced DNA-repair activity (Kohno et al. 2005). The involvement of p38α in DNA-repair signaling suggests that after certain stimuli p38α might, instead of triggering apoptosis, induce cell cycle arrest and DNA repair, thus, protecting cancer cells from apoptosis. In agreement with this idea, p38α activation has been associated with apoptosis-protective effects in hematopoietic cancer cells exposed to DNA-damaging agents such as etoposide (Kurosu et al. 2005) and photodynamic therapy-induced ROS (Cappellini et al. 2005). Furthermore, p38α mediates cell survival in U1810 non-small-cell lung cancer cells exposed to γ-irradiation by increasing the expression and activity of the DNA-repair proteins Ku86 and Ku70 (Cosaceanu et al. 2006).

In summary, the implication of p38α in DNA repair and cell differentiation (see Section 4) illustrates how tumor-suppressive mechanisms of normal cells can sometimes be switched to promote survival in cancer cells.

3.3 Reconciling pro- and anti-apoptotic functions

In the light of the above considerations, the bottom-line question is why the activation of p38α leads to apoptosis induction in some cases whereas it enhances survival in others. A popular explanation is that cell type-specific differences may account in most cases for the variability observed. However, other hypothesis may be proposed, such as the influence of the signal intensity/duration or the cross-talk with other signaling pathways.

The biological output of p38α activation might be determined by the intensity/duration of the signal. Hence, high activation of p38α following chemotherapeutic drugs would normally lead to apoptosis, whereas milder stimuli such as cytokines could result in the expression of anti-apoptotic proteins (i.e. IL-6) and the subsequent enhancement of cell survival. Indeed, cisplatin is known to require high and sustained activation of p38α for apoptosis induction, whereas apoptotic resistance correlates with transient activation of p38α (Brozovic et al. 2004; Losa et al. 2003; Mansouri et al. 2003). In agreement with this, whereas acute exposure to cisplatin leads to p38α-mediated apoptosis in a wide range of tumor cell lines (Table 3), chronic exposure of HeLa or MCF7 cells to increasing doses of cisplatin or TNF-α, respectively, leads to the development of p38α-dependent drug resistance (Brozovic et al. 2004; Weldon et al. 2004). These two antagonistic effects of p38α on cell survival could be explained by the differential regulation of p38α target recruitment depending on p38α activity levels, as previously proposed for other proteins such as p53 (Bensaad and Vousden 2005), MKK6 (Alonso et al. 2000), and ERK (Murphy and Blenis 2006). Thus, low levels of p38α activity may trigger the expression of pro-survival or differentiation-inducing proteins (Lluis et al. 2006; Saklatvala 2004), whereas higher p38α levels might additionally induce pro-apoptotic signaling that, due to its higher intensity or faster induction, could override the concomitant pro-survival signals. Accordingly, several p38α-mediated apoptotic mechanisms have been shown to function at the posttranslational level (i.e. Bcl-2/Bcl_{xL} inactivation by phosphorylation), arguing in favor of a faster kinetic of induction than, for example, the translational-dependent regulation of survival cytokines by p38α (Saklatvala 2004).

Another factor that may regulate the biological output of p38α activation could be the cross-talk with other signaling pathways. For instance, apart from inducing p38α activation, cisplatin activates the pro-survival PI3K/Akt pathway in breast cancer cells, which compensates for the p38α-mediated apoptotic signal (Winograd-Katz and Levitzki 2006). Akt has also been reported to inhibit the activation of p38α by MEKK3 and ASK1 in epithelial and endothelial cells (Gratton et al. 2001; Yuan et al. 2003), hence providing protection against the pro-apoptotic activity of p38α. Consequently, the interplay between the p38α and PI3K/Akt pathways may determine the extent of p38α-mediated apoptosis in response to chemotherapy treatments. In agreement with this, the chemotherapeutic agent taxol can activate pro-survival proteins such as NF-*k*B and p21-activated kinase (PAK) in parallel to p38α, which also results in a reduced apoptotic effect (Olson and Hallahan 2004). The JNK pathway, which is activated by most p38α-activating stimuli due to the sharing of several upstream regulators by both MAPK pathways (Ichijo 1999), can also sometimes have opposite effects to p38α on apoptosis-inducing proteins, such as BAD (Tourian et al. 2004). In conclusion, the activation of p38α may not necessarily always lead to apoptosis induction, due to the opposing effects of pro-survival pathways that are concomitantly activated by the same stimuli.

Finally, the relative activity levels of different p38 MAPK family members may also explain the controversial effect of p38α activation on cell survival. In particular, p38β has been proposed to have anti-apoptotic effects in various cell lines (Kaiser et al. 2004; Nemoto et al. 1998; Silva et al. 2006).

4 Regulation of cell differentiation

The first link between cancer and deregulated cell differentiation was the identification of embryonic biochemical markers in hepatomas, which led to the suggestion that lack of differentiation might contribute to tumor formation (Potter 1978). Subsequently, numerous reports have confirmed the link between cancer and cellular de-differentiation. For instance, the ability of several oncogenes to block cell differentiation has been proposed to mediate their transforming activity, for example Notch in breast, pancreatic, and lymphoid cancers (Sjolund et al. 2005) and Bcr-Abl in lymphoblastic leukemia (Klein et al. 2006). Furthermore, the early observation that most cells within a tumor are differentiated and weakly tumorigenic, has lead to the formulation of a recent theory that postulates that a few undifferentiated cancer stem cells within a tumor (i.e. $< 1\%$) are the true initiators and sustainers of cancer (Houghton et al. 2006). This has been supported by the isolation of cancer cells with stem cell-like properties from several human tumors and cancer cell lines (Dean et al. 2005). Consequently, treatments that specifically induce stem cell differentiation might represent an attractive approach for cancer therapy (Edsjo et al. 2006).

Along these lines, p38α is emerging as an important regulator of differentiation in several cell types, including adipocytes, neurons, myocytes, and hematopoietic cells (Lluis et al. 2006; Nebreda and Porras 2000; Uddin et al. 2004). Indeed, p38α can orchestrate the cellular differentiation process by multiple mechanisms, such as by activating differentiation-inducing transcription factors (i.e. MyoD and MEF2), by promoting cell cycle exit prior to the onset of differentiation (Perdiguero et al. 2007) or by targeting chromatin-remodeling enzymes to specific loci, thereby inducing the transcription of differentiation-specific genes (for further details see the chapters by Perdiguero and Muñoz-Cánoves and by Crump et al.). *In vivo* studies with mice have also recently shown the importance of p38α for myoblast differentiation (Perdiguero et al. 2007) as well as for maintenance of the differentiated state in adult cardiomyocytes (Engel et al. 2005), although in both cases the primary effect of p38α seems to be at the level of the cell cycle arrest required for differentiation.

Interestingly, forced activation of p38α in human cancer cell lines such as muscle rhabdomyosarcoma (Puri et al. 2000) and renal carcinoma A-498 cells (Finn et al. 2004) triggers a more differentiated and less transformed phenotype. We have also recently found that p38α-deficient mice are highly sensitized to K-Ras^{G12V}-induced lung tumorigenesis, which is mostly due to the inability of lung epithelial cells to undergo accurate differentiation in the absence of p38α (Ventura et al., submitted). These results provide *in vivo* evidence for the relevance of the differentiation-inducing activity of p38α for tumor suppression. p38α has also been shown to induce the *in vitro* differentiation of embryonic stem cells into cardiomyocytes (Aouadi et al. 2006; Schmelter et al. 2006), as well as to mediate a proliferation arrest, which was concomitant with the onset of differentiation, in hepatocyte growth factor-treated mesenchymal stem cells (Forte et al. 2006). Whether p38α might also induce the differentiation of cancer stem cells needs further investigation.

5 Inflammation

The roles of p38α in the production of pro-inflammatory cytokines and in the signal rely from cytokine receptors have been intensively studied. This work has led to the development of several p38α inhibitors currently undergoing clinical trials for inflammatory diseases (Kumar et al. 2003; O'Neill 2006; Saklatvala 2004). However, there is some concern that inhibition of p38α to ameliorate chronic inflammation might result in a higher predisposition to cancer, given the evidence in support of p38α as a tumor suppressor (Bulavin and Fornace 2004; Dolado et al. 2007). On the other hand, chronic inflammation is also a potent cancer promoter (Baniyash 2006; Karin 2006; Philip et al. 2004), which has been linked to enhanced cancer cell survival as well as to the induction of DNA damage, angiogenesis and invasion. Thus, the pro-inflammatory role of p38α may not only contribute to cancer progression but also compromise cancer treatment by increasing the sensitivity of normal tissues to chemotherapeutic drugs (Li et al. 2006; Ramesh and Reeves 2005).

p38α can induce the expression of the pro-inflammatory protein cyclooxygenase (COX)-2 (Park et al. 2002b), which has been correlated with bad prognosis in breast cancer as well as with the development of drug resistance in bladder

(T24) and cervix (HeLa) carcinoma cell lines. Indeed, COX-2 may promote cancer progression by enhancing both cell survival (Surh et al. 2001) and invasivity (Timoshenko et al. 2006). Furthermore, the pro-inflammatory cytokines TNF-α and IL-1, the pro-survival cytokines IL-2 and IL-6 and the angiogenic cytokine IL-8 have all been shown to be induced posttranscriptionally by p38α, some times indirectly through p38α-regulated kinases such as MK-2 and mitogen- and stress-activated kinase (MSK)-1 (Arthur and Darragh 2006; Kumar et al. 2003; Saklatvala 2004). In addition, p38α may induce the transcription of some of these cytokines by enhancing the activity of transcription factors such as NF-*k*B (Carter et al. 1999; Karin 2006). Thus, p38α can cooperate in cancer progression by inducing angiogenesis and invasion both via direct mRNA stabilization and indirectly by NF-*k*B-mediated transcription of pro-inflammatory proteins.

6 Cell migration and invasion

p38α may also have a direct role in tumor invasion and angiogenesis, independently of its role in inflammation. For instance, p38α has been shown to induce the expression of metalloproteinases (MMPs) such as MMP-1, MMP-3, and MMP-13 (Ono and Han 2000; Saklatvala 2004), which are key proteins for matrix remodeling and degradation by metastatic cells (Coussens et al. 2002). Vascular endothelial growth factor (VEGF), a potent inducer of tumor survival and angiogenesis (Carmeliet 2005), has been also shown to be expressed in a p38α-dependent manner in various cytokine-stimulated cellular systems (Wang et al. 2004a; Yamamoto et al. 2001; Yoshino et al. 2006). In addition, p38α may cooperate as well in the overexpression of VEGF and other angiogenic factors in hypoxic tumors in a process known as the "angiogenic switch", which correlates with enhanced cancer aggressiveness and bad prognosis (Hanahan and Folkman 1996; Zhou et al. 2006). Indeed, p38α can activate hypoxia-inducible factor (HIF)-1, a transcription factor that plays a key role in the hypoxia-driven expression of angiogenic factors, at least in part through the stabilization of its α-subunit (Emerling et al. 2005; Nakayama et al. 2007; Shemirani and Crowe 2002). In contrast, p38α has been attributed a metastasis suppressor role in human ovarian cancer (Hickson et al. 2006), although this has not been confirmed to date in other tumor types, including prostate, breast and pancreatic cancers (Vander Griend et al. 2005; Wang et al. 2004b). Furthermore, recent experiments using mouse xenografts and cultured cancer cells support a role for p38α in lung metastasis (Hiratsuka et al. 2006; Matsuo et al. 2006).

In addition to help creating the right conditions for metastasis, by enhancing tumor angiogenesis (i.e. VEGF expression) and matrix degradation (i.e. MMPs expression), p38α may also regulate cancer cell migration. Thus, p38α mediates the migration of IL-12-stimulated HeLa cells and Ras-transformed breast (MCF10A) and pancreatic (PANC-1) cell lines (Dreissigacker et al. 2006; Kim et al. 2003), as well as of other cell types stimulated by several chemotactic stimuli (McMullen et al. 2005; Rousseau et al. 2006). At the mechanistic level, the role of p38α in cell migration seems to rely mostly on its ability to induce actin polymerization and cytoskeleton remodeling through its downstream kinase MK-2, which phosphorylates the protein Hsp27 and induces its release from F-actin caps (Rousseau et al. 2006). Additionally, MK-2 may also activate the protein kinase LIMK1, which in turn phosphorylates and inactivates the actin-depolymerazing protein cofilin (Kobayashi et al. 2006). A recent report has also implicated p38α in epithelial cell migration, although via a more indirect mechanism. Namely, p38α can induce the phosphorylation of ligand-bound EGFR at tyrosine-1045, which triggers EGFR degradation. This p38α effect prevents recycling of the active EGFR receptor from the early endosomes back to the plasma membrane and seems to be associated with a proliferation-to-migration shift in the cellular response (Frey et al. 2006).

7 Concluding remarks

p38α has been proposed to have a tumor suppressor role based on its ability to negatively regulate cell proliferation and to induce cell death (Bulavin and Fornace 2004). Accordingly, a variety of anti-cancer drugs require a functional p38α pathway for efficient action (Olson and Hallahan 2004). However, most of these studies have been limited to the investigation of p38α functions in the context of the cancer cell and during tumor initiation. On the other hand, the potential implication of p38α in the interplay between tumor cells and the extracellular matrix or the immune system are largely unknown, despite the key roles of these interactions for cancer progression (Baniyash 2006; Comoglio and Trusolino 2005).

Consequently, we do not have much information on the roles of p38α in advanced cancer stages, as well as in the regulation of tumor environment-induced paracrine signaling. This might underlie the apparent controversy on the role of p38α in cancer. Thus, in spite of the many tumor-suppressive functions of p38α (see Sections 2.1, 2.2, 3.1, and 4), increased levels of phosphorylated p38α have been correlated with malignancy in follicular lymphoma (Elenitoba-Johnson et al. 2003), as well as in non-small-cell lung, thyroid and lymph node-positive breast carcinomas (Esteva et al. 2004; Greenberg et al. 2002; Pomerance et al. 2006). Furthermore, p38α has been shown to contribute to the invasiveness of breast, prostate and pancreatic cancers (Chen et al. 2004; Dreissigacker et al. 2006; Kim et al. 2003), as well as to the maintenance of the neoplastic phenotype in Ras-transformed human fibroblasts and epithelial cells (Weijzen et al. 2002). In addition, the use of chemical inhibitors such as SB203580 has shown the requirement of p38α (and maybe also p38β) for the proliferation of cancer cell lines, including chondrosarcoma, prostate carcinoma and melanoma cells (Halawani et al. 2004; Recio and Merlino 2002; Ricote et al. 2006). Thus, whereas p38α may suppress tumor initiation, it seems to serve oncogenic functions in cancer progression. Perhaps, this explains why p38α has not been found mutated or downregulated in human cancers to date, in contrast to other well-established tumor suppressor proteins (Johnstone et al. 2002).

Given the complex network of tumorigenesis-related functions coordinated by p38α, it is likely that both the type and stage of the tumor will have to be carefully taken into account in any rationale attempt to modulate p38α activity for cancer therapy. Hence, whereas p38α activation by chemotherapeutic agents may prove beneficial for first-line treatment of early-stage non-recurrent solid tumors (which are naturally sensitized to apoptosis), it might have little effect or even result hazardous in apoptosis-resistant metastatic cancers, which might in turn take advantage of p38α activity to further migrate and invade or to enter an apoptosis-resistant but reversible dormant state (Ranganathan et al. 2006). Similarly, chemotherapy-induced activation of p38α might result in therapeutic benefit in well-irrigated solid tumors, especially in combination with anti-angiogenic therapy to target for example VEGF. However, the same approach might be questionable in the case of hypoxic tumors that, apart from being less sensitive to chemotherapy due to poor blood irrigation, may instead capitalize on p38α activation by the treatment-induced inflammatory response to enhance angiogenesis and further invasion. In contrast, p38α inhibition might be of therapeutic benefit in these types of aggressive and resistant tumors (i.e. metastatic and hypoxic), if combined with simultaneous activation of p38α-independent apoptotic signaling. Another potential use of the p38α inhibitors could be in combination with drugs that induce DNA damage, because cancer cell death could be stimulated by impairing p38α-mediated cell cycle arrest and repair mechanisms (Reinhardt et al. 2007). Further work, including the use of specific mouse tumor models, should help to better define the roles of p38α in tumorigenesis and the potential value of this signaling pathway as a target for cancer therapy.

Acknowledgments

We thank members of our group for their contribution and helpful comments. Work in our laboratory is supported by grants from Ministerio de Educación y Ciencia and Fundación Científica de la Asociación Española Contra el Cáncer.

References

Aguirre-Ghiso JA, Ossowski L, Rosenbaum SK (2004) Green fluorescent protein tagging of extracellular signal-regulated kinase and p38 pathways reveals novel dynamics of pathway activation during primary and metastatic growth. Cancer Res 64:7336-7345

Alonso G, Ambrosino C, Jones M, Nebreda AR (2000) Differential activation of p38 mitogen-activated protein kinase isoforms depending on signal strength. J Biol Chem 275:40641-40648

Alvarado-Kristensson M, Melander F, Leandersson K, Ronnstrand L, Wernstedt C, Andersson T (2004) p38-MAPK signals survival by phosphorylation of caspase-8 and caspase-3 in human neutrophils. J Exp Med 199:449-458

Aouadi M, Bost F, Caron L, Laurent K, Le Marchand Brustel Y, Binetruy B (2006) p38 mitogen-activated protein kinase activity commits embryonic stem cells to either neurogenesis or cardiomyogenesis. Stem Cells 24:1399-1406

Arthur JS, Darragh J (2006) Signaling downstream of p38 in psoriasis. J Invest Dermatol 126:1689-1691

Bacus SS, Gudkov AV, Lowe M, Lyass L, Yung Y, Komarov AP, Keyomarsi K, Yarden Y, Seger R (2001) Taxol-induced apoptosis depends on MAP kinase pathways (ERK and p38) and is independent of p53. Oncogene 20:147-155

Baniyash M (2006) Chronic inflammation, immunosuppression and cancer: new insights and outlook. Semin Cancer Biol 16:80-88

Bensaad K, Vousden KH (2005) Savior and slayer: the two faces of p53. Nat Med 11:1278-1279

Brancho D, Tanaka N, Jaeschke A, Ventura JJ, Kelkar N, Tanaka Y, Kyuuma M, Takeshita T, Flavell RA, Davis RJ (2003) Mechanism of p38 MAP kinase activation *in vivo*. Genes Dev 17:1969-1978

Brozovic A, Fritz G, Christmann M, Zisowski J, Jaehde U, Osmak M, Kaina B (2004) Long-term activation of SAPK/JNK, p38 kinase and Fas-L expression by cisplatin is attenuated in human carcinoma cells that acquired drug resistance. Int J Cancer 112:974-985

Bulavin DV, Demidov ON, Saito S, Kauraniemi P, Phillips C, Amundson SA, Ambrosino C, Sauter G, Nebreda AR, Anderson CW, Kallioniemi A, Fornace AJ, Appella E (2002) Amplification of PPM1D in human tumors abrogates p53 tumor-suppressor activity. Nat Genet 31:210-215

Bulavin DV, Fornace AJ Jr (2004) p38 MAP kinase's emerging role as a tumor suppressor. Adv Cancer Res 92:95-118

Bulavin DV, Higashimoto Y, Popoff IJ, Gaarde WA, Basrur V, Potapova O, Appella E, Fornace AJ Jr (2001) Initiation of a G2/M checkpoint after ultraviolet radiation requires p38 kinase. Nature 411:102-107

Bulavin DV, Phillips C, Nannenga B, Timofeev O, Donehower LA, Anderson CW, Appella E, Fornace AJ Jr (2004) Inactivation of the Wip1 phosphatase inhibits mammary tumorigenesis through p38 MAPK-mediated activation of the p16(Ink4a)-p19(Arf) pathway. Nat Genet 36:343-350

Bulavin DV, Saito S, Hollander MC, Sakaguchi K, Anderson CW, Appella E, Fornace AJ Jr (1999) Phosphorylation of human p53 by p38 kinase coordinates N-terminal phosphorylation and apoptosis in response to UV radiation. EMBO J 18:6845-6854

Cai B, Chang SH, Becker EB, Bonni A, Xia Z (2006) p38 MAP kinase mediates apoptosis through phosphorylation of BimEL at Ser-65. J Biol Chem 281:25215-25222

Cappellini A, Tazzari PL, Mantovani I, Billi AM, Tassi C, Ricci F, Conte R, Martelli AM (2005) Antiapoptotic role of p38 mitogen activated protein kinase in Jurkat T cells and normal human T lymphocytes treated with 8-methoxypsoralen and ultraviolet-A radiation. Apoptosis 10:141-152

Carmeliet P (2005) VEGF as a key mediator of angiogenesis in cancer. Oncology 69 Suppl 3:4-10

Carter AB, Knudtson KL, Monick MM, Hunninghake GW (1999) The p38 mitogen-activated protein kinase is required for NF-kappaB-dependent gene expression. The role of TATA-binding protein (TBP). J Biol Chem 274:30858-30863

Casanovas O, Jaumot M, Paules AB, Agell N, Bachs O (2004) P38SAPK2 phosphorylates cyclin D3 at Thr-283 and targets it for proteasomal degradation. Oncogene 23:7537-7544

Casanovas O, Miro F, Estanyol JM, Itarte E, Agell N, Bachs O (2000) Osmotic stress regulates the stability of cyclin D1 in a p38SAPK2-dependent manner. J Biol Chem 275:35091-35097

Chen G, Hitomi M, Han J, Stacey DW (2000) The p38 pathway provides negative feedback for Ras proliferative signaling. J Biol Chem 275:38973-38980

Chen L, He HY, Li HM, Zheng J, Heng WJ, You JF, Fang WG (2004) ERK1/2 and p38 pathways are required for P2Y receptor-mediated prostate cancer invasion. Cancer Lett 215:239-247

Chen YC, Chen SY, Ho PS, Lin CH, Cheng YY, Wang JK, Sytwu HK (2006) Apoptosis of T-leukemia and B-myeloma cancer cells induced by hyperbaric oxygen increased phosphorylation of p38 MAPK. Leuk Res 31:805-815

Choi S-Y, Kim M-J, Kang C-M, Bae S, Cho C-K, Soh J-W, Kim J-H, Kang S, Chung HY, Lee Y-S, Lee S-J (2006) Activation of Bak and Bax through c-Abl-protein kinase Cδ-p38 MAPK signaling in response to ionizing radiation in human non-small cell lung cancer cells. J Biol Chem 281:7049-7059

Coltella N, Rasola A, Nano E, Bardella C, Fassetta M, Filigheddu N, Graziani A, Comoglio PM, Di Renzo MF (2006) p38 MAPK turns hepatocyte growth factor to a death signal that commits ovarian cancer cells to chemotherapy-induced apoptosis. Int J Cancer 118:2981-2990

Comoglio PM, Trusolino L (2005) Cancer: the matrix is now in control. Nat Med 11:1156-1159

Cosaceanu D, Budiu RA, Carapancea M, Castro J, Lewensohn R, Dricu A (2006) Ionizing radiation activates IGF-1R triggering a cytoprotective signaling by interfering with Ku-DNA binding and by modulating Ku86 expression via a p38 kinase-dependent mechanism. Oncogene 26:2423-2434

Coussens LM, Fingleton B, Matrisian LM (2002) Matrix metalloproteinase inhibitors and cancer: trials and tribulations. Science 295:2387-2392

Cowley S, Paterson H, Kemp P, Marshall CJ (1994) Activation of MAP kinase kinase is necessary and sufficient for PC12 differentiation and for transformation of NIH 3T3 cells. Cell 77:841-852

Craig R, Larkin A, Mingo AM, Thuerauf DJ, Andrews C, McDonough PM, GlEMBOtski CC (2000) p38 MAPK and NF-kappa B collaborate to induce interleukin-6 gene expression and release. Evidence for a cytoprotective autocrine signaling pathway in a cardiac myocyte model system. J Biol Chem 275:23814-23824

Crawley JB, Rawlinson L, Lali FV, Page TH, Saklatvala J, Foxwell BM (1997) T cell proliferation in response to interleukins 2 and 7 requires p38MAP kinase activation. J Biol Chem 272:15023-15027

Cuadrado A, Lafarga V, Cheung PCF, Dolado I, Llanos S, Cohen P, Nebreda AR (2007) A new p38 MAP kinase-regulated transcriptional co-activator that stimulates p53-dependent apoptosis. EMBO J 26:2115-2126

De Chiara G, Marcocci ME, Torcia M, Lucibello M, Rosini P, Bonini P, Higashimoto Y, Damonte G, Armirotti A, Amodei S, Palamara AT, Russo T, Garaci E, Cozzolino F (2006) Bcl-2 Phosphorylation by p38 MAPK: identification of target sites and biologic consequences. J Biol Chem 281:21353-21361

De Nadal E, Zapater M, Alepuz PM, Sumoy L, Mas G, Posas F (2004) The MAPK Hog1 recruits Rpd3 histone deacetylase to activate osmoresponsive genes. Nature 427:370-374

Deacon K, Mistry P, Chernoff J, Blank JL, Patel R (2003) p38 Mitogen-activated protein kinase mediates cell death and p21-activated kinase mediates cell survival during chemotherapeutic drug-induced mitotic arrest. Mol Biol Cell 14:2071-2087

Dean M, Fojo T, Bates S (2005) Tumour stem cells and drug resistance. Nat Rev Cancer 5:275-284

Dmitrieva NI, Bulavin DV, Fornace AJ Jr, Burg MB (2002) Rapid activation of G2/M checkpoint after hypertonic stress in renal inner medullary epithelial (IME) cells is protective and requires p38 kinase. Proc Natl Acad Sci USA 99:184-189

Dolado I, Swat A, Ajenjo N, De Vita G, Cuadrado A, Nebreda AR (2007) p38alpha MAP kinase as a sensor of reactive oxygen species in tumorigenesis. Cancer Cell 11:191-205

Dreissigacker U, Mueller MS, Unger M, Siegert P, Genze F, Gierschik P, Giehl K (2006) Oncogenic K-Ras down-regulates Rac1 and RhoA activity and enhances migration and invasion of pancreatic carcinoma cells through activation of p38. Cell Signal 18:1156-1168

Edsjo A, Holmquist L, Pahlman S (2006) Neuroblastoma as an experimental model for neuronal differentiation and hypoxia-induced tumor cell dedifferentiation. Semin Cancer Biol 17:248-256

Elenitoba-Johnson KS, Jenson SD, Abbott RT, Palais RA, Bohling SD, Lin Z, Tripp S, Shami PJ, Wang LY, Coupland RW, Buckstein R, Perez-Ordonez B, Perkins SL, Dube ID, Lim MS (2003) Involvement of multiple signaling pathways in follicular lymphoma transformation: p38-mitogen-activated protein kinase as a target for therapy. Proc Natl Acad Sci USA 100:7259-7264

Emerling BM, Platanias LC, Black E, Nebreda AR, Davis RJ, Chandel NS (2005) Mitochondrial reactive oxygen species activation of p38 mitogen-activated protein kinase is required for hypoxia signaling. Mol Cell Biol 25:4853-4862

Engel FB, Schebesta M, Duong MT, Lu G, Ren S, Madwed JB, Jiang H, Wang Y, Keating MT (2005) p38 MAP kinase inhibition enables proliferation of adult mammalian cardiomyocytes. Genes Dev 19:1175-1187

Engelberg D (2004) Stress-activated protein kinases-tumor suppressors or tumor initiators? Semin Cancer Biol 14:271-282

Esteva FJ, Sahin AA, Smith TL, Yang Y, Pusztai L, Nahta R, Buchholz TA, Buzdar AU, Hortobagyi GN, Bacus SS (2004) Prognostic significance of phosphorylated P38 mitogen-activated protein kinase and HER-2 expression in lymph node-positive breast carcinoma. Cancer 100:499-506

Fan L, Yang X, Du J, Marshall M, Blanchard K, Ye X (2005) A novel role of p38alpha MAPK in mitotic progression independent of its kinase activity. Cell Cycle 4:1616-1624

Fan M, Chambers TC (2001) Role of mitogen-activated protein kinases in the response of tumor cells to chemotherapy. Drug Resist Updat 4:253-267

Farley N, Pedraza-Alva, G, Serrano-Gomez D, Nagaleekar V, Aronshtam A, Krahl T, Thornton T, Rincon M (2006) p38 mitogen-activated protein kinase mediates the Fas-induced mitochondrial death pathway in CD8+ T cells. Mol Cell Biol 26:2118-2129

Fassetta M, D'Alessandro L, Coltella N, Di Renzo MF, Rasola A (2006) Hepatocyte growth factor installs a survival platform for colorectal cancer cell invasive growth and overcomes p38 MAPK-mediated apoptosis. Cell Signal 18:1967-1976

Faust D, Dolado I, Cuadrado A, Oesch F, Weiss C, Nebreda AR, Dietrich C (2005) p38alpha MAPK is required for contact inhibition. Oncogene 24:7941-7945

Finn GJ, Creaven BS, Egan DA (2004) Daphnetin induced differentiation of human renal carcinoma cells and its mediation by p38 mitogen-activated protein kinase. Biochem Pharmacol 67:1779-1788

Forte G, Minieri M, Cossa P, Antenucci D, Sala M, Gnocchi V, Fiaccavento R, Carotenuto F, De Vito P, Baldini PM, Prat M, Di Nardo P (2006) Hepatocyte growth factor effects on mesenchymal stem cells: proliferation, migration, and differentiation. Stem Cells 24:23-33

Freshney NW, Rawlinson L, Guesdon F, Jones E, Cowley S, Hsuan J, Saklatvala J (1994) Interleukin-1 activates a novel protein kinase cascade that results in the phosphorylation of Hsp27. Cell 78:1039-1049

Frey MR, Dise RS, Edelblum KL, Polk DB (2006) p38 kinase regulates epidermal growth factor receptor downregulation and cellular migration. EMBO J 25:5683-5692

Garner AP, Weston CR, Todd DE, Balmanno K, Cook SJ (2002) Delta MEKK3:ER* activation induces a p38 alpha/beta 2-dependent cell cycle arrest at the G2 checkpoint. Oncogene 21:8089-8104

Gauthier ML, Pickering CR, Miller CJ, Fordyce CA, Chew KL, Berman HK, Tlsty TD (2005) p38 regulates cyclooxygenase-2 in human mammary epithelial cells and is activated in premalignant tissue. Cancer Res 65:1792-1799

Godoy-Tundidor S, Cavarretta IT, Fuchs D, Fiechtl M, Steiner H, Friedbichler K, Bartsch G, Hobisch A, Culig Z (2005) Interleukin-6 and oncostatin M stimulation of proliferation of prostate cancer 22Rv1 cells through the signaling pathways of p38 mitogen-activated protein kinase and phosphatidylinositol 3-kinase. Prostate 64:209-216

Gollob JA, Wilhelm S, Carter C, Kelley SL (2006) Role of Raf kinase in cancer: therapeutic potential of targeting the Raf/MEK/ERK signal transduction pathway. Semin Oncol 33:392-406

Goloudina A, Yamaguchi H, Chervyakova DB, Appella E, Fornace AJ Jr, Bulavin DV (2003) Regulation of human Cdc25A stability by Serine 75 phosphorylation is not sufficient to activate a S phase checkpoint. Cell Cycle 2:473-478

Gratton JP, Morales-Ruiz M, Kureishi Y, Fulton D, Walsh K, Sessa WC (2001) Akt down-regulation of p38 signaling provides a novel mechanism of vascular endothelial growth factor-mediated cytoprotection in endothelial cells. J Biol Chem 276:30359-30365

Greenberg AK, Basu S, Hu J, Yie TA, Tchou-Wong KM, Rom WN, Lee TC (2002) Selective p38 activation in human non-small cell lung cancer. Am J Respir Cell Mol Biol 26:558-564

Grethe S, Ares MP, Andersson T, Porn-Ares MI (2004) p38 MAPK mediates TNF-induced apoptosis in endothelial cells via phosphorylation and downregulation of Bcl-x(L). Exp Cell Res 298:632-642

Grethe S, Coltella N, Di Renzo MF, Porn-Ares MI (2006) p38 MAPK downregulates phosphorylation of Bad in doxorubicin-induced endothelial apoptosis. Biochem Biophys Res Commun 347:781-790

Grethe S, Porn-Ares MI (2006) p38 MAPK regulates phosphorylation of Bad via PP2A-dependent suppression of the MEK1/2-ERK1/2 survival pathway in TNF-alpha induced endothelial apoptosis. Cell Signal 18:531-540

Halawani D, Mondeh R, Stanton LA, Beier F (2004) p38 MAP kinase signaling is necessary for rat chondrosarcoma cell proliferation. Oncogene 23:3726-3731

Han J, Lee JD, Bibbs L, Ulevitch RJ (1994) A MAP kinase targeted by endotoxin and hyperosmolarity in mammalian cells. Science 265:808-811

Hanahan D, Folkman J (1996) Patterns and emerging mechanisms of the angiogenic switch during tumorigenesis. Cell 86:353-364

Hendrickx N, Volanti C, Moens U, Seternes OM, de Witte P, Vandenheede JR, Piette J, Agostinis P (2003) Up-regulation of cyclooxygenase-2 and apoptosis resistance by p38 MAPK in hypericin-mediated photodynamic therapy of human cancer cells. J Biol Chem 278:52231-52239

Herrera B, Carracedo A, Diez-Zaera M, Guzman M, Velasco G (2005) p38 MAPK is involved in CB2 receptor-induced apoptosis of human leukaemia cells. FEBS Lett 579:5084-5088

Hickson JA, Huo D, Vander Griend DJ, Lin A, Rinker-Schaeffer CW, Yamada SD (2006) The p38 kinases MKK4 and MKK6 suppress metastatic colonization in human ovarian carcinoma. Cancer Res 66:2264-2270

Hideshima T, Akiyama M, Hayashi T, Richardson P, Schlossman R, Chauhan D, Anderson KC (2003) Targeting p38 MAPK inhibits multiple myeloma cell growth in the bone marrow milieu. Blood 101:703-705

Hiratsuka S, Watanabe A, Aburatani H, Maru Y (2006) Tumour-mediated upregulation of chemoattractants and recruitment of myeloid cells predetermines lung metastasis. Nat Cell Biol 8:1369-1375

Hirose Y, Katayama M, Stokoe D, Haas-Kogan DA, Berger MS, Pieper RO (2003) The p38 mitogen-activated protein kinase pathway links the DNA mismatch repair system to the G2 checkpoint and to resistance to chemotherapeutic DNA-methylating agents. Mol Cell Biol 23:8306-8315

Hou ST, Xie X, Baggley A, Park DS, Chen G, Walker T (2002) Activation of the Rb/E2F1 pathway by the nonproliferative p38 MAPK during Fas (APO1/CD95)-mediated neuronal apoptosis. J Biol Chem 277:48764-48770

Houghton J, Morozov A, Smirnova I, Wang TC (2006) Stem cells and cancer. Semin Cancer Biol 17:191-203

Ichijo H (1999) From receptors to stress-activated MAP kinases. Oncogene 18:6087-6093

Ito K, Hirao A, Arai F, Takubo K, Matsuoka S, Miyamoto K, Ohmura M, Naka K, Hosokawa K, Ikeda Y, Suda T (2006) Reactive oxygen species act through p38 MAPK to limit the lifespan of hematopoietic stem cells. Nat Med 12:446-451

Iyoda K, Sasaki Y, Horimoto M, Toyama T, Yakushijin T, Sakakibara M, Takehara T, Fujimoto J, Hori M, Wands JR, Hayashi N (2003) Involvement of the p38 mitogen-activated protein kinase cascade in hepatocellular carcinoma. Cancer 97:3017-3026

Johnstone RW, Ruefli AA, Lowe SW (2002) Apoptosis: a link between cancer genetics and chemotherapy. Cell 108:153-164

Kaiser RA, Bueno OF, Lips DJ, Doevendans PA, Jones F, Kimball TF, Molkentin JD (2004) Targeted inhibition of p38 mitogen-activated protein kinase antagonizes cardiac injury and cell death following ischemia-reperfusion *in vivo*. J Biol Chem 279:15524-15530

Karin M (2006) Nuclear factor-kappaB in cancer development and progression. Nature 441:431-436

Kawabe T (2004) G2 checkpoint abrogators as anticancer drugs. Mol Cancer Ther 3:513-519

Kennedy NJ, Sluss HK, Jones SN, Bar-Sagi D, Flavell RA, Davis RJ (2003) Suppression of Ras-stimulated transformation by the JNK signal transduction pathway. Genes Dev 17:629-637

Kim BJ, Ryu SW, Song BJ (2006) JNK- and p38 kinase-mediated phosphorylation of Bax leads to its activation and mitochondrial translocation and to apoptosis of human hepatoma HepG2 cells. J Biol Chem 281:21256-21265

Kim GY, Mercer SE, Ewton DZ, Yan Z, Jin K, Friedman E (2002a) The stress-activated protein kinases p38 alpha and JNK1 stabilize p21(Cip1) by phosphorylation. J Biol Chem 277:29792-29802

Kim JY, Choi JA, Kim TH, Yoo YD, Kim JI, Lee YJ, Yoo SY, Cho CK, Lee YS, Lee SJ (2002b) Involvement of p38 mitogen-activated protein kinase in the cell growth inhibition by sodium arsenite. J Cell Physiol 190:29-37

Kim MS, Lee EJ, Kim HR, Moon A (2003) p38 kinase is a key signaling molecule for H-Ras-induced cell motility and invasive phenotype in human breast epithelial cells. Cancer Res 63:5454-5461

Klein F, Feldhahn N, Herzog S, Sprangers M, Mooster JL, Jumaa H, Muschen M (2006) BCR-ABL1 induces aberrant splicing of IKAROS and lineage infidelity in pre-B lymphoblastic leukemia cells. Oncogene 25:1118-1124

Kobayashi M, Nishita M, Mishima T, Ohashi K, Mizuno K (2006) MAPKAPK-2-mediated LIM-kinase activation is critical for VEGF-induced actin remodeling and cell migration. EMBO J 25:713-726

Kohno K, Uchiumi T, Niina I, Wakasugi T, Igarashi T, Momii Y, Yoshida T, Matsuo K, Miyamoto N, Izumi H (2005) Transcription factors and drug resistance. Eur J Cancer 41:2577-2586

Kumar P, Miller AI, Polverini PJ (2004) p38 MAPK mediates -irradiation-induced endothelial cell apoptosis, and vascular endothelial growth factor protects endothelial cells through the phosphoinositide-3-kinase-Akt-Bcl-2 pathway. J Biol Chem 279:43352-43360

Kumar S, Boehm J, Lee JC (2003) p38 MAP kinases: key signalling molecules as therapeutic targets for inflammatory diseases. Nat Rev Drug Discov 2:717-726

Kurosu T, Takahashi Y, Fukuda T, Koyama T, Miki T, Miura O (2005) p38 MAP kinase plays a role in G2 checkpoint activation and inhibits apoptosis of human B cell lymphoma cells treated with etoposide. Apoptosis 10:1111-1120

Kwon Y-W, Ueda S, Ueno M, Yodoi J, Masutani H (2002) Mechanism of p53-dependent apoptosis induced by 3-methylcholanthrene. J Biol Chem 277:1837-1844

Lavoie JN, L'Allemain G, Brunet A, Muller R, Pouyssegur J (1996) Cyclin D1 expression is regulated positively by the p42/p44MAPK and negatively by the p38/HOGMAPK pathway. J Biol Chem 271:20608-20616

Lee JC, Laydon JT, McDonnell PC, Gallagher TF, Kumar S, Green D, McNulty D, Blumenthal MJ, Keys JR, Landvatter SW, Strickler JE, McLaughlin MM, Siemens IR, Fisher SM, Livi GP, White JR, Adams JL, Young PR (1994) A protein kinase involved in the regulation of inflammatory cytokine biosynthesis. Nature 372:739-746

Lee RJ, Albanese C, Stenge RJ, Watanab G, Inghiram G, Haines GK 3rd, Webster M, Muller WJ, Brugge JS, Davis RJ, Pestell RG (1999) pp60(v-src) induction of cyclin D1 requires collaborative interactions between the extracellular signal-regulated kinase, p38, and Jun kinase pathways. A role for cAMP response element-binding protein and activating transcription factor-2 in pp60(v-src) signaling in breast cancer cells. J Biol Chem 274:7341-7350

Lee S-K, Jang H-J, Lee HJ, Lee J, Jeon B-H, Jun C-D, Lee S-K, Kim E-C (2006) p38 and ERK MAP kinase mediate iron chelator-induced apoptosis and suppressed differentiation of immortalized and malignant human oral keratinocytes. Life Sci 79:1419-1427

Lemaire M, Froment C, Boutros R, Mondesert O, Nebreda AR, Monsarrat B, Ducommun B (2006) CDC25B Phosphorylation by p38 and MK-2. Cell Cycle 5:1649-1653

Lewis JS, Vijayanathan V, Thomas TJ, Pestell RG, Albanese C, Gallo MA, Thomas T (2005) Activation of cyclin D1 by estradiol and spermine in MCF-7 breast cancer cells: a mechanism involving the p38 MAP kinase and phosphorylation of ATF-2. Oncol Res 15:113-128

Li J, Campanale NV, Liang RJ, Deane JA, Bertram JF, Ricardo SD (2006) Inhibition of p38 mitogen-activated protein kinase and transforming growth factor-{beta}1/Smad signaling pathways modulates the development of fibrosis in adriamycin-induced nephropathy. Am J Pathol 169:1527-1540

Li J, Yang Y, Peng Y, Austin RJ, van Eyndhoven WG, Nguyen KC, Gabriele T, McCurrach ME, Marks JR, Hoey T, Lowe SW, Powers S (2002) Oncogenic properties of PPM1D located within a breast cancer amplification epicenter at 17q23. Nat Genet 31:133-134

Li SP, Junttila MR, Han J, Kahari VM, Westermarck J (2003) p38 Mitogen-activated protein kinase pathway suppresses cell survival by inducing dephosphorylation of mitogen-activated protein/extracellular signal-regulated kinase kinase1,2. Cancer Res 63:3473-3477

Liu Q, Hofmann PA (2004) Protein phosphatase 2A-mediated cross-talk between p38 MAPK and ERK in apoptosis of cardiac myocytes. Am J Physiol Heart Circ Physiol 286:H2204-2212

Liu RY, Fan C, Liu G, Olashaw NE, Zuckerman KS (2000) Activation of p38 mitogen-activated protein kinase is required for tumor necrosis factor-alpha -supported proliferation of leukemia and lymphoma cell lines. J Biol Chem 275:21086-21093

Lluis F, Perdiguero E, Nebreda AR, Munoz-Canoves P (2006) Regulation of skeletal muscle gene expression by p38 MAP kinases. Trends Cell Biol 16:36-44

Losa JH, Parada Cobo C, Viniegra JG, Sanchez-Arevalo Lobo VJ, Ramon y Cajal S, Sanchez-Prieto R (2003) Role of the p38 MAPK pathway in cisplatin-based therapy. Oncogene 22:3998-4006

Lu J, Quearry B, Harada H (2006) p38-MAP kinase activation followed by BIM induction is essential for glucocorticoid-induced apoptosis in lymphoblastic leukemia cells. FEBS Lett 580:3539-3544

Manke IA, Nguyen A, Lim D, Stewart MQ, Elia AE, Yaffe MB (2005) MAPKAP kinase-2 is a cell cycle checkpoint kinase that regulates the G2/M transition and S phase progression in response to UV irradiation. Mol Cell 17:37-48

Manning AM, Davis RJ (2003) Targeting JNK for therapeutic benefit: from junk to gold? Nat Rev Drug Discov 2:554-565

Mansour SJ, Matten WT, Hermann AS, Candia JM, Rong S, Fukasawa K, Vande Woude GF, Ahn NG (1994) Transformation of mammalian cells by constitutively active MAP kinase kinase. Science 265:966-970

Mansouri A, Ridgway LD, Korapati AL, Zhang Q, Tian L, Wang Y, Siddik ZH, Mills GB, Claret FX (2003) Sustained activation of JNK/p38 MAPK pathways in response to cisplatin leads to Fas ligand induction and cell death in ovarian carcinoma cells. J Biol Chem 278:19245-19256

Mao Z, Bonni A, Xia F, Nadal-Vicens M, Greenberg ME (1999) Neuronal activity-dependent cell survival mediated by transcription factor MEF2. Science 286:785-790
Matsukawa J, Matsuzawa A, Takeda K, Ichijo H (2004) The ASK1-MAP Kinase cascades in mammalian stress response. J Biochem 136:261-265
Matsuo Y, Amano S, Furuya M, Namiki K, Sakurai K, Nishiyama M, Sudo T, Tatsumi K, Kuriyama T, Kimura S, Kasuya Y (2006) Involvement of p38{alpha} Mitogen-activated protein kinase in lung metastasis of tumor cells. J Biol Chem 281:36767-36775
McMullen ME, Bryant PW, GlEMBOtski CC, Vincent PA, Pumiglia KM (2005) Activation of p38 has opposing effects on the proliferation and migration of endothelial cells. J Biol Chem 280:20995-21003
Merritt C, Enslen H, Diehl N, Conze D, Davis RJ, Rincon M (2000) Activation of p38 mitogen-activated protein kinase *in vivo* selectively induces apoptosis of CD8(+) but not CD4(+) T cells. Mol Cell Biol 20:936-946
Mikhailov A, Shinohara M, Rieder CL (2004) Topoisomerase II and histone deacetylase inhibitors delay the G2/M transition by triggering the p38 MAPK checkpoint pathway. J Cell Biol 166:517-526
Mikhailov A, Shinohara M, Rieder CL (2005) The p38-mediated stress-activated checkpoint. A rapid response system for delaying progression through antephase and entry into mitosis. Cell Cycle 4:57-62
Miller AL, Webb MS, Copik AJ, Wang Y, Johnson BH, Kumar R, Thompson EB (2005) p38 Mitogen-activated protein kinase (MAPK) is a key mediator in glucocorticoid-induced apoptosis of lymphoid cells: correlation between p38 MAPK activation and site-specific phosphorylation of the human glucocorticoid receptor at serine 211. Mol Endocrinol 19:1569-1583
Molnar A, Theodoras AM, Zon LI, Kyriakis JM (1997) Cdc42Hs, but not Rac1, inhibits serum-stimulated cell cycle progression at G1/S through a mechanism requiring p38/RK. J Biol Chem 272:13229-13235
Murphy LO, Blenis J (2006) MAPK signal specificity: the right place at the right time. Trends Biochem Sci 31:268-275
Nakayama K, Gazdoiu S, Abraham R, Pan ZQ, Ronai Z (2007) Hypoxia-induced assembly of prolyl-hydroxylase, PHD3 into complexes: implications for its activity and susceptibility for degradation by the E3 ligase Siah2. Biochem J 401:217-226
Nebreda AR, Porras A (2000) p38 MAP kinases: beyond the stress response. Trends Biochem Sci 25:257-260
Nemoto S, Xiang J, Huang S, Lin A (1998) Induction of apoptosis by SB202190 through inhibition of p38beta mitogen-activated protein kinase. J Biol Chem 273:16415-16420
Neve RM, Holbro T, Hynes NE (2002) Distinct roles for phosphoinositide 3-kinase, mitogen-activated protein kinase and p38 MAPK in mediating cell cycle progression of breast cancer cells. Oncogene 21:4567-4576
Nicke B, Bastien J, Khanna SJ, Warne PH, Cowling V, Cook SJ, Peters G, Delpuech O, Schulze A, Berns K, Mullenders J, Beijersbergen RL, Bernards R, Ganesan TS, Downward J, Hancock DC (2005) Involvement of MINK, a Ste20 family kinase, in Ras oncogene-induced growth arrest in human ovarian surface epithelial cells. Mol Cell 20:673-685
O'Neill LA (2006) Targeting signal transduction as a strategy to treat inflammatory diseases. Nat Rev Drug Discov 5:549-563

Olson JM, Hallahan AR (2004) p38 MAP kinase: a convergence point in cancer therapy. Trends Mol Med 10:125-129

Ono K, Han J (2000) The p38 signal transduction pathway: activation and function. Cell Signal 12:1-13

Park JG, Yuk Y, Rhim H, Yi SY, Yoo YS (2002a) Role of p38 MAPK in the regulation of apoptosis signaling induced by TNF-alpha in differentiated PC12 cells. J Biochem Mol Biol 35:267-272

Park JM, Greten FR, Li ZW, Karin M (2002b) Macrophage apoptosis by anthrax lethal factor through p38 MAP kinase inhibition. Science 297:2048-2051

Pedraza-Alva G, Koulnis M, Charland C, Thornton T, Clements JL, Schlissel MS, Rincon M (2006) Activation of p38 MAP kinase by DNA double-strand breaks in V(D)J recombination induces a G2/M cell cycle checkpoint. EMBO J 25:763-773

Perdiguero E, Ruiz-Bonilla V, Gresh L, Hui L, Ballestar E, Sousa-Victor P, Baeza-Raja B, Jardi M, Bosch-Comas A, Esteller M, Caelles C, Serrano AL, Wagner EF, Munoz-Canoves P (2007) Genetic analysis of p38 MAP kinases in myogenesis: fundamental role of p38alpha in abrogating myoblast proliferation. EMBO J 26:1245-1256

Philip M, Rowley DA, Schreiber H (2004) Inflammation as a tumor promoter in cancer induction. Semin Cancer Biol 14:433-439

Platanias LC (2003) Map kinase signaling pathways and hematologic malignancies. Blood 101:4667-4679

Pomerance M, Quillard J, Chantoux F, Young J, Blondeau JP (2006) High-level expression, activation, and subcellular localization of p38-MAP kinase in thyroid neoplasms. J Pathol 209:298-306

Porras A, Zuluaga S, Black E, Valladares A, Alvarez AM, Ambrosino C, Benito M, Nebreda AR (2004) p38 alpha mitogen-activated protein kinase sensitizes cells to apoptosis induced by different stimuli. Mol Biol Cell 15:922-933

Potter VR (1978) Phenotypic diversity in experimental hepatomas: the concept of partially blocked ontogeny. The 10th Walter Hubert Lecture. Br J Cancer 38:1-23

Proft M, Mas G, de Nadal E, Vendrell A, Noriega N, Struhl K, Posas F (2006) The stress-activated Hog1 kinase is a selective transcriptional elongation factor for genes responding to osmotic stress. Mol Cell 23:241-250

Pruitt K, Pruitt WM, Bilter GK, Westwick JK, Der CJ (2002) Raf-independent deregulation of p38 and JNK mitogen-activated protein kinases are critical for Ras transformation. J Biol Chem 277:31808-31817

Puri PL, Wu Z, Zhang P, Wood LD, Bhakta KS, Han J, Feramisco JR, Karin M, Wang JY (2000) Induction of terminal differentiation by constitutive activation of p38 MAP kinase in human rhabdomyosarcoma cells. Genes Dev 14:574-584

Ramesh G, Reeves WB (2005) p38 MAP kinase inhibition ameliorates cisplatin nephrotoxicity in mice. Am J Physiol Renal Physiol 289:F166-174

Ranganathan AC, Adam AP, Aguirre-Ghiso JA (2006) Opposing roles of mitogenic and stress signaling pathways in the induction of cancer dormancy. Cell Cycle 5:1799-1807

Rausch O, Marshall CJ (1999) Cooperation of p38 and extracellular signal-regulated kinase mitogen-activated protein kinase pathways during granulocyte colony-stimulating factor-induced hemopoietic cell proliferation. J Biol Chem 274:4096-4105

Recio JA, Merlino G (2002) Hepatocyte growth factor/scatter factor activates proliferation in melanoma cells through p38 MAPK, ATF-2 and cyclin D1. Oncogene 21:1000-1008

Reinhardt HC, Aslanian AS, Lees JA, Yaffe MB (2007) p53-deficient cells rely on ATM- and ATR-mediated checkpoint signaling through the p38MAPK/MK2 pathway for survival after DNA damage. Cancer Cell 11:175-189

Rennefahrt U, Janakiraman M, Ollinger R, Troppmair J (2005) Stress kinase signaling in cancer: fact or fiction? Cancer Lett 217:1-9

Ricote M, Garcia-Tuñon I, Bethencourt F, Fraile B, Onsurbe P, Paniagua R, Royuela M (2006) The p38 transduction pathway in prostatic neoplasia. J Pathol 208:401-407

Ringshausen I, Dechow T, Schneller F, Weick K, Oelsner M, Peschel C, Decker T (2004) Constitutive activation of the MAP kinase p38 is critical for MMP-9 production and survival of B-CLL cells on bone marrow stromal cells. Leukemia 18:1964-1970

Rouse J, Cohen P, Trigon S, Morange M, Alonso-Llamazares A, Zamanillo D, Hunt T, Nebreda AR (1994) A novel kinase cascade triggered by stress and heat shock that stimulates MAPKAP kinase-2 and phosphorylation of the small heat shock proteins. Cell 78:1027-1037

Rousseau S, Dolado I, Beardmore V, Shpiro N, Marquez R, Nebreda AR, Arthur JS, Case LM, Tessier-Lavigne M, Gaestel M, Cuenda A, Cohen P (2006) CXCL12 and C5a trigger cell migration via a PAK1/2-p38alpha MAPK-MAPKAP-K2-HSP27 pathway. Cell Signal 18:1897-1905

Sabbah M, Courilleau D, Mester J, Redeuilh G (1999) Estrogen induction of the cyclin D1 promoter: involvement of a cAMP response-like element. Proc Natl Acad Sci USA 96:11217-11222

Sainz-Perez A, Gary-Gouy H, Portier A, Davi F, Merle-Beral H, Galanaud P, Dalloul A (2006) High Mda-7 expression promotes malignant cell survival and p38 MAP kinase activation in chronic lymphocytic leukemia. Leukemia 20:498-504

Saklatvala J (2004) The p38 MAP kinase pathway as a therapeutic target in inflammatory disease. Curr Opin Pharmacol 4:372-377

Sanchez-Prieto R, Rojas JM, Taya Y, Gutkind JS (2000) A role for the p38 mitogen-acitvated protein kinase pathway in the transcriptional activation of p53 on genotoxic stress by chemotherapeutic agents. Cancer Res 60:2464-2472

Sarkar D, Su Z-Z, Lebedeva IV, Sauane M, Gopalkrishnan RV, Valerie K, Dent P, Fisher PB (2002) mda-7 (IL-24) mediates selective apoptosis in human melanoma cells by inducing the coordinated overexpression of the GADD family of genes by means of p38 MAPK. Proc Natl Acad Sci USA 99:10054-10059

Schmelter M, Ateghang B, Helmig S, Wartenberg M, Sauer H (2006) Embryonic stem cells utilize reactive oxygen species as transducers of mechanical strain-induced cardiovascular differentiation. FASEB J 20:1182-1184

Sebolt-Leopold JS, Herrera R (2004) Targeting the mitogen-activated protein kinase cascade to treat cancer. Nat Rev Cancer 4:937-947

Shemirani B, Crowe DL (2002) Hypoxic induction of HIF-1alpha and VEGF expression in head and neck squamous cell carcinoma lines is mediated by stress activated protein kinases. Oral Oncol 38:251-257

Silva G, Cunha A, Gregoire IP, Seldon MP, Soares MP (2006) The antiapoptotic effect of heme oxygenase-1 in endothelial cells involves the degradation of p38 alpha MAPK isoform. J Immunol 177:1894-1903

Sjolund J, Manetopoulos C, Stockhausen MT, Axelson H (2005) The Notch pathway in cancer: differentiation gone awry. Eur J Cancer 41:2620-2629

Srinivasa SP, Doshi PD (2002) Extracellular signal-regulated kinase and p38 mitogen-activated protein kinase pathways cooperate in mediating cytokine-induced proliferation of a leukemic cell line. Leukemia 16:244-253

Stadheim TA, Xiao H, Eastman A (2001) Inhibition of extracellular-signal regulated kinase (ERK) mediates cell cycle phase independent apoptosis in vinblastine-treated ML-1 cells. Cancer Res 61:1533-1540

Surh YJ, Chun KS, Cha HH, Han SS, Keum YS, Park KK, Lee SS (2001) Molecular mechanisms underlying chemopreventive activities of anti-inflammatory phytochemicals: down-regulation of COX-2 and iNOS through suppression of NF-kappa B activation. Mutat Res 480-481:243-268

Timoshenko AV, Chakraborty C, Wagner GF, Lala PK (2006) COX-2-mediated stimulation of the lymphangiogenic factor VEGF-C in human breast cancer. Br J Cancer 94:1154-1163

Todd DE, Densham RM, Molton SA, Balmanno K, Newson C, Weston CR, Garner AP, Scott L, Cook SJ (2004) ERK1/2 and p38 cooperate to induce a p21CIP1-dependent G1 cell cycle arrest. Oncogene 23:3284-3295

Tourian L Jr, Zhao H, Srikant CB (2004) p38alpha, but not p38beta, inhibits the phosphorylation and presence of c-FLIPS in DISC to potentiate Fas-mediated caspase-8 activation and type I apoptotic signaling. J Cell Sci 117:6459-6471

Tront JS, Hoffman B, Liebermann DA (2006) Gadd45a suppresses Ras-driven mammary tumorigenesis by activation of c-Jun NH2-terminal kinase and p38 stress signaling resulting in apoptosis and senescence. Cancer Res 66:8448-8454

Uddin S, Ah-Kang J, Ulaszek J, Mahmud D, Wickrema A (2004) Differentiation stage-specific activation of p38 mitogen-activated protein kinase isoforms in primary human erythroid cells. Proc Natl Acad Sci USA 101:147-152

Vander Griend DJ, Kocherginsky M, Hickson JA, Stadler WM, Lin A, Rinker-Schaeffer CW (2005) Suppression of metastatic colonization by the context-dependent activation of the c-Jun NH2-terminal kinase kinases JNKK1/MKK4 and MKK7. Cancer Res 65:10984-10991

Ventura JJ, Hubner A, Zhang C, Flavell RA, Shokat KM, Davis RJ (2006) Chemical genetic analysis of the time course of signal transduction by JNK. Mol Cell 21:701-710

Wada T, Penninger JM (2004) Mitogen-activated protein kinases in apoptosis regulation. Oncogene 23:2838-2849

Wang L, Kwak JH, Kim SI, He Y, Choi ME (2004a) Transforming growth factor-β1 stimulates vascular endothelial growth factor 164 via mitogen-activated protein kinase kinase 3-p38α and p38δ mitogen-activated protein kinase-dependent pathway in murine mesangial cells. J Biol Chem 279:33213-33219

Wang L, Pan Y, Dai JL (2004b) Evidence of MKK4 pro-oncogenic activity in breast and pancreatic tumors. Oncogene 23:5978-5985

Wang W, Chen JX, Liao R, Deng Q, Zhou JJ, Huang S, Sun P (2002) Sequential activation of the MEK-extracellular signal-regulated kinase and MKK3/6-p38 mitogen-activated protein kinase pathways mediates oncogenic ras-induced premature senescence. Mol Cell Biol 22:3389-3403

Weber HO, Ludwig RL, Morrison D, Kotlyarov A, Gaestel M, Vousden KH (2005) HDM2 phosphorylation by MAPKAP kinase 2. Oncogene 24:1965-1972

Weijzen S, Rizzo P, Braid M, Vaishnav R, Jonkheer SM, Zlobin A, Osborne BA, Gottipati S, Aster JC, Hahn WC, Rudolf M, Siziopikou K, Kast WM, Miele L (2002) Activation of Notch-1 signaling maintains the neoplastic phenotype in human Ras-transformed cells. Nat Med 8:979-986

Weldon CB, Parker AP, Patten D, Elliott S, Tang Y, Frigo DE, Dugan CM, Coakley EL, Butler NN, Clayton JL, Alam J, Curiel TJ, Beckman BS, Jaffe BM, Burow ME (2004) Sensitization of apoptotically-resistant breast carcinoma cells to TNF and TRAIL by inhibition of p38 mitogen-activated protein kinase signaling. Int J Oncol 24:1473-1480

Winograd-Katz SE, Levitzki A (2006) Cisplatin induces PKB/Akt activation and p38(MAPK) phosphorylation of the EGF receptor. Oncogene 25:7382-7390

Wu G, Luo J, Rana JS, Laham R, Sellke FW, Li J (2006) Involvement of COX-2 in VEGF-induced angiogenesis via P38 and JNK pathways in vascular endothelial cells. Cardiovasc Res 69:512-519

Yamamoto T, Kozawa O, Tanabe K, Akamatsu S, Matsuno H, Doh, S, Uemats T (2001) Involvement of p38 MAP kinase in TGF-beta-stimulated VEGF synthesis in aortic smooth muscle cells. J Cell Biochem 82:591-598

Yang YA, Zhan GM, Feigenbau L, Zhang YE (2006) Smad3 reduces susceptibility to hepatocarcinoma by sensitizing hepatocytes to apoptosis through downregulation of Bcl-2. Cancer Cell 9:445-457

Yee AS, Paulso EK, McDevit MA, Rieger-Chris K, Summerhaye I, Beras SP, Kim J, Huang CY, Zhang X (2004) The HBP1 transcriptional repressor and the p38 MAP kinase: unlikely partners in G1 regulation and tumor suppression. Gene 336:1-13

Yoshino Y, Aoyagi M, Tamaki M, Duan L, Morimoto T, Ohno K (2006) Activation of p38 MAPK and/or JNK contributes to increased levels of VEGF secretion in human malignant glioma cells. Int J Oncol 29:981-987

Yosimichi G, Nakanishi T, Nishida T, Hattori T, Takano-Yamamoto T, Takigawa M (2001) CTGF/Hcs24 induces chondrocyte differentiation through a p38 mitogen-activated protein kinase (p38MAPK), and proliferation through a p44/42 MAPK/extracellular-signal regulated kinase (ERK). Eur J Biochem 268:6058-6065

Yu C, Wang S, Dent P, Grant S (2001) Sequence-dependent potentiation of paclitaxel-mediated apoptosis in human leukemia cells by inhibitors of the mitogen-activated protein kinase kinase/mitogen-activated protein kinase pathway. Mol Pharmacol 60:143-154

Yu W, Imoto I, Inoue J, Onda M, Emi M, Inazawa J (2006) A novel amplification target, DUSP26, promotes anaplastic thyroid cancer cell growth by inhibiting p38 MAPK activity. Oncogene 26:1178-1187

Yuan ZQ, Feldman RI, Sussman GE, Coppola D, Nicosia SV, Cheng JQ (2003) AKT2 inhibition of cisplatin-induced JNK/p38 and Bax activation by phosphorylation of ASK1: implication of AKT2 in chemoresistance. J Biol Chem 278:23432-23440

Zhang Q-X, Feng R, Zhang W, Ding Y, Yang J-Y, Liu G-H (2005) Role of stress-activated MAP kinase p38 in cisplatin- and DTT-induced apoptosis of the esophageal carcinoma cell line Eca109. World J Gastroenterol 11:4451-4456

Zhou J, Schmid T, Schnitzer S, Brune B (2006) Tumor hypoxia and cancer progression. Cancer Lett 237:10-21

Zhuang ZH, Zhou Y, Yu MC, Silverman N, Ge BX (2006) Regulation of *Drosophila* p38 activation by specific MAP2 kinase and MAP3 kinase in response to different stimuli. Cell Signal 18:441-448

Zuluaga S, Alvarez-Barrientos A, Gutierrez-Uzquiza A, Benito M, Nebreda AR, Porras A (2007) Negative regulation of Akt activity by p38alpha MAP kinase in cardiomyocytes

involves membrane localization of PP2A through interaction with caveolin-1. Cell Signal 19:62-74
Zwang Y, Yarden Y (2006) p38 MAP kinase mediates stress-induced internalization of EGFR: implications for cancer chemotherapy. EMBO J 25:4195-4206

Dolado, Ignacio
CNIO (Spanish National Cancer Center), Melchor Fernández Almagro 3, 28029 Madrid, Spain

Nebreda, Angel R
CNIO (Spanish National Cancer Center), Melchor Fernández Almagro 3, 28029 Madrid, Spain
anebreda@cnio.es

List of abbreviations

CDK: cyclin-dependent kinase
CREB: cAMP-responsive element binding protein
COX: cyclooxygenase
EGFR: epidermal growth factor receptor
ERK: extracellular-signal regulated kinase
GCSF: granulocyte colony stimulating factor
HDAC: histone deacetylase
HIF: hypoxia-inducible factor
IL: interleukin
JNK: c-Jun-N-terminal kinase
MAPK: mitogen-activated protein kinase
MDA-7: melanoma differentiation-associated 7
MEFs: mouse embryo fibroblasts
MK: MAPK-activated protein kinase
MMP: matrix metalloproteinase
MSK: mitogen- and stress-activated kinase
NF-*k*B: nuclear factor *k*B
PAK: p21-activated kinase
ROS: reactive oxygen species
TNF: tumor necrosis factor
UV: ultraviolet
VEGF: vascular endothelial growth factor

Control of cell cycle by SAPKs in budding and fission yeast

Sandra Lopez-Aviles and Rosa M. Aligue

Abstract

In yeast cells as well as in higher eukaryotic organisms, the response to environmental stress is through the activation of the MAP kinases pathway, which induces the expression of genes involved in maintaining the cellular homeostasis. This pathway is activated after a variety of cellular stimuli and regulates numerous physiological processes, particularly the cell division cycle. Progression through the cell cycle is critically dependent on the presence of environmental growth factors and stress stimuli, and failure to correctly integrate such signals into the cell cycle machinery can lead to accumulation of genetic damage and genomic instability. Here, we considered the molecular mechanism by which cell cycle control is regulated by stress-activated protein kinase (SAPK) signalling pathway in yeast, *Saccharomyces cerevisiae* and *Schizosaccharomyces pombe*.

1 Introduction

Although both yeast are evolutionarily divergent, common mechanisms control their cell cycles that are conserved throughout eukaryotes (Nurse 1990). A highly conserved class of molecules termed cyclin dependent kinases (CDKs) plays a central role in coordinating the cell cycle in all eukaryotes. In both, fission and budding yeast, the cell cycle is controlled at G1-S transition and G2-M transitions by a single highly conserved CDK named Cdc28 and Cdc2 of *S. cerevisiae* and *S. pombe*, respectively (Morgan 1997; review in Humphrey and Pearce 2005). Formation of the active CDK requires its association with a cyclin molecule, its phosphorylation by a CDK-activating kinase and dephosphorylation by members of the Cdc25 family of phosphatases. Passage through cell cycle transitions depends on the periodic activity of CDKs, which is mainly regulated by the abundance of the cyclin molecules that fluctuate through the cell cycle. In addition, Cdk-cyclin complexes can be inhibited by the binding of CDK inhibitors (CKIs) (Humphrey and Pearce 2005).

Topics in Current Genetics, Vol. 20
F. Posas, A. R. Nebreda (Eds.): Stress-Activated Protein Kinases
DOI 10.1007/4735_2007_0246 / Published online: 1 August 2007

2 Cell cycle control by SAPKs in *Saccharomyces cerevisiae*

2.1 Cell cycle regulation

In the budding yeast *Saccharomyces cerevisiae*, the primary CDK involved in cell cycle control is the Cdc28 kinase. Late G1 progression and passage through START are controlled by the association of Cdc28 with three G1 cyclins: Cln1, 2, and 3, which also target the Clb-Cdc28 inhibitor Sic1 for degradation (Wittenberg et al. 1990; Schwob and Bohm et al. 1994). Subsequently, the correct timing of S phase entry requires the association of Cdc28 with the B-type cyclin Clb5 and 6 whereas G2 progression and the onset of M-phase require its association with Clbs 1-4 (Nasmyth 1993). Entry into M-phase is specifically controlled by the Cdc28-Clb2 complex, which is held in check by the protein kinase Swee1 (Wee1 in *S. pombe*). Swe1 is thought to delay entry into mitosis until critical cell size has been reached (Rupes 2002; Kellogg 2003; Harvey et al. 2005) or in response to defects in bud formation or cytoskeletal function, which are monitored by the morphogenesis checkpoint (Lew 2003). The checkpoint kinase Hsl1 monitors cytoskeleton alterations, delays bud formation and regulates Swe1 stability (Cid et al. 2001; Lew 2003). Under normal conditions, Hsl1 binds to the septin ring at the bud neck, which triggers the recruitment of Hsl7 and phosphorylation and degradation of Swe1 prior entry into mitosis (Versele and Thorner 2005; Keaton and Lew 2006).

Cell cycle progression by such Cyclin-CDK complexes is often inhibited via signal transduction pathways as a result of either intracellular stimuli.

2.2 Stress-activated protein kinase pathway and cell cycle control

To maximize the probability of survival and proliferation, cells coordinate various intracellular activities in response to changes in the extracellular environment.

In response to an increase in extracellular osmolarity, cells induce an elaborate program that includes changes in transcription and translation as well as in cell cycle progression to allow cells to adapt. Central to this response is a stress-activated protein kinase (SAPK) signalling pathway known as the high osmolarity glycerol (HOG) pathway in budding yeast (de Nadal et al. 2002).

The HOG pathway is composed by the MAPK Hog1, its MAP2K Pbs2 and a set of three MAP3K Ssk2/Ssk22 and Ste11, which are activated by two independent upstream regulatory branches that are defined by two putative osmosensors, Sln1 and Sho1 (Hohmann 2002).

Activation of the HOG pathway regulates the expression of genes involved in adaptation to high osmolarity by modulating the activity of several transcription factors as well as through chromatin remodelling (Proft and Struhl 2002; de Nadal et al. 2003). Activation of the HOG pathway by exposure to high osmolarity or mutations that lead to activation of the Hog1 SAPK also results in a transient cell

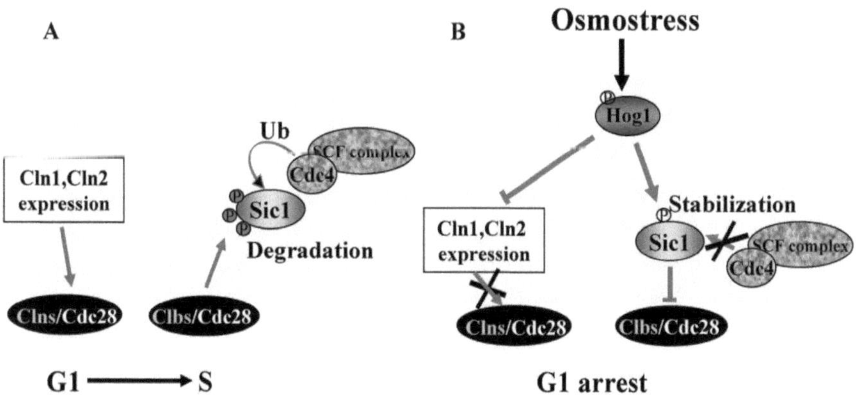

Fig. 1. Regulation of the G1-S transition by the SAPK Hog1 in budding yeast. (A) Cell Cycle progression into S-phase is driven by the formation of Clns-Cdc28 complex and degradation of Sic1 inhibitor by the SCF complex. At G1, Sic1 stability is reduced by phosphorylation on its N-terminal domain by Cln1, 2-Cdc28 complexes. Phosphorylated Sic1 is then recognized by the Cdc4 ubiquitin ligase of the SCF complex and targeted for degradation. (B) Hog1 activation promotes Sic1 stabilization and prevents entry into S-phase by inhibiting Clbs-Cdc28 complexes. Activation of Hog1 also inhibits CLN1 and CLN2 expression that results in a decrease of Cln-Cdc28 activity required for Sic1 phosphorylation and degradation. Hog1 directly phosphorylates Sic1, which interferes with the binding of Cdc4 and subsequently prevents its degradation.

cycle arrest in G1 or G2 (Alexander et al. 2001; Belli et al. 2001; Escote et al. 2004; Zapater et al. 2005; Clotet et al. 2006).

Activated Hog1 leads to G1 arrest by a dual mechanism that involves down regulation of cyclin expression and direct phosphorylation of the CDK-inhibitor Sic1 (Fig. 1). This combination results in Sic1 stabilization and inhibition of cell cycle progression to prevent premature entry into S phase before proper cell adaptation (Escote et al. 2004; Zapater et al. 2005). In response to osmotic stress, budding yeast cells also arrest at G2 by a mechanism involving the product of SWE1 gene (Alexander et al. 2001). Recently, it has been shown that in response to stress, Hog1 controls G2 progression by down regulating cycling B (Clb2) levels as well as through direct phosphorylation of the Hsl1 kinase, which leads to the stabilization of the Swe1 kinase and subsequently the decrease in Cdc28-Clb2 activity (Fig. 2) (Clotet et al. 2006).

In budding yeast, osmotic stress also regulates exit from mitosis, the cell cycle transition from mitosis to G1. Exit from mitosis is triggered by the protein phosphatase Cdc14 that acts antagonizing Cdc28-Clbs function (reviewed in Stegmeier and Amon 2004). The Cdc14 phosphatase is tightly regulated by a competitive inhibitor Cfi1/Net1. The inhibitor holds Cdc14 inactive in the nucleolus during G1, S and early M phases. During anaphase, Cdc14 is released from its inhibitor by the mitotic exit network (MEN) and spreads through the nucleus and cytoplasm to induce exit from mitosis.

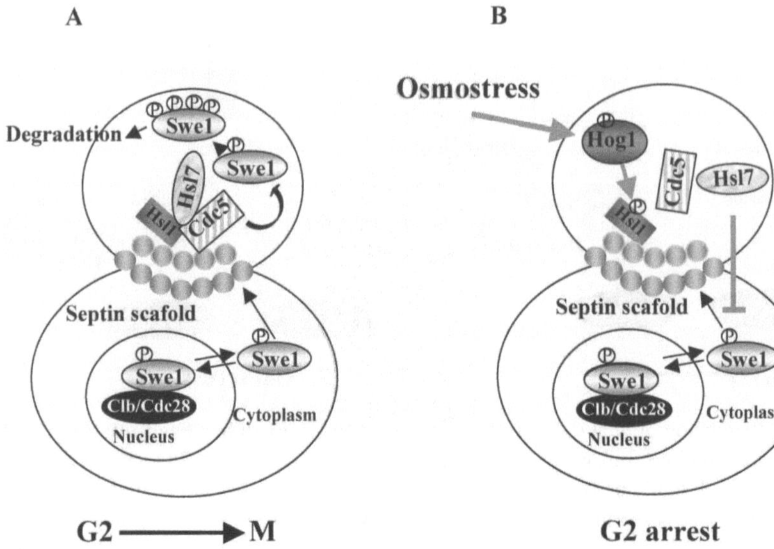

Fig. 2. Regulation of the G2-M transition by the SAPK Hog1 in budding yeast. (A) At G2, Swe1 first accumulates on the nucleus of unbudded cell. Following bud emergence, Hsl1 becomes activated at the septin scaffold tethers Hsl7, which in turn recruits Swe1 and Cdc5 (Polo kinase) to the septin collar. Phosphorylation of Swe1 at the bud neck leads to Swe1 degradation and allows progression to mitosis. (B) Upon stress, Hog1 phosphorylates a residue within the Hsl7 site of the Hsl1 kinase, which results in delocalization of Hsl7 from the septin collar and leads to Swe1 accumulation and cell cycle arrest.

Initially, it was observed that the failure of certain MEN mutants to exit from mitosis was suppressed by high osmotic conditions (Grandin et al. 1998). Recent work has shown that the suppression of the mitotic defects of MEN mutants by high osmolarity is accompanied by the release of Cdc14 from the nucleolus and is mediated by the HOG pathway, indicating that the HOG pathway brings about exit from mitosis in MEN mutants by promoting the release of Cdc14 from its inhibitor in the nucleolus (Reiser et al. 2006). Interestingly, this work also shows that only one of the two branches of the HOG pathway, the Sho1 branch, is required for promoting exit from mitosis in MEN mutants, which represents up to now the first specific function for the Sho1 branch of the HOG pathway.

3 Cell cycle control by SAPKs in *Schizosaccharomyces pombe*

3.1 Cell cycle regulation

In the fission yeast *Schizosaccharomyces pombe*, the major point of control over the cell division cycle occurs not at the G1/S transition but immediately prior to the onset of M-phase (Egel 2005). A signal transduction network regulating the activity of a homologous Cyclin-CDK pair, the Cdc13-Cdc2 kinase complex, governs the timing of mitotic initiation. Moreover, since the components of this network exist in humans, G2/M control in fission yeast has served as a paradigm for this evolutionary conserved process (Nurse 1990; Hayles et al. 1994). Phosphorylation of Cdc2 on tyrosine-15, which is critical for the inhibition of the complex, is catalysed by Wee1 and Mik1, whereas dephosphorylation of this residue by the Cdc25 phosphatase is the key event driving the initiation of mitosis (Gould and Nurse 1989; Aligue et al. 1997; Egel 2005). The activity of Cdc25 is tightly regulated throughout the cell cycle, being low during interphase and activated only as cells enter mitosis, in part because of phosphorylation of Cdc25 by Cdc2-Cdc13 and polo-like kinases at multiples sites within its N-terminal regulatory domain (Egel 2005). Cdc25 is also subject to negative regulatory phosphorylation when the genome is either damaged or not completely replicated. In these situations, Cdc25 is phosphorylated by the Chk1 and Cds1 kinases (review in Karlsson-Rosenthal and Millar 2006). In fission yeast, this occurs on at least nine phosphorylation sites in the N-terminal regulatory domain of the protein, and the phosphorylation of these sites is necessary for checkpoint arrest (Furnari et al. 1999; Zeng and Piwnica-Worms 1999).

3.2 Stress-activated protein kinase pathway and cell cycle control

Genetic analysis in fission yeast has identified a number of pathways that influence the timing of mitosis. In particular, the stress activated Sty1 (also known as Spc1) MAP kinase, a member of the stress-activated protein kinase (SAPK) signalling pathway, has been found to link the cell cycle to the extracellular environment (review in Pearce and Humphrey 2001). The Sty1/Spc1 pathway, like the HOG pathway, is stimulated by osmotic shock and it controls expression of genes important for osmoadaptation. However, regarding the pathway control there are some differences. Most significantly, while HOG pathway is apparently mainly responsive to osmotic shock, Sty1/Spc1 activity can be stimulated by a whole range of stress conditions (Hohmann 2002). The central elements of the stress-activated MAP kinases pathway in fission yeast are the MAPK Sty1/Spc1, the MAP2K Wis1 and the MAP3K Wak1 and Win1 (review in Hohmann 2002). Cells deleted for Sty1/Spc1 or Wis1 are highly elongated as a consequence of a delay in the timing of mitotic initiation, which is exacerbated in response to stress. Although such mutants still undergo cell cycle arrest in response to stress, they are unable to resume proliferation and die (review in Pearce and Humphrey 2001), suggesting that Sty1/Spc1 pathway is required for recovery from stress-induced cell cycle arrest.

A number of effectors of the Sty1/Spc1 MAP kinase have been identified including the Atf1 and Pap1 transcription factors, which are homologues to mammalian ATF-2 and c-Jun, respectively (Shiozaki and Russell 1996; Wilkinson et al. 1996). In addition, Sty1/Spc1 binds and phosphorylates two downstream kinases, Cmk2 and Srk1 (Sty1-regulated kinase), which are related to the mammalian calmodulin-dependent and MAPKAP kinases and to the budding yeast RCK1 and RCK2 kinases, which were identified as suppressors of checkpoint mutants of DNA replication (Dahlkvist et al. 1995; Alemany et al. 2002; Sanchez-Piris et al. 2002; Smith et al. 2002). Recently, a molecular link between the SAPK pathway and the G2-M transition has been described through the Srk1 kinase (Lopez-Aviles et al. 2005) and the Plo1 kinase (fission yeast Polo kinase) (Petersen and Hagan 2005). The Srk1 kinase associates with Sty1/Spc1 MAP kinase and regulates the onset of mitosis by phosphorylating and inhibiting the Cdc25 phosphatase (Lopez-Aviles et al. 2005). Phosphorylation by Srk1 causes Cdc25 to bind to Rad24, a 14-3-3 protein family member, promoting accumulation of Cdc25 in the cytoplasm and thus inducing cell cycle arrest in G2 phase (Fig. 3A). Moreover, Srk1 phosphorylates and inhibits Cdc25 in the same sites that are phosphorylated by Chk1 and Cds1, the DNA-damage and replication checkpoint kinases. However, unlike Chk1 and Cds1, Srk1 does not regulate Cdc25 in response to DNA damage but during normal cell cycle and in response to non-genotoxic environmental-stress, such as high osmolarity (Fig. 3A). Thus, the Srk1 kinase is considered as a non-genotoxic checkpoint kinase (Lopez-Aviles et al. 2005). The Plo1 kinase has been shown to be involved in mitotic commitment and recovery from stress (Petersen and Hagan 2005). During recovery from a heat stress-induced cell cycle arrest, phosphorylation of Plo1 kinase in serine 402 (Ser402) is necessary to promote association of Plo1 to the poles of the cell and re-initiation of mitosis (Fig. 3B) (Petersen and Hagan 2005). Phosphorylation of Plo1 on Ser402 requires the Sty1/Spc1 kinase, although this must be indirect because this site lacks the canonical proline +1 in the consensus site for SAP kinases (Manke et al. 2005) and also because phosphorylation is delayed relative to Sty1/Spc1 activation. At the present, it is not known whether phosphorylation of Plo1 at Ser402 is necessary to control re-entry into mitosis under these conditions by regulating Wee1, Cdc25, or Mrc1 (the fission yeast homologue of Claspin) as described in other organisms (van Vugt et al. 2004; review in Mamely et al. 2006). However, these results indicate that activation of SAP kinase induces cell cycle arrest and adds a new role to the SAP kinase pathway, which is to prepare the cell for checkpoint recovery (Petersen and Hagan 2005).

The Sty1/Spc1 pathway has also a role during mitosis regulating the spindle orientation checkpoint. Perturbation of the actin cytoskeleton delays entry into anaphase in fission yeast cells. Cells treated with the actin depolymerising agent latrunculin or cells carrying mutations in the actin gene enter mitosis and form a short, misorientated spindle (review in Gachet et al. 2006). Although, the control of the spindle orientation by the astral microtubules has been controversial (Zimmerman et al. 2004; Gachet et al. 2006), it has been shown that the spindle orientation checkpoint is regulated by Sty1/Spc1 kinase (Gachet et al. 2006). The Sty1/Spc1 is activated and localized to the nucleus by addition of low doses of latrunculin, but not by benomyl or thiabendazol (TBZ), which are microtubule inhibitors that activate the spindle association checkpoint. Moreover, it has been also reported that deletion of the Sty1/Spc1 substrate Atf1 rescues the mitotic delay imposed by the spindle assembly checkpoint activated in microtubule formation mutants (Oliferenko and Balasubramanian 2002). Recent work by Kawasaki

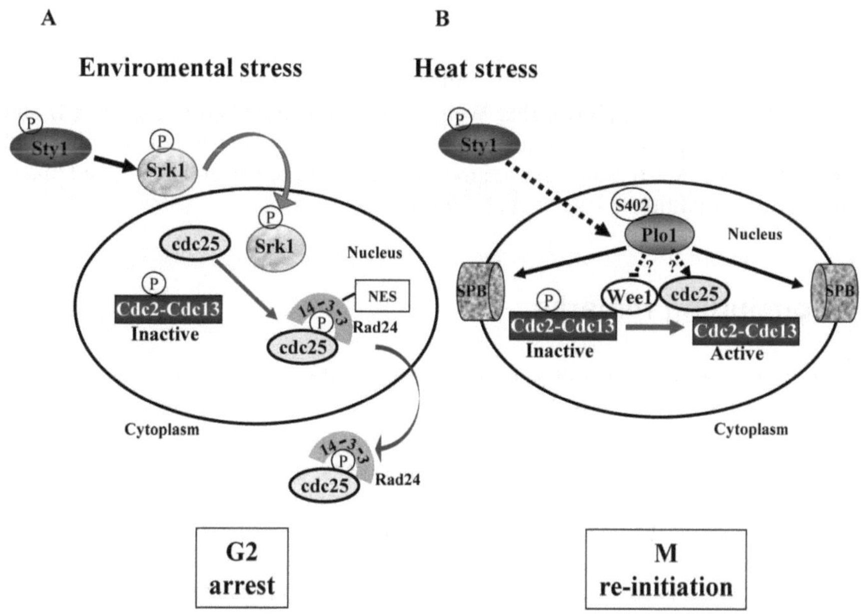

Fig. 3. Regulation of the G2-M transition by the SAPK Sty1/Spc1 in fission yeast. (A) Upon stress, Sty1/Spc1 phosphorylates and activates Srk1 kinase, which in turn phosphorylates Cdc25 phosphatase and inhibits progression to mitosis. Phosphorylation by Srk1 causes Cdc25 to bind to Rad24, a 14-3-3 protein family member, and accumulation of Cdc25 in the cytoplasm. (B) The Sty1/Spc1 activation modulates recovery from a heat stress-induced cell cycle arrest through the Plo1 kinase. Phosphorylation of Plo1 on Ser402 promotes its recruitment to SPB (spindle pole body) and thus re-initiation of mitosis during recovery from the stress. Although, SerS402 phosphorylation requires activation of Sty1/Spc1, it is not known the molecular mechanism which Plo1 activates progression to mitosis.

et al. (2006) has described an additional role for the MAPK Sty1 in chromosome segregation. Activation of the MAPK Sty1 pathway by high osmotic conditions suppressed the *ts* (temperature sensitive) mutant phenotypes of separase and securin (Cut2 and Cut1, respectively in fission yeast) which are essential for sister chromatid separation. Prior to anaphase, securin inactivates separase by forming a complex, which is essential for cell viability (Kawasaki et al. 2006). Degradation of securin occurs at the onset of anaphase and subsequently activation of separase, which specifically cleaves the subunit Scc1/Rad21 of the cohesion complex leading to chromosome separation. The authors showed that genetic suppression of securin and separase mutants under stresses occurs through a Sty1-dependent temporal arrest of the cell division, where the level of securin dramatically increased together with the enhanced complex formation between securin and the mutant separase (Kawaski et al. 2006).

All these data suggest that the SAPK pathway has an integral role in mitosis to prevent sister chromatid separation in response to perturbation of actin, microtubules attachment to chromosomes and securin degradation.

Recent studies have shown that Wsh3/Tea4, a novel cell-end protein in fission with a role in bipolar growth during normal cell cycle, contributes together with the Win1 MAP3K from the stress-signalling pathway, to cell-polarity maintenance under stress conditions (Tatebe et al. 2005).

4 Concluding remarks

In the last years, several studies in mammalian and yeast cells point to a considerable role for SAP kinase pathways in regulating different cell cycle stages in response to environmental stress (review in Pearce and Humphrey 2001). The regulation of multiple cell cycle control components by the SAPK is required to transiently arrest cell division and generate the required response for cellular adaptation and prevent the accumulation of genetic damage and genomic instability. In the same way, cells can be subjected to stress at any stage of cell division. Therefore, in response to stress the SAPK pathway is able to regulate G1-S, G2-M transitions, mitosis progression and exit, through the components that govern the cell cycle control.

As observed in yeast, mammalian cells also respond to stress by modulating cell cycle progression. Similarly as reported in budding yeast, the control of G1-S is achieved in mammalian cells by the regulation of cyclin A or D1 levels, as well as the phosphorylation of critical cell cycle regulators such as pRB, p53, p21, HBP1, or Cdc25A (Wilkinson and Millar 2000; Ambrosino and Nebreda 2001; Barnouin et al. 2002; Zapater et al. 2005). In fission yeast, a homologous mechanism regulating G1-S transition remains to be described.

Studies in fission yeast have been focused on the regulation of G2-M transition by the stress pathway. As mentioned above, the G2-phase arrest and recovery upon environmental stress are regulated by Srk1 and Plo1 kinases, respectively. In mammalian cells, they have been described as conserved mechanisms in response to DNA damage, but not to environmental stress. In response to UV-induced DNA damage, the MAPKAP-K2 (MAPK-activated protein kinase 2), which is structurally related to Srk1, associates to and is activated by the p38 SAPK (Manke et al. 2005). Human Plk1 is required for checkpoint recovery in response to DNA damage, primarily by inactivating Wee1 (van Vugt et al. 2004). Therefore, the cell cycle role of MAKAP-K2 and Plk1 under environmental stress conditions has to be further analysed.

In budding yeast, the regulation of the G2-M transition by the SAPK pathway is governed by different components. The SAPK Hog1 induces G2 arrest by direct control of the Cdc28-Clb2 activity, downregulating Clb2 cyclin levels and phosphorylating the Hsl1 kinase, a component of the morphogenesis checkpoint.

Mitosis progression monitored by the spindle orientation, assembly checkpoints as well as chromosome segregation, is also modulated by SAPK pathway in fission yeast. It will be interesting to determine whether analogous SAPK-mediated responses occur in other organisms.

The SAPK pathway also seems to regulate the exit of mitosis through controlling the Cdc14 activity (Reiser et al. 2006). While, high osmolarity is the first extracellular signal shown to affect the Cdc14 phosphatase, authors have also shown the first positive role of the SAP kinase in cell cycle progression in yeast. This observation is interesting because while cells delay commitment to mitosis after stress in order to avoid incorrect mitosis execution, it will make sense that cells in M-phase activate the resumption of M-phase to adapt and recover from the stress insult in interphase.

Finally, in budding and fission yeast, great strides have been made in dissecting the molecular mechanisms that coordinate the stress response and the cell cycle events for proper cellular adaptation and correct cell division. The identification of additional substrates of SAPK pathway will provide new insights.

Acknowledgements

We thank Jonathan Millar (University of Warwick) and Elena Hidalgo (Universitat Pompeu Fabra, Barcelona) for discussion and Eva Lambea for critical reading of the manuscript. We apologize to those researchers whose work could not be cited for length considerations. Sandra Lopez-Aviles is the recipient of a predoctoral fellowship from the Generalitat de Catalunya, Spain. Rosa Aligue is funded by the Ministerio de Educación y Ciencia and the Distinció de la Generalitat de Catalunya per a la promoció de la Recerca Universitaria. Joves investigadors.

References

Alemany V, Sanchez-Piris M, Bachs O, Aligue R (2002) Cmk2, a novel serine/threonine kinase in fission yeast. FEBS Lett 1-3:79-86

Alexander MR, Tyers M, Perret M, Craig BM, Fang KS, Gustin MC (2001) Regulation of cell cycle progression by Swe1p and Hog1p following hypertonic stress. Mol Biol Cell 1:53-62

Aligue R, Wu L, Russell P (1997) Regulation of *Schizosaccharomyces pombe* Wee1 Tyrosine Kinase. J Biol Chem 20:13320-13325

Ambrosino C, Nebreda AR (2001) Cell cycle regulation by p38 MAP kinases. Biol Cell 1-2:47-51

Barnouin K, Dubuisson ML, Child ES, Fernandez de Mattos S, Glassford J, Medema RH, Mann DJ, Lam EW (2002) H2O2 induces a transient multi-phase cell cycle arrest in mouse fibroblasts through modulating cyclin D and p21Cip1 expression. J Biol Chem 16:13761-13770

Belli G, Gari E, Aldea M, Herrero E (2001) Osmotic stress causes a G1 cell cycle delay and downregulation of Cln3/Cdc28 activity in *Saccharomyces cerevisiae*. Mol Microbiol 4:1022-1035

Cid VJ, Shulewitz MJ, McDonald KL, Thorner J (2001) Dynamic localization of the Swe1 regulator Hsl7 during the *Saccharomyces cerevisiae* cell cycle. Mol Biol Cell 6:1645-1669

Clotet J, Escote X, Adrover MA, Yaakov G, Gari E, Aldea M, de Nadal E, Posas F (2006) Phosphorylation of Hsl1 by Hog1 leads to a G2 arrest essential for cell survival at high osmolarity. EMBO J 25:2338-2346

Dahlkvist A, Kanter-Smoler G, Sunnerhagen P (1995) The RCK1 and RCK2 protein kinase genes from *Saccharomyces cerevisiae* suppress cell cycle checkpoint mutations in *Schizosaccharomyces pombe*. Mol Gen Genet 3:316-326

de Nadal E, Alepuz PM, Posas F (2002) Dealing with osmostress through MAP kinase activation. EMBO Rep 8:735-740

de Nadal E, Casadome L, Posas F (2003) Targeting the MEF2-like transcription factor Smp1 by the stress-activated Hog1 mitogen-activated protein kinase. Mol Cell Biol 1:229-237

Egel R (2005) The molecular biology of *Schizosaccharomyces pombe*. Springer Denmark.

Escote X, Zapater M, Clotet J, Posas F (2004) Hog1 mediates cell-cycle arrest in G1 phase by the dual targeting of Sic1. Nat Cell Biol 10:997-1002

Furnari B, Blasina A, Boddy MN, McGowan CH, Russell P (1999) Cdc25 inhibited *in vivo* and *in vitro* by checkpoint kinases Cds1 and Chk1. Mol Biol Cell 4:833-845

Gachet Y, Reyes C, Goldstone S, Tournier S (2006) The fission yeast spindle orientation checkpoint: a model that generates tension? Yeast 13:1015-1029

Gould KL, Nurse P (1989) Tyrosine phosphorylation of the fission yeast cdc2+ protein kinase regulates entry into mitosis. Nature 6245:39-45

Grandin N, de Almeida A, Charbonneau M (1998) The Cdc14 phosphatase is functionally associated with the Dbf2 protein kinase in *Saccharomyces cerevisiae*. Mol Gen Genet 1-2:104-116

Harvey SL, Charlet A, Haas W, Gygi SP, Kellogg DR (2005) Cdk1-dependent regulation of the mitotic inhibitor Wee1. Cell 3:407-420

Hayles J, Fisher D, Woollard A, Nurse P (1994) Temporal order of S phase and mitosis in fission yeast is determined by the state of the p34cdc2-mitotic B cyclin complex. Cell 5:813-822

Hohmann S (2002) Osmotic stress signaling and osmoadaptation in yeasts. Microbiol Mol Biol Rev 2:300-372

Humphrey T, Pearce A (2005) Cell cycle molecules and mechanisms of the budding and fission yeasts. Methods Mol Biol 3-29

Karlsson-Rosenthal C, Millar JB (2006) Cdc25: mechanisms of checkpoint inhibition and recovery. Trends Cell Biol 6:285-292

Kawasaki Y, Nagao K, Nakamura T, Yanagida M (2006) Fission yeast MAP kinase is required for the increased securing-separase interaction that rescues separase mutants under stresses. Cell Cycle 16:1831-1839

Keaton MA, Lew DJ (2006) Eavesdropping on the cytoskeleton: progress and controversy in the yeast morphogenesis checkpoint. Curr Opin Microbiol 6:540-546

Kellogg R (2003) Wee1-dependent mechanisms required for coordination of cell growth and cell division. J Cell Sci 24:4883-4890

Lew DJ (2003) The morphogenesis checkpoint: how yeast cells watch their figures. Curr Opin Cell Biol 6:648-653

Lopez-Aviles S, Grande M, Gonzalez M, Helgesen AL, Alemany V, Sanchez-Piris M, Bachs O, Millar JB, Aligue R (2005) Inactivation of the Cdc25 phosphatase by the stress-activated Srk1 kinase in fission yeast. Mol Cell 1: 49-59

Mamely I, van Vugt MA, Smits VA, Semple JI, Lemmens B, Perrakis A, Medema RH, Freire R (2006) Polo-like kinase-1 controls proteasome-dependent degradation of Claspin during checkpoint recovery. Curr Biol 19:1950-1955

Manke IA, Nguyen A, Lim D, Stewart MQ, Elia AEH, Yaffe MB (2005) MAPKAP Kinase-2 Is a cell cycle checkpoint kinase that regulates the G2/M transition and S phase progression in response to UV irradiation. Molecular Cell 1:37-48

Morgan DO (1997) Cyclin-dependent kinases: engines, clocks, and microprocessors. Annu Rev Cell Dev Biol 13:261-291

Nasmyth K (1993) Control of the yeast cell cycle by the Cdc28 protein kinase. Curr Opin Cell Biol 2:166-179

Nurse P (1990) Universal control mechanism regulating onset of M-phase. Nature 6266: 503-508

Oliferenko S, Balasubramanian MK (2002) Astral microtubules monitor metaphase spindle alignment in fission yeast. Nat Cell Biol 10:816-820

Pearce AK, Humphrey TC (2001) Integrating stress-response and cell-cycle checkpoint pathways. Trends Cell Biol 10:426-433

Petersen J, Hagan IM (2005) Polo kinase links the stress pathway to cell cycle control and tip growth in fission yeast. Nature 7041:507-512

Proft M, Struhl K (2002) Hog1 kinase converts the Sko1-Cyc8-Tup1 repressor complex into an activator that recruits SAGA and SWI/SNF in response to osmotic stress. Mol Cell 6:1307-1317

Reiser V, D'Aquino KE, Ee LS, Amon A (2006) The stress-activated mitogen-activated protein kinase signaling cascade promotes exit from mitosis. Mol Biol Cell 7:3136-3146

Rupes I (2002) Checking cell size in yeast. Trends Genet 9:479-485

Sanchez-Piris M, Posas F, Alemany V, Winge I, Hidalgo E, Bachs O, Aligue R (2002) The serine/threonine kinase Cmk2 is required for oxidative stress response in fission yeast. J Biol Chem 20:17722-17727

Schwob E, Bohm T, Mendenhall MD, Nasmyth K (1994) The B-type cyclin kinase inhibitor p40SIC1 controls the G1 to S transition in *S. cerevisiae*. Cell 2:233-244

Shiozaki K, Russell P (1996) Conjugation, meiosis, and the osmotic stress response are regulated by Spc1 kinase through Atf1 transcription factor in fission yeast. Genes Dev 18:2276-2288

Smith DA, Toone WM, Chen D, Bahler J, Jones N, Morgan BA, Quinn J (2002) The Srk1 protein kinase is a target for the Sty1 stress-activated MAPK in fission yeast. J Biol Chem 36:33411-33421

Stegmeier F, Amon A (2004) Closing mitosis: the functions of the Cdc14 phosphatase and its regulation. Annu Rev Genet 203-232

Tatebe H, Shimada K, Uzawa S, Morigasaki S, Shiozaki K (2005) Wsh3/Tea4 is a novel cell-end factor essential for bipolar distribution of Tea1 and protects cell polarity under environmental stress in *S. pombe*. Curr Biol 11:1006-1015

van Vugt MA, Bras A, Medema RH (2004) Polo-like kinase-1 controls recovery from a G2 DNA damage-induced arrest in mammalian cells. Mol Cell 5:799-811

Versele M, Thorner J (2005) Some assembly required: yeast septins provide the instruction manual. Trends Cell Biol 8:414-424

Wilkinson MG, Millar JB (2000) Control of the eukaryotic cell cycle by MAP kinase signaling pathways. FASEB J 14:2147-2157

Wilkinson MG, Samuels M, Takeda T, Toone WM, Shieh JC, Toda T, Millar JB, Jones N (1996) The Atf1 transcription factor is a target for the Sty1 stress-activated MAP kinase pathway in fission yeast. Genes Dev 18:2289-2301

Wittenberg C, Sugimoto K, Reed SI (1990) G1-specific cyclins of *S. cerevisiae*: cell cycle periodicity, regulation by mating pheromone, and association with the p34CDC28 protein kinase. Cell 2:225-237

Zapater M, Clotet J, Escote X, Posas F (2005) Control of cell cycle progression by the stress-activated Hog1 MAPK. Cell Cycle 1:6-7

Zeng Y, Piwnica-Worms H (1999) DNA damage and replication checkpoints in fission yeast require nuclear exclusion of the Cdc25 phosphatase via 14-3-3 binding. Mol Cell Biol 11:7410-7419

Zimmerman S, Daga RR, Chang F (2004) Intra-nuclear microtubules and a mitotic spindle orientation checkpoint. Nat Cell Biol 12:1245-1246

Aligue, Rosa
Department of Cell Biology, University of Barcelona, 08036 Barcelona, Spain.
aliguerosa@ub.edu

Lopez-Aviles, Sandra
Department of Cell Biology, University of Barcelona, 08036 Barcelona, Spain.

Hog1-mediated metabolic adjustments following hyperosmotic shock in the yeast *Saccharomyces cerevisiae*

Bodil Nordlander, Marcus Krantz, Stefan Hohmann

Abstract

Yeast cells exposed to hyperosmotic conditions adjust their metabolism in order to increase the production of the compatible solute glycerol. The MAPK Hog1 seems to affect glycerol production rapidly by stimulating flux through glycolysis and long-term via the transcriptional upregulation of genes encoding enzymes in glycerol formation. In addition, the glycerol channel Fps1 rapidly closes after a hyperosmotic shock to ensure efficient glycerol accumulation. Hog1 seems to modulate basal Fps1 activity and the MAPK is needed to allow complete closure of Fps1. Moreover, the expression of a number of metabolic genes is affected in both Hog1-dependent and independent ways. How those changes contribute to osmotic adaptation of yeast cells is not completely understood. To separate and analyze the different roles of Hog1 in the adjustment of metabolism will probably need a time-resolved holistic view on all components involved and a combination of theoretical modelling and experimentation. In addition, there is a need for more detailed analysis of direct and indirect Hog1-targets to elucidate the impact of such regulatory interactions.

1 Yeast osmoregulation and carbon metabolism

Maintenance of the cellular water balance is fundamental for life. The process of controlling the water balance is constantly active in living cells to maintain an appropriate intracellular environment for biochemical processes as well as turgor of cells and organisms. In the laboratory, the osmoregulatory system is most conveniently studied as a response to osmotic shock. Rapid and dramatic changes in the extracellular water activity cause either water efflux, and hence cell shrinkage (hyperosmotic shock), or influx and concomitant cell swelling (hypo-osmotic shock). The yeast *Saccharomyces cerevisiae*, as a free-living organism experiencing both slow and rapid changes in extracellular water activity, has proven a suitable and genetically tractable experimental system to study the underlying signalling pathways and regulatory processes governing osmoregulation. Although far from complete, the present picture of yeast osmoregulation is both extensive and detailed (de Nadal et al. 2002; Hohmann 2002; Klipp et al. 2005).

Topics in Current Genetics, Vol. 20
F. Posas, A. R. Nebreda (Eds.): Stress-Activated Protein Kinases
DOI 10.1007/4735_2007_0247 / Published online: 1 August 2007

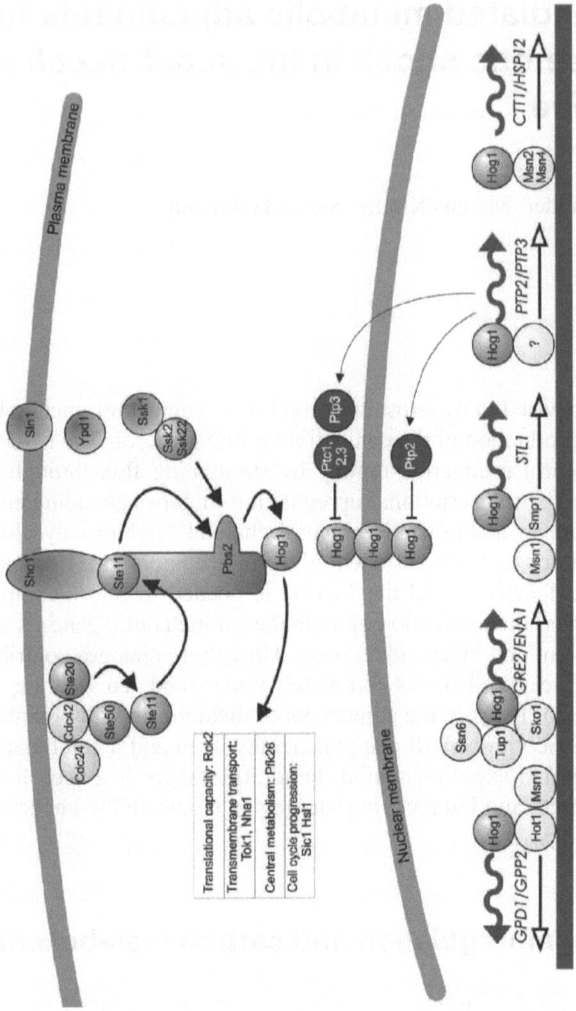

Fig. 1. The High Osmolarity Glycerol (HOG) pathway. The Sln1-branch and the Sho1-branch converge at the MAPKK Pbs2. Activated Pbs2 phosphorylates and thus activates the MAPK Hog1. Hog1 regulates events, which affect translation capacity, metabolism, cell-cycle progression, transmembrane transport, and the expression of genes such as *STL1*, *GPD1*, *GPP2*, and *GRE2*.

Following a hyperosmotic shock, yeast cells shrink within seconds and the HOG (High Osmolarity Glycerol) signal transduction system becomes activated within less than one minute (Fig. 1) (de Nadal et al. 2002; Hohmann 2002). First responses mediated by active Hog1 appear to encompass stimulation of ion export (Proft and Struhl 2004), an arrest of the cell division cycle (Escote et al. 2004; Clotet et al. 2006) and a diminished translational capacity (Bilsland-Marchesan et al. 2000; Teige et al. 2001). The glycerol channel Fps1 rapidly closes (Luyten et al. 1995; Tamás et al. 1999). In addition, it has been reported that the Hog1-dependent activation of the Pfk26 enzyme causes activation of the glycolysis to enhance production of the osmolyte glycerol (Dihazi et al. 2004). Glycerol accumulation is essential for the adaptation to hyperosmotic conditions (Albertyn et al. 1994a; Ansell et al. 1997). A significant portion of active Hog1 accumulates in the nucleus and localises to target promoters where it controls gene expression (Gasch et al. 2000; Posas et al. 2000; Rep et al. 2000; O'Rourke and Herskowitz 2004; Pokholok et al. 2006). Genes encoding enzymes in glycerol production and active glycerol uptake are among those upregulated in a Hog1-dependent manner. A mathematical model of osmoregulation explains that once the cell commences reswelling due to glycerol accumulation, the HOG pathway signalling input is terminated, Hog1 becomes dephosphorylated at a higher rate than it becomes phosphorylated and the pathway signalling is decaying. Finally, the osmotic response adapts to a level required for growth at higher osmolarity (Klipp et al. 2005).

Adjustment of cellular metabolism, especially glycolysis, is crucial for osmotic adaptation. Most significantly, yeast cells re-direct carbon towards enhanced production of glycerol to restore turgor pressure. Hence, central yeast carbon metabolism plays a significant role in osmotic adaptation (Fig. 2). Glucose (GLC-D) and fructose (FRU) are the preferred carbon sources. They are transported into yeast cells via the sugar transporters Hxt1-4, 6, and 7. Following uptake, glucose and fructose are phosphorylated by the hexokinase isoform Hxk2; expression of the isoform Hxk1 as well as the glucose-specific glucokinase Glk1 is upregulated under stress and starvation conditions (Gasch et al. 2000). Glucose-6-phosphate (G6P) is converted to fructose-6-phosphate (F6P) by phoshoglucose isomerase (Pgi1). Fructose-6-phosphate is phosphorylated by phoshofructokinase to 1,6-bisphosphate (FDP), and from this point onwards catabolism follows the standard scheme of glycolysis to pyruvate (PYR). Under conditions of glucose excess, yeast cells metabolise most pyruvate via pyruvate decarboxylase (Pdc1, 5, and 6) and alcohol dehydrogenase (Adh1 and 2) to ethanol (ETOH) and CO_2. Only when glucose is limited and oxygen available *S. cerevisiae* oxidises most pyruvate completely to CO_2 via the pyruvate dehydrogenase complex and the TCA cycle. The unusual behaviour to ferment excess sugar to ethanol even in the presence of oxygen is though to provide yeast with a competitive advantage in its natural environment, since yeast is highly tolerant to ethanol (Goncalves and Planta 1998; Hohmann and Meacock 1998; Ozcan and Johnston 1999; Gramser 2005).

Trehalose (TRE) and glycerol (GLYC) metabolism branch off from central glycolysis (Thevelein and Hohmann 1995; Hohmann 2002). The disaccharide trehalose is produced from two molecules of glucose-6-phosphate (G6P). One molecule of glucose-6-phosphate is first converted to UDP-glucose (UDPG) in two reactions catalysed by phosphoglucose isomerase (Pgm1 and Pgm2) and UDP-glucose pyrophosphorylase (Ugp1). UDP-glucose and glucose-6-phosphate are then combined by trehalose-6-phosphate synthase (Tps1) to trehalose-6-phosphate (TRE6P), which is dephosphorylated by trehalose-6-phosphatase (Tps2) to trehalose. For degradation trehalose is hydrolysed to two glucose residues by a cytoplasmic trehalase (Nth1 and 2). Trehalose can be transported across the plasma membrane under certain conditions but this does not seem to be a physiologically relevant process (Francois and Parrou 2001).

Glycerol is produced in two steps from the glycolytic intermediate dihydroxyacetone phosphate (DHAP). Those steps are catalysed by NAD-dependent glycerol-3-phosphate dehydrogenase (Gpd1 and 2) and glycerol-3-phosphatase (Gpp1 and 2) (Hohmann 2002). Glycerol can also be utilised as a carbon and energy source. Glycerol is taken up by proton symport mediated by Stl1 while glycerol efflux from the cell is passive and mediated by Fps1 (Luyten et al. 1995; Oliveira et al. 2003; Ferreira et al. 2005). Glycerol is phosphorylated to glycerol-3-phosphate (GLYC3P) by glycerol kinase (Gut1) and converted to dihydroxyacetonephosphate by the FAD-dependent glycerol-3-phosphate dehydrogenase (Hohmann 2002).

Interestingly, mutations blocking trehalose or glycerol metabolism lead to complete deregulation and arrest of the main glycolytic pathway. Hence, these two branches have important regulatory and/or metabolic buffering roles. Knockout of *TPS1* (but not *TPS2*) results in an inability of yeast cells to grow on glucose due to an imbalance in metabolite levels between the upper and lower part of glycolysis. This phenomenon has been explained by a feedback on the hexokinases via trehalose-6-phosphate, although the regulatory details are not fully explained (Thevelein and Hohmann 1995). Similarly, a complete block of glycerol production (by deletion of *GPD1* and *GPD2*) causes an inability of yeast cells to grow with glucose as carbon source in the absence of oxygen. This is because glycerol production seems to be the only way for re-oxidising access NADH when the respiratory chain is inactive (Ansell et al. 1997).

2 Osmolytes: glycerol and trehalose as cell protectants

During steady-state growth, the cell keeps its intracellular environment at a slightly higher osmotic pressure than the external environment. This difference is due to the high cytosolic concentration of solutes and/or ions as compared to the external concentration causing a force for water inflow. As the cell wall is less elastic than the plasma membrane, it resists the expansion of the cell and creates turgor pressure, which is defined as the difference in hydrostatic pressure between the inside and the outside of the cell (Marechal et al. 1995; Martinez de Maranon et al. 1996; Gervais and Beney 2001). The cell wall, which provides cell shape and physical shelter but also a protection against bursting upon an instant hypo-osmotic shock, allows the cell to survive in environments where water availability continuously changes (Gervais and Beney 2001; Harold 2002). When yeast cells are exposed to an increased external osmotic pressure, two distinct phases of osmoregulation are observed. First, the passive phase, which corresponds to rapid water flow across the plasma membrane, loss of turgor and a volume decrease. The final volume depends on stress intensity. Second, the cell responds by actively triggering an increased accumulation of glycerol to lower the osmotic potential, thereby allowing passive water uptake and re-established volume and turgor (Marechal et al. 1995; Martinez de Maranon et al. 1996; Gervais and Beney 2001).

The accumulation of osmotically active compounds, more commonly named "compatible solutes", is a common strategy to protect cells from prolonged dehydration and death (Yancey et al. 1982). The molecules known to serve as compatible solutes fall into a limited number of chemical categories, such as polyols (glycerol, sorbitol, inositols, etc), small carbohydrates (e.g. trehalose), amino acids and derivatives (glycine, proline, taurine, etc), and methylamines (e.g. glycine betaine). These compounds have very little impact on cellular functions. In addition, most organic osmolytes are believed to be highly soluble, neutral at physiological pH, but also effective in stabilising proteins and other cell components (Yancey et al. 1982; Kempf and Bremer 1998; Yancey 2005). A number of these organic osmolytes seems to have unique properties besides maintaining cell volume, for example, in antioxidation, in providing redox-balance, in detoxifying sulphide as well as in stabilising cellular proteins and structures mentioned above (Yancey 2005).

In response to hyperosmotic stress, *S. cerevisiae* mainly produces the three-carbon polyol glycerol as well as the disaccharide trehalose (Blomberg and Adler 1992). Although accumulated to high concentrations, glycerol does not disturb vital cellular processes, but rather protects macromolecular structure and function under dehydration. According to the preferential exclusion model, glycerol is believed to stabilise and promote native conformations of proteins and subunit aggregations. This could be explained by the fact that glycerol itself is excluded from the vicinal hydration sphere of proteins which would minimize protein-solvent interactions (Yancey et al. 1982; Blomberg and Adler 1992).

While glucose-growing yeast cells accumulate glycerol under hyperosmotic stress, cells growing on galactose or ethanol seem to accumulate trehalose in addition to glycerol (Andre et al. 1991; Hounsa et al. 1998). Moreover, glycerol is essential for growth under moderate osmotic stress whereas trehalose contributes to survival under severe osmotic stress (Hounsa et al. 1998; Hohmann 2002). While exponentially growing cells are susceptible to dehydration, stationary phase cells contain increased levels of trehalose and an enhanced ability to survive dehydration. Such a correlation between trehalose accumulation and yeast stress tolerance is observed under a number of different growth conditions, which has been interpreted as indication for a role of trehalose in stress protection. This may be true under certain conditions while significant acquisition of stress tolerance has been observed even in mutants unable to accumulate trehalose (De Virgilio et al. 1994; Hottiger et al. 1994; Hottiger 1987; Thevelein and Hohmann 1995; Hounsa et al. 1998; Singer and Lindquist 1998; Francois and Parrou 2001).

Trehalose is a widespread disaccharide throughout the biological world (Elbein et al. 2003) and believed to have a number of important functions. During almost complete dehydration, such as spore formation, the yeast cell accumulates trehalose. Desiccation causes destabilisation of the lipid bilayer through fusion and lipid phase transitions and it has been shown that trehalose inhibits both these transitions, thus protecting and preserving membrane structure. Trehalose also preserves labile proteins during drying, protects proteins from denaturation, and

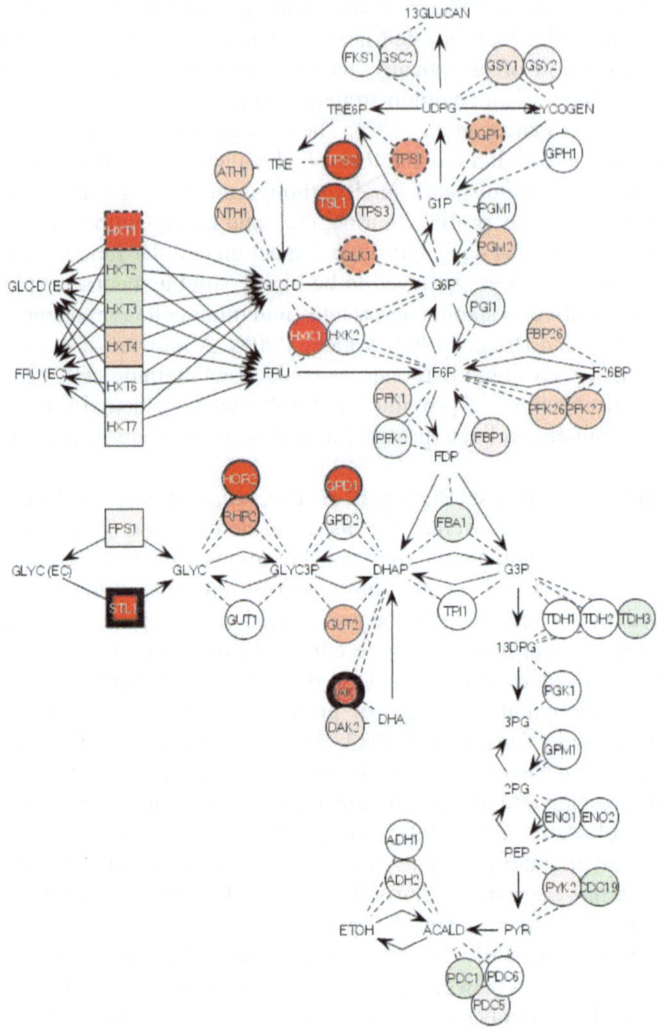

Fig. 2. Glycolysis and its branches. Osmotic stress leads to transcriptional upregulation of genes encoding enzymes in glycerol production and trehalose turnover. Enzymes are indicated by spheres and coloured according to osmotic induction of their corresponding genes. Enzymes are connected with their substrates and products by dashed lines. Metabolic conversions are indicated by solid arrows. Induction and repression are derived from three independent time course experiments, from which the lowest peak change has been selected (Gasch et al. 2000; Krantz et al. 2004). Red and green colours indicate induction and repression, respectively, ranging between +3 and -3 on a log2 scale for genes showing a consistent expression pattern. Transcripts with ambiguous changes are shown as white spheres. Hog1 dependence is indicated with increasing line thickness from -2 to -5 on a log2 scale, and dashed lines for an effect between -1 and -2 (Rep et al. 2000).

prevents the aggregation of denaturated proteins to promote a subsequent refolding by molecular chaperons (Thevelein 1984; Singer and Lindquist 1998; Francois and Parrou 2001; Elbein et al. 2003). Still, the precise roles of trehalose in yeast physiology are not fully understood.

3 Flux control and potential direct effects on metabolism by Hog1: PFK2

Following hyperosmotic shock glycerol accumulation starts within less than one minute resulting in substantial glycerol levels already after twenty minutes (Klipp et al. 2005). This rapid increase seems to be mainly due to basal glycerol production and closure of the glycerol channel Fps1, but there are also indications of a more direct regulation of glycolytic flux by Hog1 to increase glycerol production (Tamas et al. 1999; Dihazi et al. 2004; Krantz et al. 2004). In addition, Hog1 promotes increased expression of genes involved in glycerol metabolism to further enhance and maintain glycerol production over a longer period of time (Albertyn et al. 1994a; 1994b; Ansell et al. 1997; Påhlman et al. 2001). The different events known so far contributing to glycerol accumulation are shown in Figure 4.

Cells regulate their glycolytic flux in order to survive under different stress conditions. In the past, one regulatory concept was that of rate-limiting steps, which would control the pace of the throughput of the pathway. However, Metabolic Control Analysis (MCA) led to the view that glycolytic flux is regulated more subtly and that regulation is shared between all steps (Westerhoff 1995). Several enzymes in glycolysis are allosterically regulated, such as hexokinase (by trehalose-6-phosphate *in vitro* (Blazquez et al. 1993)), phosphofructokinase (see further), or pyruvate kinase (by fructose-1,6-bisphosphate (Murcott et al. 1992; Jurica et al. 1998)) suggesting that allosteric effectors as well as global substrates such as ATP, ADP, NAD+, and NADH could potentially have pronounced roles in regulating the flux (Gancedo 1989; Sierkstra et al. 1992). After a hyperosmotic stress, there are changes in the expression of a large number of genes encoding metabolic enzymes. These changes vastly exceed those that could be contributed to Hog1 directly (Fig. 2, 3). The question is whether these changes are required for, or rather an effect of, adaptation to high glycerol production and growth under the stress condition. In summary, the regulation of the flux seems to depend on the coordinated triggering of multiple events, such as protein modification and modulation of gene expression, but also the allosteric regulation of enzymatic activities (Goncalves and Planta 1998), an effect which might be the most immediate one when the cell is experiencing acute stress conditions like hyperosmotic stress.

Phosphofructokinase was for a long time thought to be the main control point for steady-state glycolytic flux. Today it is no longer considered to have a major control of the flux under steady-state conditions (Westerhoff 1995). However, *in vitro* studies have shown that the enzyme is sensitive to a number of regulators, such as fructose-6-phosphate, fructose-1,6-bisphosphate, ATP, AMP, NH_4^+, and

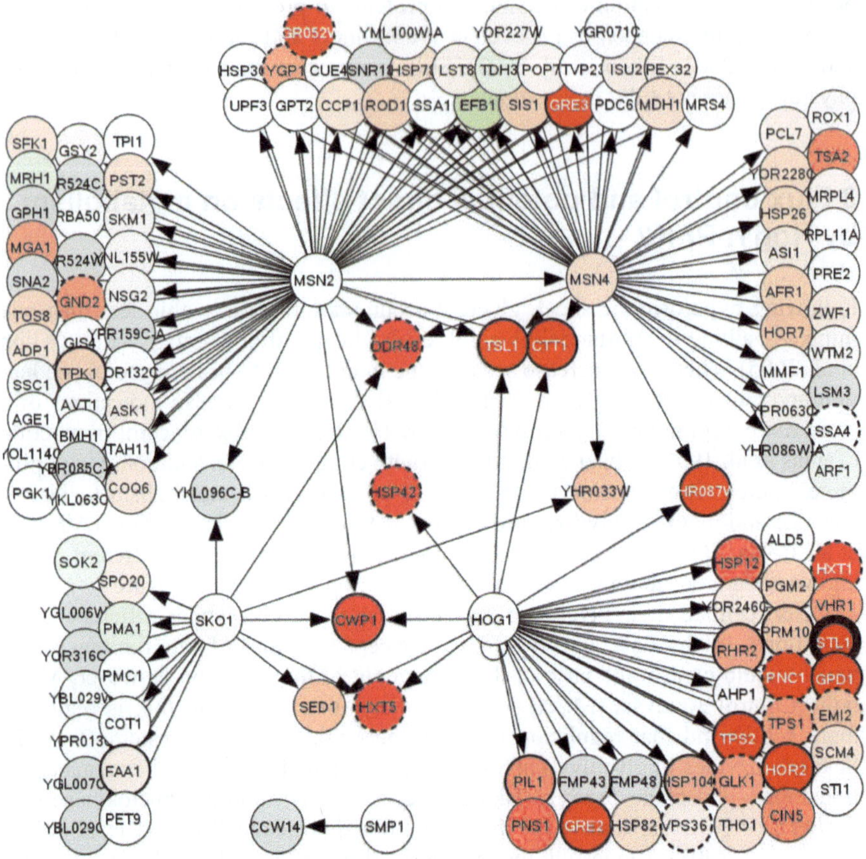

Fig. 3. Transcriptional response. The promoters to which Hog1, Msn2, Msn4, Sko1, and Smp1 localise are indicated with solid arrows. Each of these factors, as well as Hot1 and Msn1, has been implicated in the transcriptional response to osmotic stress. While osmotic stress leads to an altered expression of hundreds of genes (Gasch et al. 2000; Rep et al. 2000) a large part of this response is not stress-specific and includes for example transient repression of many ribosomal protein genes. Consistently, the targets of the transcriptional regulators known to be involved in the osmotic stress response are clearly not enough to account for these extensive changes, even though their targets include other transcription factors such as Sok2 and Cin5. Which physiological alterations cause these widespread transcriptional changes remains unknown. The regulatory network is derived from global localisation studies where Msn1 had no high confidence targets and data for Hot1 was missing (MacIsaac et al. 2006; Pokholok et al. 2006). Colouring is as in Figure 2, with gray ORFs indicating that information is missing.

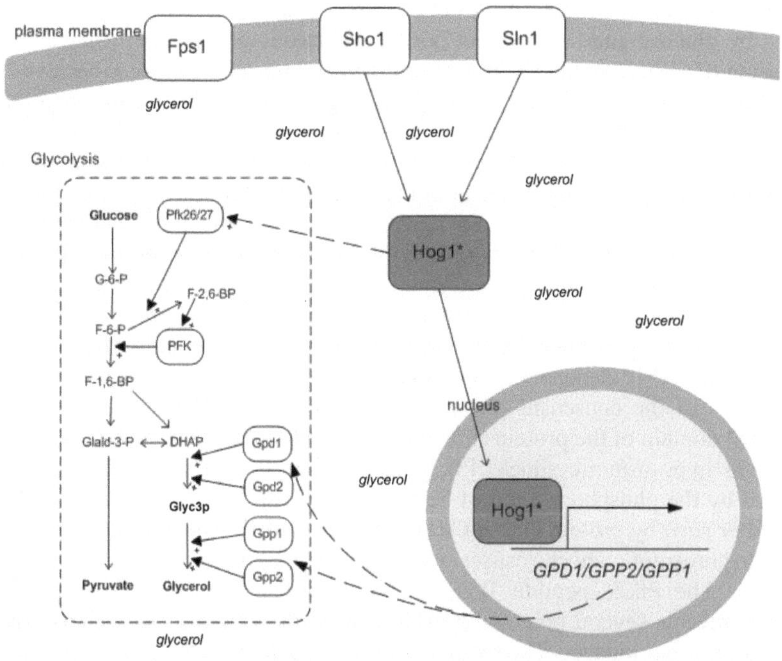

Fig. 4. Glycerol accumulation. Expression of the *GPD1* and *GPP2,* which encode enzymes in glycerol synthesis, is upregulated in a HOG-dependent manner. Gpd1 exerts a major control on glycerol production and hence an increase of Gpd-proteins increases glycerol formation. Hog1 appears to be involved in the regulation of the PFK2-enzyme (Pfk26/Pfk27). An increased activity of PFK2 enhances the production of the allosteric activator F26BP, which activates phosphofructokinase. Such stimulation might increase the pool of DHAP, the substrate for Gpd. The glycerol export channel Fps1 closes quickly following hyperosmotic stress. Hog1 might be involved in the closure of Fps1 since a *hog1Δ* shows a slighter higher glycerol leakage under basal as well as hyperosmotic conditions as compared to wild type.

fructose-2,6-bisphosphate (Reibstein et al. 1986; Gancedo 1989). Fructose-2,6-bisphosphate (F26BP) was considered as the most significant activator of phosphofructokinase because of its strong *in vitro* activating effect (Bartrons et al. 1982; Schaftingen 1982). However, a simultaneous overproduction of phosphofructokinase and F26BP *in vivo* did not increase the glycolytic flux and a strain devoid of F26BP did not significantly affect the steady-state glycolytic flux (Boles et al. 1996; Muller et al. 1997; Goncalves and Planta 1998). However, when transferring respiring F26BP -deficient mutant cells to conditions where glucose is fermented, glucose consumption, and ethanol production rates were initially diminished as compared to wild type. This led to the interesting idea that F26BP activation of phosphofructokinase is important during transitions between different metabolic conditions (Boles et al. 1996; Goncalves and Planta 1998).

The two genes *PFK26* and *PFK27* encode the 6-phosphofructo-2-kinase, PFK2. PFK2 catalyses the conversion of fructose-6-phosphate to the allosteric activator

F26BP (Aragon et al. 1987; Boles et al. 1996). The expression of *PFK27* is stimulated by glucose and sucrose, but not by galactose and maltose, whereas the expression of *PFK26* is higher in respiring cells (Boles et al. 1996; Goncalves et al. 1997; Goncalves and Planta 1998). During growth on ethanol, the major PFK2 isoform Pfk26 is believed to be present in an inactive form, which can be readily activated by protein kinase A upon the transition to the fermentative mode of metabolism. Thus, the rapid increase in the level of cAMP after the addition of, for example, glucose was shown to be responsible for the rapid activation of 6-phosphofructo-2-kinase activity (Francois et al. 1984; Goncalves and Planta 1998; Dihazi et al. 2003). The Pfk27 enzyme is assumed to be constitutively active. The activity of the Pfk26 enzyme seems to be modulated by post-translational modifications, such as phosphorylation and acetylation (Dihazi et al. 2001, 2003, 2004, 2005). The cAMP-dependent protein kinase A phosphorylates and activates Pfk26 at Ser644 of the consensus phosphorylation site (RRYS) located within the C-terminal domain of the protein (Francois et al. 1984; Dihazi et al. 2003).

Upon hypo-osmotic shock, Pfk26 is inactivated *in vivo*, which probably is caused by the phosphorylation of Ser652; this phosphorylation appears to be mediated *in vitro* by protein kinase C (Dihazi et al. 2001). The opposite stress condition, hyperosmotic shock, causes activation of the enzyme in a Hog1-dependent manner. The Pfk26 peptide $T_{67\text{-}101}$ is phosphorylated on four sites and the sequence reveals several potential MAPK consensus sequences as well as potential targets for other kinases. The removal of this peptide from the protein resulted in a catalytically inactive enzyme (Dihazi et al. 2004). Although different mutants in the HOG pathway, such as *hog1Δ* and *pbs2Δ* mutants, do not show the osmoshock-dependent activation of PFK2 normally observed in wild type cells (Dihazi et al. 2004), a direct involvement of Hog1 in the phosphorylation of the enzyme has not been demonstrated yet.

Following osmotic shifts there are rapid changes in the amount of intracellular glucose-6-phosphate and glycerol illustrating that metabolism is redirected (Dihazi et al. 2001, 2004). Survival after a hypo-osmotic shock requires a rapid decrease in the intracellular glycerol content which depends on glycerol outflow through the Fps1 channel protein and (probably) a decrease in the production of glycerol (Tamás et al. 1999; Dihazi et al. 2001). The accumulation of glucose-6-phosphate, which occurs after a hypo-osmotic shock, results in an increased availability of essential precursors for cell wall material such as glucans (Boada et al. 2000; Shahinian and Bussey 2000). Following hyperosmotic shock, increased glucose-6-phosphate consumption and enhanced glycerol production are observed. Moreover, it was shown that the overproduction of PFK2 increased glycerol accumulation dramatically compared to cells lacking this enzyme (Dihazi et al. 2004). Potentially, stimulation of phosphofructokinase by F26BP increases the pool of the glycolytic intermediate dihydroxyacetonephosphate (DHAP). In glycerol producing cells, the regeneration of NADH is continuously required to support the massive glycerol production while maintaining a proper NADH/NAD+ ratio and hence redox balance in the cell. Blomberg et al. demonstrated an increase in acetate production in cells exposed to 0.7M NaCl which could enhance NADH regeneration (Blomberg and Adler 1989). Metabolic Control Analysis showed that the Gpd enzymes exerts a major control on glycerol production, but also revealed that the optimal strategy to improve glycerol production would be to simultaneously increase the levels of DHAP, the Gpd enzyme level as well as the NADH/NAD+ ratio (Cronwright et al. 2002). This strategy seems to be employed by the yeast cell.

4 Glycerol export and import

The aquaporin family of transmembrane proteins encompasses water-specific aquaporins as well as aquaglyceroporins with permeability for glycerol and other small uncharged compounds (Hohmann 2001; Pettersson et al. 2006). Fps1 is characterised by six transmembrane domains where loops B and E dip back into the membrane and meet in the central plane of the membrane bilayer. These two loops are part of the transmembrane pore constriction and contain the two NPA motifs, which are characteristics for the aquaporin family. Fps1, like many other fungal aquaglyceroporins, has long extensions at the N- and the C-terminus (Pettersson et al. 2005).

The yeast plasma membrane is rather impermeable for glycerol (Oliveira et al. 2003) and hence employing a regulated glycerol export channel allows yeast cells to rapidly control and fine-tune their glycerol content. Mutants lacking Fps1 display a higher intracellular glycerol level (Tao et al. 1999) but under hyperosmotic conditions accumulate glycerol with a similar profile as wild type cells (Tamas et al. 1999). When yeast cells are then exposed to a hypo-osmotic shock, i.e. by transfer into distilled water, most of the intracellular glycerol is released to the surrounding medium within less than three minutes. Rapid glycerol export is fully dependent on Fps1 (Tamas et al. 1999). Mutants lacking Fps1 are sensitive to a hypo-osmotic shock and only a fraction of the cells survives such a treatment (Tamas et al. 1999). Deletion of *FPS1* is lethal when combined with mutations that affect the strength of the cell wall, illustrating the role of Fps1 as an osmotic pressure safety valve (Tamás et al. 1999; Tong et al. 2004).

Fps1 is probably regulated through a gating mechanism. Constitutively open Fps1 mutants exist and those mediate glycerol efflux even under hyperosmotic conditions. As a consequence, such yeast cells are unable to properly accumulate glycerol, which they try to compensate by enhanced glycerol production (Tamás et al. 1999, 2003; Hedfalk et al. 2004; Karlgren et al. 2004). Mutants expressing open Fps1 grow more poorly under hyperosmotic conditions. Hence, mutants lacking Fps1 are sensitive to hypo-osmotic stress while mutants with overactive Fps1 are sensitive to hyper-osmotic stress, illustrating the importance of Fps1 for yeast osmoregulation. While proteins with similar functions in regulated osmolyte efflux exist in probably most organisms, the precise type of aquaglyceroporin represented by *S. cerevisiae* Fps1 appears to be a yeast-specific invention in evolution (Pettersson et al. 2005).

The control mechanism of Fps1 has been investigated by targeted and random mutagenesis. It appears that three regions play a role in Fps1 gating:

1. The B-loop. A random genetic screen for constitutively open Fps1 (Karlgren et al. 2004) identified residues G348 and H350, both facing the intracellular mouth of the pore in the B-loop. Both residues are located just adjacent to the conserved NPA motif (residues 352-354). It is possible that the mouth of the cytoplasmic side of the pore serves as the target for the closing mechanism.
2. The region of about 40 amino acids proximal of the first transmembrane domain was first identified by truncation analysis and then by targeted and random point mutations (Tamas et al. 1999, 2003; Karlgren et al. 2004). Within this domain 15/18 residues are conserved in all yeast orthologues. Also the spacing of this region to the first transmembrane domain is conserved (18 residues), and several mutations in this region affecting gating have been identified. We assume that this proximal domain serves as a "lid" that interacts with the B-loop, but direct evidence for such a scenario is presently missing.
3. The ten amino acids immediately distal of the sixth transmembrane domain are also highly conserved (8/10 identical in all yeast sequences). Truncation of the C-terminus within this sequence, but not further downstream, renders Fps1 constitutively active (Hedfalk et al. 2004; Karlgren et al. 2004). The C-terminal extension may have a more indirect role in gating, for instance by supporting a specific configuration of the N-terminal extension and/or oligomerisation. Usually, aquaporins are tetramers.

A link between the Fps1 N-terminus and Hog1 signalling was recently discovered in an entirely different context. Fps1 mediates the influx of toxic trivalent arsenite into yeast cells (Wysocki et al. 2001), a feature it shares with other aquaglyceroporins. Hydrated arsenite resembles glycerol. Mutants lacking Fps1 are highly tolerant to arsenite while expression of constitutively open Fps1 causes arsenite hypersensitivity (Wysocki et al. 2001). Mutants lacking Hog1 are sensitive to arsenite. It appears that the Hog1 target in the acquisition of arsenite tolerance is Fps1. Hog1 phosphorylates T231 in the highly conserved N-terminal domain and this effect seems to cause inactivation of Fps1 (Thorsen et al. 2006). Mutation of this residue causes constitutively open Fps1, arsenite sensitivity and also hyper-osmosensitivity (Tamás et al. 2003; Karlgren et al. 2004; Thorsen et al. 2006). However, whether the same Hog1-dependent mechanism controls Fps1 also following hyper-osmotic shock is presently unclear. Rather, the Hog1 effect may represent a long-term control of Fps1 activity.

Measuring the expression of the Sugar Transporter-Like gene *STL1* is frequently done to investigate HOG pathway activity since the *STL1* expression in response to hyperosmotic shock is strong and completely Hog1-dependent (Rep et al. 2000). Stl1 represents a glycerol uptake system in yeast that transports glycerol by proton symport during conditions where glycerol symport is functional. Expression of *STL1* is subject to glucose-induced inactivation and stimulated upon the diauxic shift in a Cat8-dependent manner (Haurie et al. 2001; Ferreira et al. 2005). This suggests that the physiological role of Stl1 is in glycerol uptake for its utilisation as a carbon source. Under hyperosmotic stress, Stl1 may support glycerol accumulation by taking up glycerol from the environment. Active glycerol uptake under osmotic stress is observed in many other yeasts (e.g. Prista et al. 2005).

5 Integration: a potential timeline of adjustments under osmo-stress

Establishing a detailed timeline of the relative quantitative importance of different control mechanisms acting on glycerol accumulation will require a combination of mathematical modelling and experimental studies. Such a model could integrate HOG signalling, transcription, translation, metabolism as well as the biophysical changes that occur due to the water movement over the membrane. An integrative view of the system should facilitate our understanding of how certain processes, often studied in isolation, are interconnected. In addition, as turgor maintenance is performed independently by each cell, it will be necessary to analyse single cells to capture cell-to-cell variations and thereby important system parameters such as noise and thresholds. Such an approach will allow predicting cellular events in a more precise way.

Acknowledgements

Relevant work in the Hohmann lab is supported by grants from the Swedish Research Council and the European Commission (QUASI: LSHG-CT2003-503230; SystemsBiology: MEST2-CT2004-514169). VINNOVA supports the collaboration between the Hohmann lab and the Systems Biology Institute (Interdisciplinary BIO Sweden-Japan). Marcus Krantz is a postdoctoral fellow supported by the Japan Society for the Promotion of Science.

References

Albertyn J, Hohmann S, Prior BA (1994a) Characterization of the osmotic-stress response in *Saccharomyces cerevisiae*: osmotic stress and glucose repression regulate glycerol-3-phosphate dehydrogenase independently. Curr Genet 25:12-18

Albertyn J, Hohmann S, Thevelein JM, Prior BA (1994b) *GPD1*, which encodes glycerol-3-phosphate dehydrogenase, is essential for growth under osmotic stress in *Saccharomyces cerevisiae*, and its expression is regulated by the high-osmolarity glycerol response pathway. Mol Cell Biol 14:4135-4144

Andre L, Hemming A, Adler L (1991) Osmoregulation in *Saccharomyces cerevisiae*. Studies on the osmotic induction of glycerol production and glycerol-3-phosphate dehydrogenase (NAD+). FEBS Lett 286:13-17

Ansell R, Granath K, Hohmann S, Thevelein JM, Adler L (1997) The two isoenzymes for yeast NAD+-dependent glycerol 3-phosphate dehydrogenase encoded by *GPD1* and

GPD2 have distinct roles in osmoadaptation and redox regulation. EMBO J 16:2179-2187

Aragon JJ, Gomez ME, Gancedo C (1987) Identification of two forms of 6-phosphofructo-2-kinase in yeast. FEBS Lett 226:121-124

Bartrons R, Van Schaftingen E, Vissers S, Hers HG (1982) The stimulation of yeast phosphofructokinase by fructose 2,6-bisphosphate. FEBS Lett 143:137-140

Bilsland-Marchesan E, Arino J, Saito H, Sunnerhagen P, Posas F (2000) Rck2 kinase is a substrate for the osmotic stress-activated mitogen-activated protein kinase Hog1. Mol Cell Biol 20:3887-3895

Blazquez MA, Lagunas R, Gancedo C, Gancedo JM (1993) Trehalose-6-phosphate, a new regulator of yeast glycolysis that inhibits hexokinases. FEBS Lett 329:51-54

Blomberg A, Adler L (1989) Roles of glycerol and glycerol-3-phosphate dehydrogenase (NAD+) in acquired osmotolerance of *Saccharomyces cerevisiae*. J Bacteriol 171:1087-1092

Blomberg A, Adler L (1992) Physiology of osmotolerance in fungi. Adv Microb Physiol 33:145-212

Boada J, Roig T, Perez X, Gamez A, Bartrons R, Cascante M, Bermudez J (2000) Cells overexpressing fructose-2,6-bisphosphatase showed enhanced pentose phosphate pathway flux and resistance to oxidative stress. FEBS Lett 480:261-264

Boles E, Gohlmann HW, Zimmermann FK (1996) Cloning of a second gene encoding 6-phosphofructo-2-kinase in yeast, and characterization of mutant strains without fructose-2,6-bisphosphate. Mol Microbiol 20:65-76

Clotet J, Escote X, Adrover MA, Yaakov G, Gari E, Aldea M, de Nadal E, Posas F (2006) Phosphorylation of Hsl1 by Hog1 leads to a G2 arrest essential for cell survival at high osmolarity. EMBO J 25:2338-2346

Cronwright GR, Rohwer JM, Prior BA (2002) Metabolic control analysis of glycerol synthesis in *Saccharomyces cerevisiae*. Appl Environ Microbiol 68:4448-4456

de Nadal E, Alepuz PM, Posas F (2002) Dealing with osmostress through MAP kinase activation. EMBO Rep 3:735-740

De Virgilio C, Hottiger T, Dominguez J, Boller T, Wiemken A (1994) The role of trehalose synthesis for the acquisition of thermotolerance in yeast. I. Genetic evidence that trehalose is a thermoprotectant. Eur J Biochem 219:179-186

Dihazi H, Kessler R, Eschrich K (2001) Phosphorylation and inactivation of yeast 6-phosphofructo-2-kinase contribute to the regulation of glycolysis under hypotonic stress. Biochemistry 40:14669-14678

Dihazi H, Kessler R, Eschrich K (2003) Glucose-induced stimulation of the Ras-cAMP pathway in yeast leads to multiple phosphorylations and activation of 6-phosphofructo-2-kinase. Biochemistry 42:6275-6282

Dihazi H, Kessler R, Eschrich K (2004) High osmolarity glycerol (HOG) pathway-induced phosphorylation and activation of 6-phosphofructo-2-kinase are essential for glycerol accumulation and yeast cell proliferation under hyperosmotic stress. J Biol Chem 279:23961-23968

Dihazi H, Kessler R, Muller GA, Eschrich K (2005) Lysine 3 acetylation regulates the phosphorylation of yeast 6-phosphofructo-2-kinase under hypo-osmotic stress. Biol Chem 386:895-900

Elbein AD, Pan YT, Pastuszak I, Carroll D (2003) New insights on trehalose: a multifunctional molecule. Glycobiology 13:17R-27R

Escote X, Zapater M, Clotet J, Posas F (2004) Hog1 mediates cell-cycle arrest in G1 phase by the dual targeting of Sic1. Nat Cell Biol 6:997-1002

Ferreira C, van Voorst F, Martins A, Neves L, Oliveira R, Kielland-Brandt MC, Lucas C, Brandt A (2005) A member of the sugar transporter family, Stl1p is the glycerol/H+ symporter in *Saccharomyces cerevisiae*. Mol Biol Cell 16:2068-2076

Francois J, Parrou JL (2001) Reserve carbohydrates metabolism in the yeast *Saccharomyces cerevisiae*. FEMS Microbiol Rev 25:125-145

Francois J, Van Schaftingen E, Hers HG (1984) The mechanism by which glucose increases fructose 2,6-bisphosphate concentration in *Saccharomyces cerevisiae*. A cyclic-AMP-dependent activation of phosphofructokinase 2. Eur J Biochem 145:187-193

Gancedo C, Serrano R (1989) Energy-yielding metabolism. In: Rose AH, Harrison JS (eds) The yeasts. 2nd ed. Vol. 3. London, England: Academic Press, pp 206–259

Gasch AP, Spellman PT, Kao CM, Carmel-Harel O, Eisen MB, Storz G, Botstein D, Brown PO (2000) Genomic expression programs in the response of yeast cells to environmental changes. Mol Biol Cell 11:4241-4257

Gervais P, Beney L (2001) Osmotic mass transfer in the yeast *Saccharomyces cerevisiae*. Cell Mol Biol (Noisy-le-grand) 47:831-839

Goncalves P, Planta RJ (1998) Starting up yeast glycolysis. Trends Microbiol 6:314-319

Goncalves PM, Griffioen G, Bebelman JP, Planta RJ (1997) Signalling pathways leading to transcriptional regulation of genes involved in the activation of glycolysis in yeast. Mol Microbiol 25:483-493

Gramser S (2005) Alcohol and science: the party gene. Nature 438:1068-1069

Harold FM (2002) Force and compliance: rethinking morphogenesis in walled cells. Fungal Genet Biol 37:271-282

Haurie V, Perrot M, Mini T, Jeno P, Sagliocco F, Boucherie H (2001) The transcriptional activator Cat8p provides a major contribution to the reprogramming of carbon metabolism during the diauxic shift in *Saccharomyces cerevisiae*. J Biol Chem 276:76-85

Hedfalk K, Bill RM, Mullins JG, Karlgren S, Filipsson C, Bergstrom J, Tamás MJ, Rydstrom J, Hohmann S (2004) A regulatory domain in the C-terminal extension of the yeast glycerol channel Fps1p. J Biol Chem 279:14954-14960

Hohmann S (2002) Osmotic stress signaling and osmoadaptation in yeasts. Microbiol Mol Biol Rev 66:300-372

Hohmann S, Meacock PA (1998) Thiamin metabolism and thiamin diphosphate-dependent enzymes in the yeast *Saccharomyces cerevisiae*: genetic regulation. Biochim Biophys Acta 1385:201-219

Hohmann S, Nielsen S, Agre P (2001) Aquaporins. San Diego, CA: Academic Press

Hottiger T, De Virgilio C, Hall MN, Boller T, Wiemken A (1994) The role of trehalose synthesis for the acquisition of thermotolerance in yeast. II. Physiological concentrations of trehalose increase the thermal stability of proteins *in vitro*. Eur J Biochem 219:187-193

Hounsa CG, Brandt EV, Thevelein J, Hohmann S, Prior BA (1998) Role of trehalose in survival of *Saccharomyces cerevisiae* under osmotic stress. Microbiology 144:671-680

Jurica MS, Mesecar A, Heath PJ, Shi W, Nowak T, Stoddard BL (1998) The allosteric regulation of pyruvate kinase by fructose-1,6-bisphosphate. Structure 6:195-210

Karlgren S, Filipsson C, Mullins JG, Bill RM, Tamás MJ, Hohmann S (2004) Identification of residues controlling transport through the yeast aquaglyceroporin Fps1 using a genetic screen. Eur J Biochem 271:771-779

Kempf B, Bremer E (1998) Uptake and synthesis of compatible solutes as microbial stress responses to high-osmolality environments. Arch Microbiol 170:319-330

Klipp E, Nordlander B, Kruger R, Gennemark P, Hohmann S (2005) Integrative model of the response of yeast to osmotic shock. Nat Biotechnol 23:975-982

Krantz M, Nordlander B, Valadi H, Johansson M, Gustafsson L, Hohmann S (2004) Anaerobicity prepares *Saccharomyces cerevisiae* cells for faster adaptation to osmotic shock. Eukaryot Cell 3:1381-1390

Luyten K, Albertyn J, Skibbe WF, Prior BA, Ramos J, Thevelein JM, Hohmann S (1995) Fps1, a yeast member of the MIP family of channel proteins, is a facilitator for glycerol uptake and efflux and is inactive under osmotic stress. EMBO J 14:1360-1371

MacIsaac KD, Wang T, Gordon DB, Gifford DK, Stormo GD, Fraenkel E (2006) An improved map of conserved regulatory sites for *Saccharomyces cerevisiae*. BMC Bioinformatics 7:113

Marechal PA, de Maranon IM, Molin P, Gervais P (1995) Yeast cell responses to water potential variations. Int J Food Microbiol 28:277-287

Martinez de Maranon I, Marechal PA, Gervais P (1996) Passive response of *Saccharomyces cerevisiae* to osmotic shifts: cell volume variations depending on the physiological state. Biochem Biophys Res Commun 227:519-523

Muller S, Zimmermann FK, Boles E (1997) Mutant studies of phosphofructo-2-kinases do not reveal an essential role of fructose-2,6-bisphosphate in the regulation of carbon fluxes in yeast cells. Microbiology 143:3055-3061

Murcott TH, Gutfreund H, Muirhead H (1992) The cooperative binding of fructose-1,6-bisphosphate to yeast pyruvate kinase. EMBO J 11:3811-3814

O'Rourke SM, Herskowitz I (2004) Unique and redundant roles for HOG MAPK pathway components as revealed by whole-genome expression analysis. Mol Biol Cell 15:532-542

Oliveira R, Lages F, Silva-Graca M, Lucas C (2003) Fps1p channel is the mediator of the major part of glycerol passive diffusion in *Saccharomyces cerevisiae*: artefacts and redefinitions. Biochim Biophys Acta 1613:57-71

Ozcan S, Johnston M (1999) Function and regulation of yeast hexose transporters. Microbiol Mol Biol Rev 63:554-569

Påhlman AK, Granath K, Ansell R, Hohmann S, Adler L (2001) The yeast glycerol 3-phosphatases Gpp1p and Gpp2p are required for glycerol biosynthesis and differentially involved in the cellular responses to osmotic, anaerobic, and oxidative stress. J Biol Chem 276:3555-3563

Pettersson N, Filipsson C, Becit E, Brive L, Hohmann S (2005) Aquaporins in yeasts and filamentous fungi. Biol Cell 97:487-500

Pettersson N, Hagstrom J, Bill RM, Hohmann S (2006) Expression of heterologous aquaporins for functional analysis in *Saccharomyces cerevisiae*. Curr Genet 50:247-255

Pokholok DK, Zeitlinger J, Hannett NM, Reynolds DB, Young RA (2006) Activated signal transduction kinases frequently occupy target genes. Science 313:533-536

Posas F, Chambers JR, Heyman JA, Hoeffler JP, de Nadal E, Arino J (2000) The transcriptional response of yeast to saline stress. J Biol Chem 275:17249-17255

Prista C, Loureiro-Dias MC, Montiel V, Garcia R, Ramos J (2005) Mechanisms underlying the halotolerant way of *Debaryomyces hansenii*. FEMS Yeast Res 5:693-701

Proft M, Struhl K (2004) MAP kinase-mediated stress relief that precedes and regulates the timing of transcriptional induction. Cell 118:351-361

Reibstein D, den Hollander JA, Pilkis SJ, Shulman RG (1986) Studies on the regulation of yeast phosphofructo-1-kinase: its role in aerobic and anaerobic glycolysis. Biochemistry 25:219-227

Rep M, Krantz M, Thevelein JM, Hohmann S (2000) The transcriptional response of *Saccharomyces cerevisiae* to osmotic shock. Hot1p and Msn2p/Msn4p are required for the induction of subsets of high osmolarity glycerol pathway-dependent genes. J Biol Chem 275:8290-8300

Schaftingen H-GHaEV (1982) Fructose 2,6-bisphosphate 2 years after its discovery. Biochem J 206:1-12

Shahinian S, Bussey H (2000) beta-1,6-Glucan synthesis in *Saccharomyces cerevisiae*. Mol Microbiol 35:477-489

Sierkstra LN, Verbakel JM, Verrips CT (1992) Analysis of transcription and translation of glycolytic enzymes in glucose-limited continuous cultures of *Saccharomyces cerevisiae*. J Gen Microbiol 138:2559-2566

Singer MA, Lindquist S (1998) Multiple effects of trehalose on protein folding *in vitro* and *in vivo*. Mol Cell 1:639-648

Tamas MJ, Karlgren S, Bill RM, Hedfalk K, Allegri L, Ferreira M, Thevelein JM, Rydstrom J, Mullins JG, Hohmann S (2003) A short regulatory domain restricts glycerol transport through yeast Fps1p. J Biol Chem 278:6337-6345

Tamás MJ, Karlgren S, Bill RM, Hedfalk K, Allegri L, Ferreira M, Thevelein JM, Rydstrom J, Mullins JG, Hohmann S (2003) A short regulatory domain restricts glycerol transport through yeast Fps1p. J Biol Chem 278:6337-6345

Tamás MJ, Luyten K, Sutherland FC, Hernandez A, Albertyn J, Valadi H, Li H, Prior BA, Kilian SG, Ramos J, Gustafsson L, Thevelein JM, Hohmann S (1999) Fps1p controls the accumulation and release of the compatible solute glycerol in yeast osmoregulation. Mol Microbiol 31:1087-1104

Tao W, Deschenes RJ, Fassler JS (1999) Intracellular glycerol levels modulate the activity of Sln1p, a *Saccharomyces cerevisiae* two-component regulator. J Biol Chem 274:360-367

Teige M, Scheikl E, Reiser V, Ruis H, Ammerer G (2001) Rck2, a member of the calmodulin-protein kinase family, links protein synthesis to high osmolarity MAP kinase signaling in budding yeast. Proc Natl Acad Sci USA 98:5625-5630

Thevelein JM (1984) Regulation of trehalose mobilization in fungi. Microbiol Rev 48:42-59

Thevelein JM, Hohmann S (1995) Trehalose synthase: guard to the gate of glycolysis in yeast? Trends Biochem Sci 20:3-10

Thorsen M, Di Y, Tangemo C, Morillas M, Ahmadpour D, Van der Does C, Wagner A, Johansson E, Boman J, Posas F, Wysocki R, Tamás MJ (2006) The MAPK Hog1p modulates Fps1p-dependent arsenite uptake and tolerance in yeast. Mol Biol Cell 17:4400-4410

Tong AH, Lesage G, Bader GD, Ding H, Xu H, Xin X, Young J, Berriz GF, Brost RL, Chang M, Chen Y, Cheng X, Chua G, Friesen H, Goldberg DS, Haynes J, Humphries C, He G, Hussein S, Ke L, Krogan N, Li Z, Levinson JN, Lu H, Menard P, Munyana C, Parsons AB, Ryan O, Tonikian R, Roberts T, Sdicu AM, Shapiro J, Sheikh B, Suter B, Wong SL, Zhang LV, Zhu H, Burd CG, Munro S, Sander C, Rine J, Greenblatt J, Peter M, Bretscher A, Bell G, Roth FP, Brown GW, Andrews B, Bussey H, Boone C (2004) Global mapping of the yeast genetic interaction network. Science 303:808-813

Westerhoff HV (1995) Subtlety in control--metabolic pathway engineering. Trends Biotechnol 13:242-244

Wysocki R, Chery CC, Wawrzycka D, Van Hulle M, Cornelis R, Thevelein JM, Tamás MJ (2001) The glycerol channel Fps1p mediates the uptake of arsenite and antimonite in *Saccharomyces cerevisiae*. Mol Microbiol 40:1391-1401

Yancey PH (2005) Organic osmolytes as compatible, metabolic and counteracting cytoprotectants in high osmolarity and other stresses. J Exp Biol 208:2819-2830

Yancey PH, Clark ME, Hand SC, Bowlus RD, Somero GN (1982) Living with water stress: evolution of osmolyte systems. Science 217:1214-1222

Nordlander, Bodil
Department for Cell and Molecular Biology, Göteborg University, Box 462, 40530 Göteborg, Sweden
bodil.nordlander@cmb.gu.se

Krantz, Marcus
The Systems Biology Institute, 6-31-15 Jingumae, M-31 Suite 6A, Shibuya-ku, Tokyo 150-0001, Japan

Hohmann, Stefan
Department for Cell and Molecular Biology, Göteborg University, Box 462, 40530 Göteborg, Sweden

Control of mRNA stability by SAPKs

Miguel A. Rodríguez-Gabriel and Paul Russell

Abstract

Control of mRNA turnover is an essential step in the regulation of gene expression in eukaryotes. The concerted action of many enzymes regulates the way each mRNA is degraded. Moreover, the degradation of each mRNA is influenced by the environment surrounding the cell. The connection between the environment and changes in the half-lives of mRNAs is regulated by the activity of stress activated MAP kinases (SAPKs) and their substrates. Here, we will describe some of those mechanisms, and how SAPKs regulate mRNA stability in eukaryotic cells.

1 Introduction

Stress activated MAP kinases (SAPKs) are kinases activated in response to external stimuli, like stress, hormones, growth factors, etc. There are two families of SAPKs: the c-Jun N-terminal kinases (JNKs) and the p38 kinases. After activation, they coordinate the activation of gene transcription, protein synthesis, cell cycle machinery, cell death, and differentiation exerting a profound effect in many physiological processes including cell proliferation, inflammation, and development (Kyriakis and Avruch 2001). SAPKs are highly conserved in eukaryotes from yeast to human (e.g. mammalian p38, p38-like *S. cerevisiae* Hog1 and *S. pombe* Spc1/Sty1), and because of their importance in cell physiology and disease, they are the subject of intense study.

One of the main functions of SAPKs is the regulation of transcription factors that bind DNA promoters and regulate transcription of different genes. This regulation is achieved through the phosphorylation of transcription factors like ATF2 or c-Jun (Kyriakis and Avruch 2001). However, it is becoming clear that there is another important function of SAPKs that contributes to the correct and timely expression of genes: the control of mRNA stability. Changes in turnover rates of individual mRNAs have great effects in their steady-state levels (Ross 1995; Caponigro and Parker 1996; Wilusz et al. 2001). The advantages of this additional level of regulation are several:

1. Speed: translation and degradation occur with pre-existing mRNAs, saving time in the production of additional mRNAs when necessary, or eliminating mRNAs when the translational machinery needs to be dedicated to other functions.

Topics in Current Genetics, Vol. 20
F. Posas, A. R. Nebreda (Eds.): Stress-Activated Protein Kinases
DOI 10.1007/4735_2007_0248 / Published online: 1 August 2007

2. Economy: for similar reasons, stabilization of mRNAs can save precious energy used in transcription, giving an important advantage to the always energy-efficient cell machinery.
3. Control: control of mRNA stability is an additional layer of post-transcriptional regulation that helps to finely adjust the appropriate levels of each mRNA to the environmental conditions.

We are going to discuss some of the mechanisms that control mRNA decay and how SAPKS relate to them.

2. Eukaryotic mRNA turnover

2.1. Proteins involved

Through the study of yeast and mammals, two general pathways of mRNA decay have been identified in eukaryotic cells (Parker and Song 2004)(Fig. 1). In the first pathway, the mRNA poly(A) tail is first shortened by mRNA deadenylases. After deadenylation, a decapping enzyme formed by Dcp1 and Dcp2, removes the cap structure in 5', allowing the 5'-3' digestion of the mRNA by Xrn1 exonuclease (Parker and Song 2004). In the second mechanism, deadenylation is followed by 3'-5' degradation of the mRNA by the cytoplasmic exosome, a multisubunit complex formed by many exonucleases (Anderson and Parker 1998; Chen et al. 2001; Wang and Kiledjian 2001; Mukherjee et al. 2002).

There are other pathways of mRNA degradation that are more specialized:

A. Nuclear degradation of pre-mRNAs that failed to complete processing (Moore 2002).
B. Degradation of mRNAs with early stop codons in a process referred to as nonsense-mediated decay (NMD). Those mRNAs are degraded by decapping independent of deadenylation or by accelerated 3'-5' exonucleolytic digestion (Muhlrad and Parker 1994; Cao and Parker 2003; Lejeune et al. 2003; Mitchell and Tollervey 2003; Takahashi et al. 2003).
C. mRNAs lacking stop codon are degraded in a processes named non-stop decay (NSD), by the cytoplasmic exosome (Frischmeyer et al. 2002; van Hoof et al. 2002).

As mentioned above, the first step in the general degradation of mRNA in eukaryotes is deadenylation. At least three different complexes with deadenylation capacity have been identified in eukaryotes: the Ccr4/Pop2 complex (Tucker et al. 2001), the Pan2/Pan3 (PAN) complex (Tucker et al. 2001), and poly(A)-specific exonuclease referred to as PARN (Astrom et al. 1992; Korner and Wahle 1997). In mammalian cells, PARN possesses the major deadenylase activity *in vitro* (Gao et al. 2000). The poly(A)-binding protein (Pab1) can have opposing effects on the different deadenylating activities, being able to inhibit Ccr4/Pop2 and PARN activity and stimulating PAN. As proposed by Parker and Song (Parker and Song 2004), the composition of the mRNA-protein complex (mRNP) would have important implications for the regulation of mRNA turnover. First, the changes in

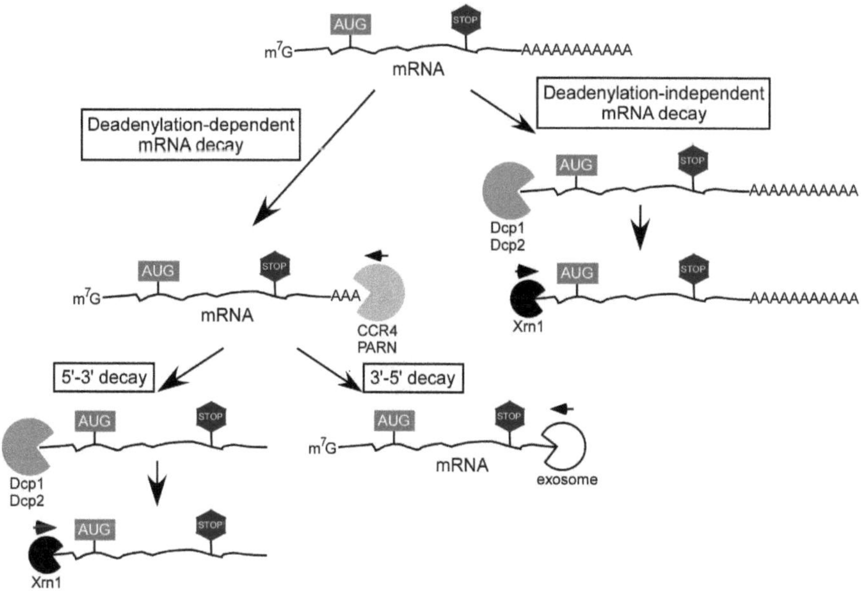

Fig. 1. Main mRNA degradation pathways in eukaryotes. Deadenylation-dependent and deadenylation-independent mechanisms of mRNA decay are depicted with the main enzymes involved.

mRNP composition will alter the susceptibility to different deadenylases. Second, the alteration of one deadenylase would affect a specific set of transcripts. Third, mRNA binding proteins could function by attracting or repelling deadenylase activities to the 3' end of the mRNA.

2.2 Localization

Several studies have shown that mRNA turnover could take place in cytoplasmic bodies. In yeast, many enzymes involved in mRNA degradation like decapping enzymes or exonucleases have been localize to discrete cytoplasmic bodies, were mRNA degradation intermediates have also been found (Sheth and Parker 2003). A different structure, the stress granules, are also cytoplasmic structures that contain translationally repressed mRNAs and mRNAs targeted for degradation (Kedersha and Anderson 2002; Anderson and Kedersha 2006). They may function as a triage system to select mRNAs that need to be translated or degraded, depending of the physiological conditions of the cell.

3 mRNA *cis* acting elements

The decay rates of specific mRNAs can be determined by their sequence elements. The best studied are AU-riche elements (AREs), which are generally found in the 3' untranslated region (UTR) of mRNAS. AREs modulate stability at the level of deadenylation, decapping or 3'-5' degradation, and can also control translational efficiency (Shaw and Kamen 1986; Chen and Shyu 1995; Bakheet et al. 2001; Wilusz and Wilusz 2004). Proteins that associate with these AREs can either enhance or inhibit mRNA decay rates. Several of these proteins have been described, such as AUF1, Tristetraprolin (TTP), HuR, KSRP, NF90, and TIA-1 (Wilusz and Wilusz 2004). They belong to different protein families, they have different effects on mRNA stability (stabilizing or destabilizing), and modulate different aspects of mRNA metabolism like deadenilation, exosome binding, or decapping.

AU-binding proteins recruit the exosome to degrade ARE-containing mRNAs (Chen et al. 2001). Regulated ARE-mediated mRNA decay is also conserved in *Saccharomyces cerevisiae* (Vasudevan and Peltz 2001), and possibly in other eukaryotic microorganisms, opening the possibility of studying these processes in model organisms.

Although the exosome does not exhibit ARE-binding activity, it is recruited to ARE-RNAs through interactions with several ARE-binding proteins such us KSRP and TTP, which in turn recruit the exosome to affect 3'-5' mRNA degradation. ARE-mediated mRNA stability is also regulated by microRNAs in mammalian cells (Jing et al. 2005).

4 Control of mRNA stability by SAPKs

Mammalian SAPKs (JNK and p38) regulate mRNA stability. Stabilization of Interleukin-2 mRNA by the c-Jun NH2-Terminal Kinase Pathway (Chen et al. 1998). The main target of p38 MAP kinase is the kinase MK2, and this signaling pathway regulates the stability and translation of tumor necrosis factor(TNF) and IL-6 mRNAs through a process that involves the AU-rich elements in the 3' UTR of these mRNAs (Kotlyarov et al. 1999; Neininger et al. 2002). Microarray experiments demonstrated that many ARE-containing mRNAs are regulated by p38 activity, but many others are regulated independently of p38 (Frevel et al. 2003).

SAPK p38 signals to cytokine induced mRNA stabilization via MK2 and AU-rich elements (Winzen et al. 1999); it also stabilizes mRNA that contain cyclooxygenase-2 and TNF AREs by inhibiting deadenylation (Dean et al. 2003) and has been found to regulate both the translation and the stability of inflammatory mRNAs. The mRNAs regulated by p38 share common AU-rich elements (ARE) present in their 3'-untranslated regions. AREs act as mRNA instability determinants but also confer stabilization of the mRNA by the p38 pathway. In recent years, AREs have shown to be binding sites for numerous proteins as mentioned before. However, it is unclear which protein is responsible for mRNA stabilization by p38 (Dean et al. 2004).

4.1 RNA binding proteins involved in SAPK regulation of mRNA stability

4.1.1. Csx1

Schizosaccharomyces pombe Csx1 is a protein with three RNA recognition motifs (RRMs). Under oxidative stress, Csx1 is able to stabilize the mRNA encoding the transcription factor Atf1, being the *csx1Δ* cells sensitive to oxidation. This regulation of Atf1 mRNA half-life is specific of oxidative stress and it is not produced under other types of insult. The absence of Csx1 protein produces a decrease in Atf1 mRNA that, in turn, affect the expression of multiple mRNAs necessary for the correct response to oxidation in fission yeast (Rodriguez-Gabriel et al. 2003).

Csx1 binds to Atf1 mRNA and regulates its turnover rate, and this seems to be the way Csx1 regulates its half-life. Under oxidative stress, Csx1 is phosphorylated dependent on the activity of the p38-like kinase Spc1/Sty1, and correlating with the stabilization of Atf1 mRNA. However, the relationship between this phosphorylation and Csx1 function is unclear (Rodriguez-Gabriel et al. 2003).

Two other RRM-containing proteins have been involved in the regulation of Atf1 mRNA stability in response to oxidative stress in *S. pombe*: Cip1 and Cip2 (Csx1 Interacting Protein 1 and 2). These two proteins interact with Csx1 protein. Elimination of any of them is able to rescue the sensitivity of *csx1Δ* cells to oxidation. Moreover, the absence of Cip1 or Cip2 can override the rapid disappearance of Atf1 mRNA in *csx1Δ* cells and re-establish the correct expression of many of the Csx1-regulated mRNAs (Martin et al. 2006).

4.1.2 HuR

HuR is a mammalian RNA-binding protein that also has three RRMs. HuR binds to several mRNAs (e.g. those that encode TNFalpha, ATF3, p21WAF1) through their 3′AREs and helps to stabilize them (Wang et al. 2000; Dean et al. 2001; Giles et al. 2003; Pan et al. 2005; Xu et al. 2005). It has been reported that HuR does not alter accumulation of target mRNAs in the absence of the destabilizing function of tristetraprolin (TTP), indicating the functional relationship between both proteins (Katsanou et al. 2005). The way in which HuR functions is still controversial. Recent reports describe the function of HuR in the derepression of gene expression produced by miRNA binding to 3′UTRs. After HuR action, those mRNAs can be released from P-bodies and translated into proteins in response to stress (Bhattacharyya et al. 2006). In some cases, like urokinase and urokinase receptor mRNAs, the stabilization is linked to cytoplasmic accumulation of HuR induced by activated MK2, joining p38 pathway and mRNA stabilization (Tran et al. 2003).

4.1.3 TTP (Tristetraprolin)

TTP downregulates IL-2 gene expression through ARE-mediated mRNA decay (Ogilvie et al. 2005). It is thought that this regulation is possibly achieved by TTP promoting deadenylation of ARE-containing substrates by PARN (Lai et al. 2003). One of the most interesting effects of TTP is the delay of tumor progression. Tumor suppression can be achieved by interfering with RNA turnover (Stoecklin et al. 2003). However many of the phenotypes related with TTP deficiency are related with its regulation of TNF alpha mRNA regulation. TTP $^{-/-}$ mice have high levels of TNF alpha mRNA. These mice suffer cachexia, arthritis, and autoimmunity resulting from TTP deficiency. Many of the phenotypes of TTP $^{-/-}$ mice can be overcome by injection of TNF alpha antibodies (Taylor et al. 1996). TTP destabilizes TNF alpha mRNA (Carballo et al. 1998), binding to its AU-rich element and promoting its deadenylation (Lai et al. 1999).

TTP deficient cell lines are less sensitive to p38 inhibitors, indicating a possible role of TTP in p38 pathway (Carballo et al. 2001). However there are other reports indicating that p38 does not alter TTP function (Rigby et al. 2005). Depending of the type of stress, TTP can localize to stress granules, were it could regulate degradation of ARE-containing mRNAs along with other ARE-binding factors as TIA-1 and HuR. TTP becomes activated in a p38 dependent manner, binds to 14-3-3 and is excluded from the stress granules (Stoecklin et al. 2004). Therefore, ARE-containing mRNAs are stabilized. It is thought that exclusion of TTP from the stress granules allows HuR to exert its stabilizing effect.

Mitogen-activated protein kinase p38 controls the expression and posttranslational modification of TTP, regulating TNF alpha mRNA stability (Mahtani et al. 2001), and cyclin D and myc mRNAs (Marderosian et al. 2006). The abundance of TTP mRNA is regulated by p38 and TTP itself (Tchen et al. 2004).

MK2 regulates TNF mRNA stability and translation mainly by altering TTP expression, stability and binding to AREs (Hitti et al. 2006). MK2 phosphorylates TTP of *in vivo* sites, including Ser178, that is a site required for 14-3-3 binding (Chrestensen et al. 2004). MK2 induced TTP:14-3-3 complexes prevent stress granule association and ARE-mRNA decay (Stoecklin et al. 2004).

4.1.4 AUF1

AUF1 has several isoforms. It facilitates rapid decay of AU-rich mRNAs (Sarkar et al. 2003). It has opposite functions to HuR, and the balance of this two proteins could be important for the regulation of mRNA stability (Lu and Schneider 2004).

4.1.5 NF90

Nuclear export of NF90 is required for interleukin-2 mRNA stabilization (Shim et al. 2002), and it could have overlapping functions with HuR.

4.1.6 RHAU

The RHAU RNA helicase facilitates mRNA deadenylation and decay. It interacts with PARN and exosome and the RNA binding proteins HuR and NFAR1 (Tran et al. 2004).

4.1.7 KSRP

KSRP, a KH domain RNA binding protein, was cloned and purified as a component of a splicing complex (Min et al. 1997). It interacts with ARE, exosome and PARN to promote mRNA decay (Gherzi et al. 2004). In fact, tethering KSRP to mRNAs elicit mRNA decay (Chou et al. 2006). It was recently shown that phosphorylation of KSRP in a p38 dependent manner regulates myogenic transcripts, linking SAPK activity and mRNA stability through this RNA binding protein (Briata et al. 2005).

4.1.8 TIA-1 and TIAR

TIA-1 and TIAR show alternative forms of splicing that present several RNA recognition motifs (Beck et al. 1996). TIAR is essential for primordial germ cell development (Beck et al. 1998), and TIA-1 -/- mice develop mild arthritis, with overproduction of TNF alpha and cyclooxigenase-2 by some cell types (Phillips et al. 2004). Proteins from the TIAR/TIA-1 family bind AREs and can also regulate pre-mRNA splicing (Forch et al. 2000). During stress, TIA-1 shuttles between the cytoplasm and the stress granules, accompanying the recruitment of mRNAs to these granules (Kedersha et al. 2000; Kedersha and Anderson 2002). It seems that the main function of the TIAR/TIA-1 family members is the translational repression of mRNAs in response to stress (Mazan-Mamczarz et al. 2006). Although the relationship of these proteins with SAPKs has not been clearly demonstrated, they seem to work in the same situations and under similar stress conditions, opening the possibility of a biochemical relationship between them.

5 Concluding remarks

The role of SAPK pathways in the regulation of posttranscriptional events, and more specifically of mRNA turnover rates, has been the subject of numerous studies in recent years. The participation of SAPKs (and its substrates) in the regulation of mRNA stability, the importance of RNA binding proteins in this process, and the different *cis* elements found in mRNAs (AU-rich elements and others) are well established facts. However, there are still many open questions that lay ahead: how is the combination of RNA binding proteins regulating RNA turnover? How are they regulated? How do they interplay with deadenylation machinery, or exosome, or decapping enzymes? How is everything coordinated to function as a eukaryotic operons (Keene and Tenenbaum 2002)? Is it possible to target those mechanisms for therapeutic approaches? With the current rate of progress we can expect answers to these questions in the not too distant future.

Acknowledgement

Miguel A. Rodríguez-Gabriel is a Ramon y Cajal Investigator. Work in his group is supported by grants BFU2006-01767 from the Spanish Ministry of Education and Science and a Marie Curie International Reintegration Grant from the EU. Work in Paul Russell's lab related to this review is funded by grant number ES10337 from the National Institute of Environmental Health Sciences, NIH.

References

Anderson JS, Parker RP (1998) The 3' to 5' degradation of yeast mRNAs is a general mechanism for mRNA turnover that requires the SKI2 DEVH box protein and 3' to 5' exonucleases of the exosome complex. EMBO J 17:1497-1506

Anderson P, Kedersha N (2006) RNA granules. J Cell Biol 172:803-808

Astrom J, Astrom A, Virtanen A (1992) Properties of a HeLa cell 3' exonuclease specific for degrading poly(A) tails of mammalian mRNA. J Biol Chem 267:18154-18159

Bakheet T, Frevel M, Williams BR, Greer W, Khabar KS (2001) ARED: human AU-rich element-containing mRNA database reveals an unexpectedly diverse functional repertoire of encoded proteins. Nucleic Acids Res 29:246-254

Beck AR, Medley QG, O'Brien S, Anderson P, Streuli M (1996) Structure, tissue distribution and genomic organization of the murine RRM-type RNA binding proteins TIA-1 and TIAR. Nucleic Acids Res 24:3829-3835

Beck AR, Miller IJ, Anderson P, Streuli M (1998) RNA-binding protein TIAR is essential for primordial germ cell development. Proc Natl Acad Sci USA 95:2331-2336

Bhattacharyya SN, Habermacher R, Martine U, Closs EI, Filipowicz W (2006) Relief of microRNA-mediated translational repression in human cells subjected to stress. Cell 125:1111-1124

Briata P, Forcales SV, Ponassi M, Corte G, Chen CY, Karin M, Puri PL, Gherzi R (2005) p38-dependent phosphorylation of the mRNA decay-promoting factor KSRP controls the stability of select myogenic transcripts. Mol Cell 20:891-903

Cao D, Parker R (2003) Computational modeling and experimental analysis of nonsense-mediated decay in yeast. Cell 113:533-545

Caponigro G, Parker R (1996) Mechanisms and control of mRNA turnover in *Saccharomyces cerevisiae*. Microbiol Rev 60:233-249

Carballo E, Cao H, Lai WS, Kennington EA, Campbell D, Blackshear PJ (2001) Decreased sensitivity of tristetraprolin-deficient cells to p38 inhibitors suggests the involvement of tristetraprolin in the p38 signaling pathway. J Biol Chem 276:42580-42587

Carballo E, Lai WS, Blackshear PJ (1998) Feedback inhibition of macrophage tumor necrosis factor-alpha production by tristetraprolin. Science 281:1001-1005

Chen CY, Del Gatto-Konczak F, Wu Z, Karin M (1998) Stabilization of interleukin-2 mRNA by the c-Jun NH2-terminal kinase pathway. Science 280:1945-1949

Chen CY, Gherzi R, Ong SE, Chan EL, Raijmakers R, Pruijn GJ, Stoecklin G, Moroni C, Mann M, Karin M (2001) AU binding proteins recruit the exosome to degrade ARE-containing mRNAs. Cell 107:451-464

Chen CY, Shyu AB (1995) AU-rich elements: characterization and importance in mRNA degradation. Trends Biochem Sci 20:465-470

Chou CF, Mulky A, Maitra S, Lin WJ, Gherzi R, Kappes J, Chen CY (2006) Tethering KSRP, a decay-promoting AU-rich element-binding protein, to mRNAs elicits mRNA decay. Mol Cell Biol 26:3695-3706

Chrestensen CA, Schroeder MJ, Shabanowitz J, Hunt DF, Pelo JW, Worthington MT, Sturgill TW (2004) MAPKAP kinase 2 phosphorylates tristetraprolin on *in vivo* sites including Ser178, a site required for 14-3-3 binding. J Biol Chem 279:10176-10184

Dean JL, Sarsfield SJ, Tsounakou E, Saklatvala J (2003) p38 Mitogen-activated protein kinase stabilizes mRNAs that contain cyclooxygenase-2 and tumor necrosis factor AU-rich elements by inhibiting deadenylation. J Biol Chem 278:39470-39476

Dean JL, Sully G, Clark AR, Saklatvala J (2004) The involvement of AU-rich element-binding proteins in p38 mitogen-activated protein kinase pathway-mediated mRNA stabilisation. Cell Signal 16:1113-1121

Dean JL, Wait R, Mahtani KR, Sully G, Clark AR, Saklatvala J (2001) The 3' untranslated region of tumor necrosis factor alpha mRNA is a target of the mRNA-stabilizing factor HuR. Mol Cell Biol 21:721-730

Forch P, Puig O, Kedersha N, Martinez C, Granneman S, Seraphin B, Anderson P, Valcarcel J (2000) The apoptosis-promoting factor TIA-1 is a regulator of alternative pre-mRNA splicing. Mol Cell 6:1089-1098

Frevel MA, Bakheet T, Silva AM, Hissong JG, Khabar KS, Williams BR (2003) p38 Mitogen-activated protein kinase-dependent and -independent signaling of mRNA stability of AU-rich element-containing transcripts. Mol Cell Biol 23:425-436

Frischmeyer PA, van Hoof A, O'Donnell K, Guerrerio AL, Parker R, Dietz HC (2002) An mRNA surveillance mechanism that eliminates transcripts lacking termination codons. Science 295:2258-2261

Gao M, Fritz DT, Ford LP, Wilusz J (2000) Interaction between a poly(A)-specific ribonuclease and the 5' cap influences mRNA deadenylation rates *in vitro*. Mol Cell 5:479-488

Gherzi R, Lee KY, Briata P, Wegmuller D, Moroni C, Karin M, Chen CY (2004) A KH domain RNA binding protein, KSRP, promotes ARE-directed mRNA turnover by recruiting the degradation machinery. Mol Cell 14:571-583

Giles KM, Daly JM, Beveridge DJ, Thomson AM, Voon DC, Furneaux HM, Jazayeri JA, Leedman PJ (2003) The 3'-untranslated region of p21WAF1 mRNA is a composite cis-acting sequence bound by RNA-binding proteins from breast cancer cells, including HuR and poly(C)-binding protein. J Biol Chem 278:2937-2946

Hitti E, Iakovleva T, Brook M, Deppenmeier S, Gruber AD, Radzioch D, Clark AR, Blackshear PJ, Kotlyarov A, Gaestel M (2006) Mitogen-activated protein kinase-activated protein kinase 2 regulates tumor necrosis factor mRNA stability and translation mainly by altering tristetraprolin expression, stability, and binding to adenine/uridine-rich element. Mol Cell Biol 26:2399-2407

Jing Q, Huang S, Guth S, Zarubin T, Motoyama A, Chen J, Di Padova F, Lin SC, Gram H, Han J (2005) Involvement of microRNA in AU-rich element-mediated mRNA instability. Cell 120:623-634

Katsanou V, Papadaki O, Milatos S, Blackshear PJ, Anderson P, Kollias G, Kontoyiannis DL (2005) HuR as a negative posttranscriptional modulator in inflammation. Mol Cell 19:777-789

Kedersha N, Anderson P (2002) Stress granules: sites of mRNA triage that regulate mRNA stability and translatability. Biochem Soc Trans 30:963-969

Kedersha N, Cho MR, Li W, Yacono PW, Chen S, Gilks N, Golan DE, Anderson P (2000) Dynamic shuttling of TIA-1 accompanies the recruitment of mRNA to mammalian stress granules. J Cell Biol 151:1257-1268

Keene JD, Tenenbaum SA (2002) Eukaryotic mRNPs may represent posttranscriptional operons. Mol Cell 9:1161-1167

Korner CG, Wahle E (1997) Poly(A) tail shortening by a mammalian poly(A)-specific 3'-exoribonuclease. J Biol Chem 272:10448-10456

Kotlyarov A, Neininger A, Schubert C, Eckert R, Birchmeier C, Volk HD, Gaestel M (1999) MAPKAP kinase 2 is essential for LPS-induced TNF-alpha biosynthesis. Nat Cell Biol 1:94-97

Kyriakis JM, Avruch J (2001) Mammalian mitogen-activated protein kinase signal transduction pathways activated by stress and inflammation. Physiol Rev 81:807-869

Lai WS, Carballo E, Strum JR, Kennington EA, Phillips RS, Blackshear PJ (1999) Evidence that tristetraprolin binds to AU-rich elements and promotes the deadenylation and destabilization of tumor necrosis factor alpha mRNA. Mol Cell Biol 19:4311-4323

Lai WS, Kennington EA, Blackshear PJ (2003) Tristetraprolin and its family members can promote the cell-free deadenylation of AU-rich element-containing mRNAs by poly(A) ribonuclease. Mol Cell Biol 23:3798-3812

Lejeune F, Li X, Maquat LE (2003) Nonsense-mediated mRNA decay in mammalian cells involves decapping, deadenylating, and exonucleolytic activities. Mol Cell 12:675-687

Lu JY, Schneider RJ (2004) Tissue distribution of AU-rich mRNA-binding proteins involved in regulation of mRNA decay. J Biol Chem 279:12974-12979

Mahtani KR, Brook M, Dean JL, Sully G, Saklatvala J, Clark AR (2001) Mitogen-activated protein kinase p38 controls the expression and posttranslational modification of tristetraprolin, a regulator of tumor necrosis factor alpha mRNA stability. Mol Cell Biol 21:6461-6469

Marderosian M, Sharma A, Funk AP, Vartanian R, Masri J, Jo OD, Gera JF (2006) Tristetraprolin regulates Cyclin D1 and c-Myc mRNA stability in response to rapamycin in an Akt-dependent manner via p38 MAPK signaling. Oncogene 25:6277-90

Martin V, Rodriguez-Gabriel MA, McDonald WH, Watt S, Yates JR 3rd, Bahler J, Russell P (2006) Cip1 and Cip2 are novel RNA-recognition-motif proteins that counteract Csx1 function during oxidative stress. Mol Biol Cell 17:1176-83

Mazan-Mamczarz K, Lal A, Martindale JL, Kawai T, Gorospe M (2006) Translational repression by RNA-binding protein TIAR. Mol Cell Biol 26:2716-2727

Min H, Turck CW, Nikolic JM, Black DL (1997) A new regulatory protein, KSRP, mediates exon inclusion through an intronic splicing enhancer. Genes Dev 11:1023-1036

Mitchell P, Tollervey D (2003) An NMD pathway in yeast involving accelerated deadenylation and exosome-mediated 3'-->5' degradation. Mol Cell 11:1405-1413

Moore MJ (2002) Nuclear RNA turnover. Cell 108:431-434

Muhlrad D, Parker R (1994) Premature translational termination triggers mRNA decapping. Nature 370:578-581

Mukherjee D, Gao M, O'Connor JP, Raijmakers R, Pruijn G, Lutz CS, Wilusz J (2002) The mammalian exosome mediates the efficient degradation of mRNAs that contain AU-rich elements. EMBO J 21:165-174

Neininger A, Kontoyiannis D, Kotlyarov A, Winzen R, Eckert R, Volk HD, Holtmann H, Kollias G, Gaestel M (2002) MK2 targets AU-rich elements and regulates biosynthesis

of tumor necrosis factor and interleukin-6 independently at different post-transcriptional levels. J Biol Chem 277:3065-3068
Ogilvie RL, Abelson M, Hau HH, Vlasova I, Blackshear PJ, Bohjanen PR (2005) Tristetraprolin down-regulates IL-2 gene expression through AU-rich element-mediated mRNA decay. J Immunol 174:953-961
Pan YX, Chen H, Kilberg MS (2005) Interaction of RNA-binding proteins HuR and AUF1 with the human ATF3 mRNA 3'-untranslated region regulates its amino acid limitation-induced stabilization. J Biol Chem 280:34609-34616
Parker R, Song H (2004) The enzymes and control of eukaryotic mRNA turnover. Nat Struct Mol Biol 11:121-127
Phillips K, Kedersha N, Shen L, Blackshear PJ, Anderson P (2004) Arthritis suppressor genes TIA-1 and TTP dampen the expression of tumor necrosis factor alpha, cyclooxygenase 2, and inflammatory arthritis. Proc Natl Acad Sci USA 101:2011-2016
Rigby WF, Roy K, Collins J, Rigby S, Connolly JE, Bloch DB, Brooks SA (2005) Structure/function analysis of tristetraprolin (TTP): p38 stress-activated protein kinase and lipopolysaccharide stimulation do not alter TTP function. J Immunol 174:7883-7893
Rodriguez-Gabriel MA, Burns G, McDonald WH, Martin V, Yates JR 3rd, Bahler J, Russell P (2003) RNA-binding protein Csx1 mediates global control of gene expression in response to oxidative stress. EMBO J 22:6256-6266
Ross J (1995) mRNA stability in mammalian cells. Microbiol Rev 59:423-450
Sarkar B, Xi Q, He C, Schneider RJ (2003) Selective degradation of AU-rich mRNAs promoted by the p37 AUF1 protein isoform. Mol Cell Biol 23:6685-6693
Shaw G, Kamen R (1986) A conserved AU sequence from the 3' untranslated region of GM-CSF mRNA mediates selective mRNA degradation. Cell 46:659-667
Sheth U, Parker R (2003) Decapping and decay of messenger RNA occur in cytoplasmic processing bodies. Science 300:805-808
Shim J, Lim H, J RY, Karin M (2002) Nuclear export of NF90 is required for interleukin-2 mRNA stabilization. Mol Cell 10:1331-1344
Stoecklin G, Gross B, Ming XF, Moroni C (2003) A novel mechanism of tumor suppression by destabilizing AU-rich growth factor mRNA. Oncogene 22:3554-3561
Stoecklin G, Stubbs T, Kedersha N, Wax S, Rigby WF, Blackwell TK, Anderson P (2004) MK2-induced tristetraprolin:14-3-3 complexes prevent stress granule association and ARE-mRNA decay. EMBO J 23:1313-1324
Takahashi S, Araki Y, Sakuno T, Katada T (2003) Interaction between Ski7p and Upf1p is required for nonsense-mediated 3'-to-5' mRNA decay in yeast. EMBO J 22:3951-3959
Taylor GA, Carballo E, Lee DM, Lai WS, Thompson MJ, Patel DD, Schenkman DI, Gilkeson GS, Broxmeyer HE, Haynes BF, Blackshear PJ (1996) A pathogenetic role for TNF alpha in the syndrome of cachexia, arthritis, and autoimmunity resulting from tristetraprolin (TTP) deficiency. Immunity 4:445-454
Tchen CR, Brook M, Saklatvala J, Clark AR (2004) The stability of tristetraprolin mRNA is regulated by mitogen-activated protein kinase p38 and by tristetraprolin itself. J Biol Chem 279:32393-32400
Tran H, Maurer F, Nagamine Y (2003) Stabilization of urokinase and urokinase receptor mRNAs by HuR is linked to its cytoplasmic accumulation induced by activated mitogen-activated protein kinase-activated protein kinase 2. Mol Cell Biol 23:7177-7188
Tran H, Schilling M, Wirbelauer C, Hess D, Nagamine Y (2004) Facilitation of mRNA deadenylation and decay by the exosome-bound, DExH protein RHAU. Mol Cell 13:101-111

Tucker M, Valencia-Sanchez MA, Staples RR, Chen J, Denis CL, Parker R (2001) The transcription factor associated Ccr4 and Caf1 proteins are components of the major cytoplasmic mRNA deadenylase in *Saccharomyces cerevisiae*. Cell 104:377-386

van Hoof A, Frischmeyer PA, Dietz HC, Parker R (2002) Exosome-mediated recognition and degradation of mRNAs lacking a termination codon. Science 295:2262-2264

Vasudevan S, Peltz SW (2001) Regulated ARE-mediated mRNA decay in *Saccharomyces cerevisiae*. Mol Cell 7:1191-1200

Wang W, Furneaux H, Cheng H, Caldwell MC, Hutter D, Liu Y, Holbrook N, Gorospe M (2000) HuR regulates p21 mRNA stabilization by UV light. Mol Cell Biol 20:760-769

Wang Z, Kiledjian M (2001) Functional link between the mammalian exosome and mRNA decapping. Cell 107:751-762

Wilusz CJ, Wilusz J (2004) Bringing the role of mRNA decay in the control of gene expression into focus. Trends Genet 20:491-497

Wilusz CJ, Wormington M, Peltz SW (2001) The cap-to-tail guide to mRNA turnover. Nat Rev Mol Cell Biol 2:237-246

Winzen R, Kracht M, Ritter B, Wilhelm A, Chen CY, Shyu AB, Muller M, Gaestel M, Resch K, Holtmann H (1999) The p38 MAP kinase pathway signals for cytokine-induced mRNA stabilization via MAP kinase-activated protein kinase 2 and an AU-rich region-targeted mechanism. EMBO J 18:4969-4980

Xu YZ, Di Marco S, Gallouzi I, Rola-Pleszczynski M, Radzioch D (2005) RNA-binding protein HuR is required for stabilization of SLC11A1 mRNA and SLC11A1 protein expression. Mol Cell Biol 25:8139-8149

Rodríguez-Gabriel, Miguel A.

Depto. Microbiologia II. Facultad de Farmacia. Universidad Complutense de Madrid. 28040 Madrid, Spain

Russell, Paul

Dept. Molecular Biology. The Scripps Research Institute. La Jolla, CA 92037. U.S.A.

prussell@scripps.edu

Intrinsically active (MKK- independent) variants of SAPKs - How do they work?

Inbal Maayan and David Engelberg

Abstract

The activity of Stress activated protein kinases (SAPKs) is critical for proper development, differentiation, and survival. Several SAPKs are expressed in each eukaryotic cell and are co-activated in response to stimulation. It is difficult therefore to reveal the specific functions of each SAPK. These functions could be accurately addressed by expression of SAPKs mutants that are spontaneously active *in vivo*. We have recently produced such mutants of the yeast Hog1 and of the four isoforms of mammalian p38. Here, we summarize the current understanding of their mechanism of action. Structural studies show that the mutations cause conformational changes at the L16 domain, primarily the disruption of a hydrophobic core. These changes confer an auto-phosphorylation capability to the enzyme and consequently activation. In addition, the mutants are spontaneously recruiting specific upstream components for increasing their activity. We suggest that this process emulates the natural mechanism of p38 activation, achieved via phosphorylation of Tyr323 (the "bypass" pathway). This notion suggests that the intrinsically active mutants are legitimate for biological studies.

1 Introduction

Mitogen activated protein kinases (MAPKs) form a family comprised of signaling mediators involved in critical cellular and multi-cellular functions (Davis 2000; Ono and Han 2000; Kyriakis and Avruch 2001; Pearson et al. 2001; Rubinfeld and Seger 2005; Zarubin and Han 2005; Murphy and Blenis 2006; Yoon and Seger 2006). The MAPKs, which are conserved in all eukaryotes, are divided to subgroups according to sequence similarity and biological activity. In mammals there are at least three major subgroups, ERK, JNK, and p38. Each of these subfamilies includes a number of isoforms. Also, most of the isoforms give rise to several splicing variants (Jiang et al. 1997; Casanova et al. 2000; Faccio et al. 2000; Yung et al. 2000; Sudo et al. 2002; Aebersold et al. 2004; Shaul and Seger 2006). While the ERK subfamily is activated mainly by growth factors, the JNKs and p38s subfamilies respond to a variety of stress conditions and therefore they are also called Stress Activated Protein Kinases (SAPKs) (Kyriakis et al. 1994). It should be noted, however, that no MAPK is signal specific and many MAPKs are concomitantly induced in the same cell (to different levels) by any given stimulus.

Topics in Current Genetics, Vol. 20
F. Posas, A. R. Nebreda (Eds.): Stress-Activated Protein Kinases
DOI 10.1007/4735_2007_0249 / Published online: 1 August 2007

Once activated, MAPKs may induce different biological outcomes depending on the cell type and/or the particular signal involved. The SAPKs, for example, could stimulate in some systems, or under some conditions, apoptosis or cell cycle arrest (Eilers et al. 1998; Davis 2000; Porras et al. 2004). In other systems, they induce differentiation or proliferation (Morooka and Nishida 1998; Williams et al. 1998; Zetser et al. 1999; Keren et al. 2006). It is still unknown what are the mechanisms through which the same enzymes impose different effects in different cell types, or under different conditions.

SAPKs are associated with many diseases. For example, over activated p38 is correlated with several types of cancer and with inflammatory diseases (Schett et al. 2000; Kumar et al. 2003; Uzgare et al. 2003; Bulavin and Fornace 2004; Orchel et al. 2005; Schieven 2005; Westra and Limburg 2006). Over activation of JNK is associated with neuronal degenerative diseases (Zhu et al. 2001; Lagalwar et al. 2006). In some other types of cancer, inactivation of JNK or p38 seems to be involved (Bulavin and Fornace 2004; reviewed by Engelberg 2004). In all these cases, it is not known whether the abnormal SAPKs activity is the direct cause of the disease or maybe just one of its consequences.

The involvement of SAPKs in so many cellular and multi-cellular functions and their role in currently non-curable diseases, underlie the need to understand the precise unique activities of each SAPK. This mission is not trivial not only because many isoforms and splicing variants are co-activated, but also because there is a complex crosstalk between the MAPK cascades. Therefore, it is impossible to insulate one MAPK and study its biochemical and biological activities. As a consequence, despite many years of research the specific role of each particular SAPK, in health or in disease, is still not known.

A useful tool for accurately revealing the role of each MAPK would be intrinsically active variants of each enzyme. Such molecules are by definition independent of upstream regulation and when expressed in a biological system they spontaneously execute biochemical, molecular, and biological functions of the parental protein. As these activities are executed when cells are not exposed to stress or other stimulations, and all other MAPKs are essentially dormant, this approach accurately discloses the molecular targets and the biological effects of the given active MAPK. It is required, of course, that besides its intrinsic activity, the active variant maintains all properties of the native parental protein.

Recently, we developed such active variants of all members of the p38 family, including the yeast Hog1 (Bell et al. 2001) and all isoforms of the human p38 (Diskin et al. 2004; Avitzour et al. 2007). In this chapter, we briefly review the development of these active variants and describe in detail our current understanding of their mechanism of action. Understanding this mechanism is critical because it will reveal whether the mutants are imitating natural activating mechanisms, making them legitimate tools for biological studies.

2 Intrinsically active variants of SAPKs - bypassing their natural mode of activation

In order to be active, a MAPK has to be phosphorylated on a unique motif containing Thr and Tyr residues separated by only one amino acid (Cobb and Goldsmith 1995; Canagarajah et al. 1997; English et al. 1999). This unique dual phosphorylation is catalyzed by proteins known as MAPK kinases (MAPKKs, also called MKKs and MEKs), which are themselves activated via phosphorylation catalyzed by enzymes called MAPKK kinases (MAPKKKs, also called MKKKs or MEKKs). Although this relay race of phosphorylation is required for the activation of every MAPK in nature, there seem to be some additional cascades that activate MAPKs via alternative routes that bypass the MEKK-MEK module. One example is the activation of the mammalian MAPK p38α by the TAB1 protein in response to stimulation with TGFβ (Ge et al. 2002). TAB1 associates directly with p38α and induces an auto-phosphorylation activity. Another example exists in T cells. In these cells, ligand- bound T cell receptor (TCR) recruits the tyrosine kinase ZAP70 that in turn phosphorylates p38α on Tyr323. This phosphorylation induces auto-phosphorylation and activation of p38α (Salvador et al. 2005).

MKKs and MKKKs are not specific and can activate more than one protein so that activation of one of those kinases leads to activation of several MAPKs. Also, ZAP70 phosphotylates many other substrates in addition to p38α (Hivroz and Fischer 1994; Scielzo et al. 2006). Therefore, the best way to study the specific effects of a given MAPK is to exclusively activate it, and not upstream activators. Exclusive activation could be achieved by expressing an intrinsically active variant of the MAPK of interest. This mission is however not trivial because in spite of the comprehensive knowledge about the mechanism of MAPK activation and the structural requirements involved, it is still not known how to mimic by mutagenesis the PO_4-Thr - Xaa - PO_4-Tyr motif or the resulting structural changes (Hanks and Hunter 1995; Canagarajah et al. 1997; English et al. 1999) . As a result, researchers applied indirect and not entirely specific approaches including using active forms of upstream components (Cowley et al. 1994; Mansour et al. 1994, 1996; Gillespie-Brown et al. 1995), exposing cells to relatively specific stimuli (Han et al. 1994; Minden et al. 1994; Zetser et al. 1999), or producing MEK-MAPK chimera (Robinson et al. 1998; Zheng et al. 1999; reviewed by Askari et al. 2006).

Another approach was to try to isolate intrinsically active MAPKs using unbiased genetic screens in *Drosophila melanogaster* and in *S. cerevisiae.* These screens searched for gain-of-function mutations hoping that the mutated proteins would acquire an intrinsic unregulated activity. These screens uncovered several interesting mutations in the yeast MAPKs Kss1 and Fus3, and in the *Drosophila* Erk2 MAPK (Rolled) (Brill et al. 1994; Brunner et al. 1994; Hall et al. 1996; Madhani et al. 1997). These mutations rendered MAPKs biologically hyperactive, but none of them made the proteins intrinsically active. All of them were still dependent on their upstream kinases (Bott et al. 1994; Brill et al. 1994; reviewed by Askari et al. 2006). Therefore, in order to obtain variants of MAPKs that are absolutely independent of their upstream MAPKK, we devised a different genetic screen. The rationale was that a screen for gain-of-function mutants, that are truly independent of the upstream cascade, should not be performed in the background of wild type cells, but rather in cells lacking the upstream activator altogether (i.e. MAPKKΔ cells). For details on the rationale and the methodology see (Engelberg and Livnah 2006). We applied this rationale in the yeast Hog1 system (screening for active Hog1 in the background of *pbs2Δ* cells). This screen led to the discovery of nine *HOG1* alleles, each carrying a single point mutation (Bell et al. 2001). Each of these Hog1 mutants enabled *pbs2Δ* cells to grow under hyper-osmotic conditions and was shown biochemically to be intrinsically active. Combination of two point mutations in the same Hog1 protein caused an increase in catalytic activity and a severe growth arrest on the yeast cells (Yaakov et al. 2003). The synergism observed by integrating the two mutations may suggest that each acts via a different mechanism. This notion is also supported by the observation that the levels of the catalytic activities manifested by the mutants are different (Bell et al. 2001; Yaakov et al. 2003; Diskin et al. 2004) (Table1).

The mammalian homologs of the yeast Hog1p are members of the p38 subfamily that is composed of four isoforms (α, β, γ, and δ). Sequence alignment of these isoforms with *HOG1* shows that some of the nine point mutations, identified to render Hog1 active, occurred in residues conserved in one or more isoforms (Bell et al. 2001; Askari et al. 2007) (Table1). Three of the residues; Y68, D170, and W332, were found to be conserved in all p38 isoforms. Another residue, A314, was not conserved in p38γ (but was in all other isoforms) and Phe322 was conserved in p38β and p38α. Most importantly, Phe318 of Hog1, when mutated gave rise to the most active Hog1 allele (Bell et al. 2001) that was found to be conserved only in p38δ. All other isoforms possess a Tyr residue at this position. In p38α, it is Tyr 323 which was recently shown to be phosphorylated by ZAP70 (Salvador et al. 2005) (Table 1). This conservation led us to insert systematically the activating mutations to the conserved residues in p38s. We first inserted seven mutations into p38α. Only three of them: D176A, F327S, and F327L rendered p38α intrinsically active (Diskin et al. 2004) (Table 1). Later, we mutated Tyr323 to either Ser or Leu and found these mutants to render the kinase intrinsically active, but to a low level (Avitzour et al. 2007)(Table 1). Finally, we inserted some mutations to other p38 isoforms, and found the resulting proteins to be intrinsically active (Askari et al. 2007; Avitzour et al. 2007)(Table 1). As marked in Table 1, the spontaneous activity of all mutants was demonstrated both *in vivo* in mammalian cell lines (Askari et al. 2007) and *in vitro* using recombinant proteins purified from *E. coli* cells (Avitzour et al. 2007; Diskin et al. 2007). Thus, *bona fide* intrinsically active variants have been obtained for the yeast MAPK Hog1 as well as for all the different isoforms of the mammalian MAPK p38. Further studies confirmed that the active mutants maintained the biological and pharmacological properties of the parental protein (Diskin et al. 2004; Askari et al. 2007; Avitzour et al. 2007).

Table 1. Levels of activity of Hog1 and p38 isoforms carrying activating mutations.

Hog1			p38α			p38β			p38γ			p38δ			Position on structure
	In vivo[a]	*In vitro*		*In vivo*[a]	*In vitro*[b]		*In vivo*[a]	*In vitro*[b]		*In vivo*[a]	*In vitro*[b]		*In vivo*[a]	*In vitro*[b]	
Y68H	150%	N.D	Y69A	N.D	0.3%	Y69			Y72			Y70			C-helix
			Y69H	0%											
D170A	300%	N.D	D176A	100%	9%	D176A	5%	60%	D179A	500%	6.5%	D176A	150%	0%	Phosphorylation lip
			D177A	N.D											
A314T	N.D	N.D	A320T	0%	0.3%	A320			V324			A320			L16
F318S	300%	N.D	Y323S	N.D	<1%	Y323S	N.D	-	Y326			F324S	400%	13%	L16 (3/10 helix)
F318L	600%	N.D	Y323L	N.D	<1%	Y323L	0%	47%							
W320R	100%	N.D	Q325	N.D	N.D	E333	N.D	N.D	D328	N.D	N.D	D326	N.D	N.D	L16
F322L	250%	N.D	F327S	20%	11%	V327S	1%	5%	F330S	N.D	N.D	L328S	N.D	N.D	L16
			F327L	10%	10%	V327L	N.D	-							
W332R	200%	N.D	W337A	0%	<1%	W345			W340			W338R	N.D	N.D	L16
N391D	N.D	N.D	N.E			N.E			N.E			N.E			
D170A+ F318L	500%	N.D	D176A+ F327L	700%	25%										Phosphorylation lip + L16

[a]- Activity of proteins immunoprecipitated from *S. cerevisiea* (for Hog1), or from HEK293 cells (for p38α, p38β, p38γ, and p38δ) not exposed to any stimulus. The level of this activity is presented as percentage of maximal activated, dually phosphorylated, wild type proteins. Activation levels are estimated from the data presented in (Bell et al. 2001; Yaakov et al. 2003; Askari et al. 2007)

[b]- Activity of recombinant protein purified from *E. coli* and not treated further. Activity is presented as percentage of maximally activated, dually phosphorylated, wild type proteins. (Reported in Diskin et al. 2004; Avitzour et al. 2007)

N.D- not determined. The activity of these mutants was not tested.

N.E- not exists. This position is located in the Hog1 C terminus and does not exist in p38s that are shorter.

Gray background- Sites that were not mutated and were not tested.

3 Mechanism of activation of native SAPKs and of intrinsically active SAPKs

3.1 The role of the phosphoacceptors in the intrinsically active variants

How do the mutations render Hog1 and p38s intrinsically active? To date, all known MAPK activation mechanisms, those dependent on the MAPKKK/MAPKK systems as well as those bypassing these cascades, lead ultimately to dual phosphorylation at the MAPKs' phosphorylation lip.

As the phosphorylation of the phosphoacceptors is essential for activation of SAPKs by all activating mechanisms, it was most important to test whether the intrinsically active variants are bypassing this requirement. Two facts may suggest a model in which the mutants do not require phosphorylation for activity: 1) The mutants are biologically active in cells lacking their relevant MEKs (Bell et al. 2001; Yaakov et al. 2003; Diskin et al. 2004; Askari et al. 2007). 2) They are catalytically active when expressed in and purified from *E. coli* cells (Diskin et al. 2004; Avitzour et al. 2007). Namely, the activating mutations may be directly enforcing an active fold on the molecule, mimicking the fold naturally conferred by phosphorylation. Such a model implies that in the mutants, the Thr and Tyr residues in the phosphorylation lip are dispensable for activity. Several approaches were taken to test this model. First, we inserted into all Hog1 active variants the mutations Tyr176 Phe, Thr174Ala, or both. We found that all Hog1 active variants, mutated in the Tyr phosphoacceptor, are still biologically active, and even maintain their unusual capabilities of rescuing *pbs2*Δ cells (Bell and Engelberg 2003). On the other hand, when the Thr residue in Hog1 phosphorylation lip (Thr174) was changed to Ala, all Hog1 variants lost their capability to rescue yeast cells from hyper-osmotic stress, even in the presence of Pbs2. Kinase assays performed with immunoprecipitated proteins, demonstrated a clear correlation between the biological activity and the catalytic activity (Bell and Engelberg 2003). Namely, intrinsically active Hog1 molecules that also harbor a T174A mutation lost all catalytic activity, whereas active variants harboring the Y176F mutation were catalytically active in the absence of any stimulation. However, although maintaining their spontaneous catalytic activity, the active mutants, carrying in addition the Y176F mutation, were significantly less active than those carrying the native Y176. Also, the catalytic activity of the molecules mutated in Tyr176 did not increase upon exposure to hyperosmotic stress (Bell and Engelberg 2003). The deleterious effect of the Thr174Ala mutation on all active mutants may suggest that Thr174 phosphorylation is crucial for their activity. However, in spite of significant efforts to measure phosphorylation of Thr174 of the active variants in *pbs2*Δ cells (in which the mutants are active) the Thr phosphoacceptor was not demonstrated to be phosphorylated (Bell and Engelberg 2003; Yaakov et al. 2003). Thus, these combined results suggest that in the active mutants the phosphoacceprots are not dispensable, but there is a different role to each one of them. While the Tyr176 residue is not vital for activity, it is clearly required for enhancing the activity after exposure to osmotic stress. The Thr174 residue is essential for activity, probably not as a phosphoacceptor, but as a structural component, stabilizing the protein in its active fold imposed by the activating mutation. This idea is supported by the fact that even a Thr174Glu mutation is deleterious to Hog1 activity (Bell and Engelberg 2003). Such different roles for the phosphoacceptors in SAPKs activation are not unique to the active mutants, but reflect probably the situation in native MAPKs as well. For example, native Hog1, mutated in the Tyr phosphoacceptor, is biologically active. When such a molecule is overexpressed, it rescues *hog1Δ* cells from hyperosmotic stress showing that Tyr176 is not essential for survival (Bell and Engelberg 2003). Various roles were also suggested for mono-phosphorylated ERKs (Prowse et al. 2001; Zhou and Zhang 2002). In addition to their unusual property of intrinsic activity, which requires intact Thr174 and allows them to rescue *pbs2Δ* cells, the Hog1 active mutants acquired the property to become spontaneously dually phosphorylated if Pbs2 is present. Thus, the active Hog1 mutants, just like the native Hog1, manifest their maximal activity when dually phosphorylated.

Just like the Hog1 active mutants, the active p38 mutants require an intact TGY motif for their activity (Askari et al. data not shown). In this case, however, Thr180 is clearly crucial as a phosphoacceptor. This notion is based on the observation that all intrinsically active p38 variants were found to be phosphorylated on Thr180 but not on Tyr182, when purified from *E. coli* (Avitzour et al. 2007; Diskin et al. 2007). Their activity could be further activated after incubation with the relevant MAPKK, MKK6, most probably through additional phosphorylation of Tyr182 in the phosphorylation motif. For example, when purified from *E. coli* cells, the active double mutant $p38\alpha^{D176A,F327S}$ reaches activity level of 25% in reference to the phosphorylated $p38\alpha^{w.t.}$(Diskin et al. 2004), and exhibits an activity level of more than 100% after *in vitro* activation by MKK6 (Askari et al. 2007). Curiously, the maximal activity obtained by p38α mutated in its Tyr phosphoacceptor, (phosphorylated by MKK6) is 25% of maximum, similar to the basal activity manifested by the intrinsically active p38α variant (Diskin et al. 2007). Another indication for the crucial role of Thr180 phosphorylation is that when the mutants were purified from bacteria that were grown at 20°C they were not phosphorylated and not active. Similarly, when the intrinsically active p38α variants were synthesized in an *in vitro* expression system they were not phosphorylated and not active (Maayan et al., data not shown). In both cases, the unphosphorylated, inactive, purified proteins could be phosphorylated and activated by their native MAPKK, MKK6, indicating that some level of phosphorylation is essential for the mutants' activity. Also, similar to the Hog1 mutants, the p38 mutants are spontaneously dually phosphorylated *in vivo* (see below).

In summary, the Thr phosphoacceptor at the phosphorylation lip is critical for the intrinsic activity of both Hog1 and p38 active variants. In the p38 mutants, it must be phosphorylated. In the Hog1 mutants, we could not detect spontaneous phosphorylation of this threonine, but it may be phosphorylated at low levels, below our detection threshold. Thus, although the active mutants were isolated from $MEK^{-/-}$ cells, and although active as recombinant proteins, phosphorylation still plays an important role in their activation. In fact, the mutants manifest their maximal activity *in vivo*, in wild type cells (yeast or mammalian) when they are spontaneously dually phosphorylated.

3.2 The mutants acquired an auto-phosphorylation capability

On one hand, the activity of the p38 mutants is absolutely independent of upstream regulation, particularly that of MEKs (they are active as purified recombinant proteins). On the other hand, their activity is dependent on Thr182 phosphorylation. How are the mutants become phosphorylated on Thr182 than? It is unlikely that a bacterial protein, a MEK-like kinase, is responsible for phosphorylating the recombinant active SAPKs, because such a kinase would phosphorylate $p38^{wt.}$ as well. It seems more likely that the mutants auto-phosphorylate. This idea is not unforeseen because several studies showed that MAPKs are capable of auto-phosphorylation. Particularly, as described above, auto-phosphorylation of p38α is normally induced by either association with TAB1, or by Tyr323 phosphorylation. Other MAPKs (e.g. ERKs, JNKs) were also reported to auto-phosphorylate on both Thr and Tyr (Seger et al. 1991; Wu et al. 1991; Rossomando et al. 1992; Cui et al. 2005). We tested the potential auto-phosphorylation activity of all intrinsically active p38 isoforms. We observed that while wild type p38 molecules showed almost no such activity, the active mutants manifested significant auto-phosphorylation capabilities (Bell et al. 2001; Avitzour et al. 2007; Diskin et al. 2007). Curiously, $p38\beta^{w.t.}$ showed some auto-phosphorylation activity that was much lower than that of the $p38\beta^{D176A}$ active variant (Avitzour et al. 2007). Biochemical and biological assays revealed a strong correlation between the auto-phosphorylation capability of a p38 protein and its catalytic activity towards external substrates. Hence, $p38\alpha^{w.t.}$ and $p38\delta^{w.t.}$ that had no auto-phosphorylation ability were not catalytically active, while the $p38\alpha^{D176A}$, $p38\beta^{D176A}$, and $p38\delta^{F324S}$ active mutants, that displayed high levels of auto-phosphorylation, showed high level of catalytic activity (Avitzour et al. 2007). Accordingly, $p38\beta^{w.t.}$, which possesses some basal auto-phosphorylation activity, also manifests a measurable basal catalytic activity (Avitzour et al. 2007).

This model, suggests that the active p38 variants acquired a capability of mono-auto-phosphorylation activity (on Thr182), can also explain the fact that all active mutants, those of Hogl and those of all p38 isoforms, are much more active *in vivo* (in yeast or in mammalian cells) than *in vitro* (as purified recombinant proteins). Namely, *in vivo*, the active variants are dually phosphorylated, whereas *in vitro* they are mono-auto-phosphorylated. It is not clear how the mutants are spontaneously dually phosphorylated *in vivo*, but in yeast their dual phosphorylation is clearly dependent on Pbs2 because they are not phosphorylated in *pbs2Δ* cells (Bell et al. 2001; Yaakov et al. 2003). In mammalian cells, phosphorylation of p38 mutants is partially reduced in $MKK3/6^{-/-}$ cells (Askari et al. 2007). An open question here is how do the mutants spontaneously (i.e. in the absence of external stimulation) recruit the upstream kinases. One explanation could be that the mutants have acquired high affinity to MEKs and get phosphorylated in the absence of any external signal. However, we do not favor this model. We think that the mutants could not have acquired an increased affinity to MEK because they were initially isolated from cells in which the relevant MEK was knocked out (Bell et al. 2001). An alternative (and more preferred by us) explanation is that the active mutants initiate a positive feedback loop. Namely, downstream targets of the active SAPKs activate (for example via some exocrinic effect) an upstream component of the SAPK cascade. Further research is clearly required to reveal the exact mechanism of the mutants' spontaneous phosphorylation *in vivo*. In summary, the active SAPKs have the capability to auto-phosphorylate *in vivo* and *in vitro*. *In vivo* they are also further phosphorylated spontaneously by MEK.

3.3 Structural changes due to activation

The biochemical and biological data reviewed above suggest that the activating mutations may be mimicking the effect of the bypassing pathways. The location of many of the activating mutations at the L16 domain, adjacent to Tyr323, suggests that they may be mimicking the effect of Tyr323 phosphorylation. Do the mutations impose conformational changes similar to those imposed by ZAP70-mediated Tyr323 phosphorylation? Or they impose perhaps changes acquired by p38 following dual phosphorylation? Or do they acquire yet another conformation, unique to the mutants, which has no biological relevance? The complete answer to these questions cannot be provided now because crystal structures of Tyr323-phosphorylated or Tyr182/Thr184-phosphorylated-p38 have not been obtained yet. To date, the only available structure of active p38 is that of active p38γ and is at relatively low resolution (Bellon et al. 1999). However, the 3D structures of non-active p38α and of some of the active p38α mutants are available at high resolution (Wilson et al. 1996; Wang et al. 1997; Diskin et al. 2007), allowing important insight into the structure-function relation of the kinases.

Detailed data on the structural differences between the active and inactive forms of a MAPK is available to date only for the mammalian Erk2 MAPK. Comparing the Ekr2 and the phosphor-Erk2 crystal structures reveals a phosphorylation-dependent movement and re-orientation of the phosphorylation lip toward the L16 domain and creation of a 3/10 helix within the L16. Those changes expose the p+1 substrate binding domain and probably also lead to dimerization of Erk2 (Zhang et al. 1994; Canagarajah et al. 1997; Khokhlatchev et al. 1998; Cobb and Goldsmith 2000). Other domains in the protein remain essentially unchanged supporting the idea that the L16-phosphorylation lip association is sufficient for activation.

Most of the activating mutations (six out of nine), isolated in the Hog1 screen, are located in the L16 domain. Another mutation, D170A, is located within the phosphorylation lip. Analysis of the structure of the non-active p38α revealed a hydrophobic core that is stabilized by three aromatic residues (Tyr69, Phe327, and Trp337), two of which are located within the L16 region. Strikingly, all three residues are homologous to residues that were found mutated in the Hog1 screen (Table 1). The crystal structure of the intrinsically active mutant $p38\alpha^{F327L}$ showed that the mutated Phe327 residue can no longer participate in the stabilization of the hydrophobic core, strongly suggesting that disruption of this core imposes p38 and Hog1 activation (Diskin et al. 2007). Yet, not any type of destabilization would render the kinase active. For example, mutating Phe327 to Ala has no effect (Diskin et al. 2004). Also, mutating Tyr69 to His and Trp337 to Ala did not induce activity in p38α (Diskin et al. 2004; Askari et al. 2007), although similar mutations in Hog1 did impose activity. There is most probably a defined subset of allowed conformations in the region of the hydrophobic core that evoke activity. Systematic mutagenesis and resolution of the activity and structures of the resulting proteins would reveal these conformations. We predict that phosphorylation of Tyr 323 increases the charge and polarity in the vicinity of the hydrophobic core, affecting mainly Phe327 that is nearby. The result is a disruption of the hydrophobic core leading to auto-phosphorylation just as the activating mutations do.

Another interesting region is the phosphorylation lip. Unlike the situation in Erk2 in which this lip acquires a stable structure that associates with L16 (Canagarajah et al. 1997), in p38 this element is highly disordered and therefore can not be seen in the crystal structures. One of the activating mutations, Asp170 in Hog1, or Asp176 in p38, is located within the phosphorylation lip. This mutation is sufficient to generate an intrinsic activity and when combined with another mutation, such as Phe327Leu or Phe327Ser, it gives rise to higher activity. The p38α^{D176A}, p38β^{D176A}, and p38γ^{D179A} mutants show auto-phosphorylation activity (Avitzour et al. 2007). As in the case of all p38 structures obtained so far (Wilson et al. 1996; Wang et al. 1997), the crystal structures of the mutant p38α^{D176A} also does not disclose the fold of the phosphorylation lip, not allowing to appreciate the effect of the mutation on this region. The p38α^{D176A} structure revealed no conformational changes in any other domain in reference to the native inactive p38α structure (Diskin et al. 2007). The D176A mutation confers an auto-phosphorylation capability to the p38 molecule, similar to the effect of mutations in the L16. Perhaps, similar to the case of active ERK2, the phosphorylation lip associates with L16 that consequently acquires a new fold, implying that the D176A mutation does cause a conformational change in the L16 by the movement of the phosphorylation lip. It should be appreciated that the crystal structures of the active mutants were obtained from non-phosphorylated, non-active proteins (purified from bacteria grown at 20^{0}C) and the putative effect of D176A mutation on the L16 domain could not be seen in the crystal structure due to the inactive protein state.

The fact that the mutations D176A and F327L or F327S synergies (Diskin et al. 2004) may suggest that each one imposes some change in L16 and together they destabilize the hydrophobic core, giving rise to one of the potent allowed active folds.

4 Discussion

The identification of truly intrinsically active Hog1 and p38 variants allows for the first time examination of the unique and specific roles of a given SAPK without activating any other unwanted component in the cell. Indeed, the first application of the p38 mutants in biological systems revealed a different role for p38α and p38γ in the activation of the AP-1 transcriptional machinery. Active mutants of p38α were shown to dramatically induce AP-1 activity in human cells (HEK293) whereas active variants of p38γ suppressed this activity (Askari et al. 2007). It should be noted that such results are valid only if the active mutants maintain all properties of their wild type original protein. Indeed, it seems that the active variants created so far, maintain substrate specificity, target genes specificity and sensitivity for inhibitors, similar to these of the parental native protein (Diskin et al. 2004; Askari et al. 2007; Avitzour et al. 2007). As we outlined in this chapter, these properties are maintained because the mutations induced intrinsic activity by imposing structural changes and mechanism of activation similar to those induced by natural activating systems. It seems, therefore, that the mutants are legitimate for biological studies.

Although the exact activation mechanism imposed by the activating mutations is still not known, the information accumulated so far indicates that their mechanism of activation most probably emulates the natural "bypass" activation mechanism reported to occur in T cells. Namely, the mutations mimic the changes imposed on p38 molecules by Tyr323 phosphorylation. The original mutations that served as a basis for mutating all of the p38 family were isolated in the yeast Hog1 that was not reported to be activated via bypass pathways. Furthermore, Hog1 harbors a Phe at a position equivalent to Tyr323 in p38α. Curiously, p38δ also harbors a Phe at this position. p38β and p38γ harbor a Tyr at this position, just as p38α does. In all these SAPKs this position is one of the best targets for activating mutations although only p38α was reported to be naturally activated via phosphorylation at Tyr323 and only in T cells. The finding that mutating this position in all p38s and even in the yeast Hog1 protein induce intrinsic activity, may suggest that a bypass pathway exists not only for p38α and not only in T cells. Such pathways may exist actually from early in evolution. We have recently observed indeed Hog1 activation in response to osmotic stress in *pbs2Δ* cells (Maayan et al., unpublished data). It is easy to envisage how the putative bypass pathways activate p38α, p38β, and p38γ because they all have the target Tyr residue. How p38δ and Hog1 are activated by such a pathway remains a puzzle.

The mutations found to render Hog1 and p38 active are conserved not only in these MAPKs, but also in ERKs and JNKs. We believe, therefore, that the insights from Hog1 and p38s active variants can be used for the design of mutations to be inserted to other MAPK families in order to create intrinsically active variants. This issue should be addressed with precautions because the mutants may be mimicking a pathway that does not exist for ERKs or JNKs. These kinases do not have a conserved tyrosine, equivalent to Tyr323 of p38s. Thus, activating mutations of ERKs and JNKs may reside in different domains of the molecules. Curiously, however, ERKs harbor a Phe at this position similar to p38δ and Hog1 (Bell et al. 2001). Also, in all MAPKs, the phosphorylation lip and the L16 domains undergoes structural changes upon activation and are key targets for mutagenesis. Therefore, even if the exact mutations that render Hog1 and p38 active will not be relevant for activation of JNK and ERK, the L16 and the phosphorylation lip are clearly attractive targets for mutagenesis.

References

Aebersold DM, Shaul YD, Yung Y, Yarom N, Yao Z, Hanoch T, Seger R (2004) Extracellular signal-regulated kinase 1c (ERK1c), a novel 42-kilodalton ERK, demonstrates unique modes of regulation, localization, and function. Mol Cell Biol 24:10000-10015

Askari N, Diskin R, Avitzour M, Capone R, Livnah O, Engelberg D (2007) Hyperactive variants of p38alpha induce, whereas hyperactive variants of p38gamma suppress, activating protein 1-mediated transcription. J Biol Chem 282:91-99

Askari N, Diskin R, Avitzour M, Yaakov G, Livnah O, Engelberg D (2006) MAP-quest: could we produce constitutively active variants of MAP kinases? Mol Cell Endocrinol 252:231-240

Avitzour M, Diskin R, Raboy B, Askari N, Engelberg D, Livnah O (2007) Intrinsically active variants of all human p38 isoforms. Febs J 274:963-975

Bell M, Capone R, Pashtan I, Levitzki A, Engelberg D (2001) Isolation of hyperactive mutants of the MAPK p38/Hog1 that are independent of MAPK kinase activation. J Biol Chem 276:25351-25358

Bell M, Engelberg D (2003) Phosphorylation of Tyr-176 of the yeast MAPK Hog1/p38 is not vital for Hog1 biological activity. J Biol Chem 278:14603-14606

Bellon S, Fitzgibbon MJ, Fox T, Hsiao HM, Wilson KP (1999) The structure of phosphorylated p38gamma is monomeric and reveals a conserved activation-loop conformation. Structure 7:1057-1065

Bott CM, Thorneycroft SG, Marshall CJ (1994) The sevenmaker gain-of-function mutation in p42 MAP kinase leads to enhanced signalling and reduced sensitivity to dual specificity phosphatase action. FEBS Lett 352:201-205

Brill JA, Elion EA, Fink GR (1994) A role for autophosphorylation revealed by activated alleles of FUS3, the yeast MAP kinase homolog. Mol Biol Cell 5:297-312

Brunner D, Oellers N, Szabad J, Biggs WH 3rd, Zipursky SL, Hafen E (1994) A gain-of-function mutation in *Drosophila* MAP kinase activates multiple receptor tyrosine kinase signaling pathways. Cell 76:875-888

Bulavin DV, Fornace AJ Jr (2004) p38 MAP kinase's emerging role as a tumor suppressor. Adv Cancer Res 92:95-118

Canagarajah BJ, Khokhlatchev A, Cobb MH, Goldsmith EJ (1997) Activation mechanism of the MAP kinase ERK2 by dual phosphorylation. Cell 90:859-869

Casanova E, Callejo AI, Calvo P, Chinchetru MA (2000) Analysis of splicing of four mouse JNK/SAPKalpha variants. Neuroreport 11:305-309

Cobb MH, Goldsmith EJ (1995) How MAP kinases are regulated. J Biol Chem 270:14843-14846

Cobb MH, Goldsmith EJ (2000) Dimerization in MAP-kinase signaling. Trends Biochem Sci 25:7-9

Cowley S, Paterson H, Kemp P, Marshall CJ (1994) Activation of MAP kinase kinase is necessary and sufficient for PC12 differentiation and for transformation of NIH 3T3 cells. Cell 77:841-852

Cui J, Holgado-Madruga M, Su W, Tsuiki H, Wedegaertner P, Wong AJ (2005) Identification of a specific domain responsible for JNK2alpha2 autophosphorylation. J Biol Chem 280:9913-9920

Davis RJ (2000) Signal transduction by the JNK group of MAP kinases. Cell 103:239-252

Diskin R, Askari N, Capone R, Engelberg D, Livnah O (2004) Active mutants of the human p38alpha mitogen-activated protein kinase. J Biol Chem 279:47040-47049

Diskin R, Lebendiker M, Engelberg D, Livnah O (2007) Structures of p38alpha active mutants reveal conformational changes in L16 loop that induce autophosphorylation and activation. J Mol Biol 365:66-76

Eilers A, Whitfield J, Babij C, Rubin LL, Ham J (1998) Role of the Jun kinase pathway in the regulation of c-Jun expression and apoptosis in sympathetic neurons. J Neurosci 18:1713-1724

Engelberg D (2004) Stress-activated protein kinases-tumor suppressors or tumor initiators? Semin Cancer Biol 14:271-282

Engelberg D, Livnah O (2006) Isolation of intrinsically active mutants of MAP kinases via genetic screens in yeast. Methods 40:255-261

English J, Pearson G, Wilsbacher J, Swantek J, Karandikar M, Xu S, Cobb MH (1999) New insights into the control of MAP kinase pathways. Exp Cell Res 253:255-270

Faccio L, Chen A, Fusco C, Martinotti S, Bonventre JV, Zervos AS (2000) Mxi2, a splice variant of p38 stress-activated kinase, is a distal nephron protein regulated with kidney ischemia. Am J Physiol Cell Physiol 278:C781-790

Ge B, Gram H, Di Padova F, Huang B, New L, Ulevitch RJ, Luo Y, Han J (2002) MAPKK-independent activation of p38alpha mediated by TAB1-dependent autophosphorylation of p38alpha. Science 295:1291-1294

Gillespie-Brown J, Fuller SJ, Bogoyevitch MA, Cowley S, Sugden PH (1995) The mitogen-activated protein kinase kinase MEK1 stimulates a pattern of gene expression typical of the hypertrophic phenotype in rat ventricular cardiomyocytes. J Biol Chem 270:28092-28096

Hall JP, Cherkasova V, Elion E, Gustin MC, Winter E (1996) The osmoregulatory pathway represses mating pathway activity in *Saccharomyces cerevisiae*: isolation of a FUS3 mutant that is insensitive to the repression mechanism. Mol Cell Biol 16:6715-6723

Han J, Lee JD, Bibbs L, Ulevitch RJ (1994) A MAP kinase targeted by endotoxin and hyperosmolarity in mammalian cells. Science 265:808-811

Hanks SK, Hunter T (1995) Protein kinases 6. The eukaryotic protein kinase superfamily: kinase (catalytic) domain structure and classification. Faseb J 9:576-596

Hivroz C, Fischer A (1994) Immunodeficiency diseases. Multiple roles for ZAP-70. Curr Biol 4:731-733

Jiang Y, Gram H, Zhao M, New L, Gu J, Feng L, Di Padova F, Ulevitch RJ, Han J (1997) Characterization of the structure and function of the fourth member of p38 group mitogen-activated protein kinases, p38delta. J Biol Chem 272:30122-30128

Keren A, Tamir Y, Bengal E (2006) The p38 MAPK signaling pathway: a major regulator of skeletal muscle development. Mol Cell Endocrinol 252:224-230

Khokhlatchev AV, Canagarajah B, Wilsbacher J, Robinson M, Atkinson M, Goldsmith E, Cobb MH (1998) Phosphorylation of the MAP kinase ERK2 promotes its homodimerization and nuclear translocation. Cell 93:605-615

Kumar S, Boehm J, Lee JC (2003) p38 MAP kinases: key signalling molecules as therapeutic targets for inflammatory diseases. Nat Rev Drug Discov 2:717-726

Kyriakis JM, Avruch J (2001) Mammalian mitogen-activated protein kinase signal transduction pathways activated by stress and inflammation. Physiol Rev 81:807-869

Kyriakis JM, Banerjee P, Nikolakaki E, Dai T, Rubie EA, Ahmad MF, Avruch J, Woodgett JR (1994) The stress-activated protein kinase subfamily of c-Jun kinases. Nature 369:156-160

Lagalwar S, Guillozet-Bongaarts AL, Berry RW, Binder LI (2006) Formation of phospho-SAPK/JNK granules in the hippocampus is an early event in Alzheimer disease. J Neuropathol Exp Neurol 65:455-464

Madhani HD, Styles CA, Fink GR (1997) MAP kinases with distinct inhibitory functions impart signaling specificity during yeast differentiation. Cell 91:673-684

Mansour SJ, Candia JM, Gloor KK, Ahn NG (1996) Constitutively active mitogen-activated protein kinase kinase 1 (MAPKK1) and MAPKK2 mediate similar transcriptional and morphological responses. Cell Growth Differ 7:243-250

Mansour SJ, Matten WT, Hermann AS, Candia JM, Rong S, Fukasawa K, Vande Woude GF, Ahn NG (1994) Transformation of mammalian cells by constitutively active MAP kinase kinase. Science 265:966-970

Minden A, Lin A, McMahon M, Lange-Carter C, Derijard B, Davis RJ, Johnson GL, Karin M (1994) Differential activation of ERK and JNK mitogen-activated protein kinases by Raf-1 and MEKK. Science 266:1719-1723

Morooka T, Nishida E (1998) Requirement of p38 mitogen-activated protein kinase for neuronal differentiation in PC12 cells. J Biol Chem 273:24285-24288

Murphy LO, Blenis J (2006) MAPK signal specificity: the right place at the right time. Trends Biochem Sci 31:268-275

Ono K, Han J (2000) The p38 signal transduction pathway: activation and function. Cell Signal 12:1-13

Orchel A, Dzierzewicz Z, Parfiniewicz B, Weglarz L, Wilczok T (2005) Butyrate-induced differentiation of colon cancer cells is PKC and JNK dependent. Dig Dis Sci 50:490-498

Pearson G, Robinson F, Beers Gibson T, Xu BE, Karandikar M, Berman K, Cobb MH (2001) Mitogen-activated protein (MAP) kinase pathways: regulation and physiological functions. Endocr Rev 22:153-183

Porras A, Zuluaga S, Black E, Valladares A, Alvarez AM, Ambrosino C, Benito M, Nebreda AR (2004) P38 alpha mitogen-activated protein kinase sensitizes cells to apoptosis induced by different stimuli. Mol Biol Cell 15:922-933

Prowse CN, Deal MS, Lew J (2001) The complete pathway for catalytic activation of the mitogen-activated protein kinase, ERK2. J Biol Chem 276:40817-40823

Robinson MJ, Stippec SA, Goldsmith E, White MA, Cobb MH (1998) A constitutively active and nuclear form of the MAP kinase ERK2 is sufficient for neurite outgrowth and cell transformation. Curr Biol 8:1141-1150

Rossomando AJ, Wu J, Michel H, Shabanowitz J, Hunt DF, Weber MJ, Sturgill TW (1992) Identification of Tyr-185 as the site of tyrosine autophosphorylation of recombinant mitogen-activated protein kinase p42mapk. Proc Natl Acad Sci USA 89:5779-5783

Rubinfeld H, Seger R (2005) The ERK cascade: a prototype of MAPK signaling. Mol Biotechnol 31:151-174

Salvador JM, Mittelstadt PR, Guszczynski T, Copeland TD, Yamaguchi H, Appella E, Fornace AJ Jr, Ashwell JD (2005) Alternative p38 activation pathway mediated by T cell receptor-proximal tyrosine kinases. Nat Immunol 6:390-395

Schett G, Tohidast-Akrad M, Smolen JS, Schmid BJ, Steiner CW, Bitzan P, Zenz P, Redlich K, Xu Q, Steiner G (2000) Activation, differential localization, and regulation of the stress-activated protein kinases, extracellular signal-regulated kinase, c-JUN N-terminal kinase, and p38 mitogen-activated protein kinase, in synovial tissue and cells in rheumatoid arthritis. Arthritis Rheum 43:2501-2512

Schieven GL (2005) The biology of p38 kinase: a central role in inflammation. Curr Top Med Chem 5:921-928

Scielzo C, Camporeale A, Geuna M, Alessio M, Poggi A, Zocchi MR, Chilosi M, Caligaris-Cappio F, Ghia P (2006) ZAP-70 is expressed by normal and malignant human B-cell subsets of different maturational stage. Leukemia 20:689-695

Seger R, Ahn NG, Boulton TG, Yancopoulos GD, Panayotatos N, Radziejewska E, Ericsson L, Bratlien RL, Cobb MH, Krebs EG (1991) Microtubule-associated protein 2 kinases, ERK1 and ERK2, undergo autophosphorylation on both tyrosine and threonine residues: implications for their mechanism of activation. Proc Natl Acad Sci USA 88:6142-6146

Shaul YD, Seger R (2006) ERK1c regulates Golgi fragmentation during mitosis. J Cell Biol 172:885-897

Sudo T, Yagasaki Y, Hama H, Watanabe N, Osada H (2002) Exip, a new alternative splicing variant of p38 alpha, can induce an earlier onset of apoptosis in HeLa cells. Biochem Biophys Res Commun 291:838-843

Uzgare AR, Kaplan PJ, Greenberg NM (2003) Differential expression and/or activation of P38MAPK, erk1/2, and jnk during the initiation and progression of prostate cancer. Prostate 55:128-139

Wang Z, Harkins PC, Ulevitch RJ, Han J, Cobb MH, Goldsmith EJ (1997) The structure of mitogen-activated protein kinase p38 at 2.1-A resolution. Proc Natl Acad Sci USA 94:2327-2332

Westra J, Limburg PC (2006) p38 mitogen-activated protein kinase (MAPK) in rheumatoid arthritis. Mini Rev Med Chem 6:867-874

Williams NG, Zhong H, Minneman KP (1998) Differential coupling of alpha1-, alpha2-, and beta-adrenergic receptors to mitogen-activated protein kinase pathways and differentiation in transfected PC12 cells. J Biol Chem 273:24624-24632

Wilson KP, Fitzgibbon MJ, Caron PR, Griffith JP, Chen W, McCaffrey PG, Chambers SP, Su MS (1996) Crystal structure of p38 mitogen-activated protein kinase. J Biol Chem 271:27696-27700

Wu J, Rossomando AJ, Her JH, Del Vecchio R, Weber MJ, Sturgill TW (1991) Autophosphorylation *in vitro* of recombinant 42-kilodalton mitogen-activated protein kinase on tyrosine. Proc Natl Acad Sci USA 88:9508-9512

Yaakov G, Bell M, Hohmann S, Engelberg D (2003) Combination of two activating mutations in one HOG1 gene forms hyperactive enzymes that induce growth arrest. Mol Cell Biol 23:4826-4840

Yoon S, Seger R (2006) The extracellular signal-regulated kinase: multiple substrates regulate diverse cellular functions. Growth Factors 24:21-44

Yung Y, Yao Z, Hanoch T, Seger R (2000) ERK1b, a 46-kDa ERK isoform that is differentially regulated by MEK. J Biol Chem 275:15799-15808

Zarubin T, Han J (2005) Activation and signaling of the p38 MAP kinase pathway. Cell Res 15:11-18

Zetser A, Gredinger E, Bengal E (1999) p38 mitogen-activated protein kinase pathway promotes skeletal muscle differentiation. Participation of the Mef2c transcription factor. J Biol Chem 274:5193-5200

Zhang F, Strand A, Robbins D, Cobb MH, Goldsmith EJ (1994) Atomic structure of the MAP kinase ERK2 at 2.3 A resolution. Nature 367:704-711

Zheng C, Xiang J, Hunter T, Lin A (1999) The JNKK2-JNK1 fusion protein acts as a constitutively active c-Jun kinase that stimulates c-Jun transcription activity. J Biol Chem 274:28966-28971

Zhou B, Zhang ZY (2002) The activity of the extracellular signal-regulated kinase 2 is regulated by differential phosphorylation in the activation loop. J Biol Chem 277:13889-13899

Zhu X, Castellani RJ, Takeda A, Nunomura A, Atwood CS, Perry G, Smith MA (2001) Differential activation of neuronal ERK, JNK/SAPK and p38 in Alzheimer disease: the 'two hit' hypothesis. Mech Ageing Dev 123:39-46

Engelberg, David

Department of Biological Chemistry, The Alexander Silvberman Institute of Life Sciences, The Hebrew University of Jerusalem, Jerusalem, 91904, Israel.
engelber@cc.huji.ac.il

Maayan, Inbal

Department of Biological Chemistry, The Alexander Silvberman Institute of Life Sciences, The Hebrew University of Jerusalem, Jerusalem, 91904, Israel.

Regulation of MAPK signaling in yeast

Fabian Rudolf, Serge Pelet, Matthias Peter

Abstract

Mitogen activated protein kinase (MAPK) pathways orchestrate the response of cells to their changing extracellular environment. Budding yeast has proven to be a useful model system to elucidate the basic principles of these complex signal transduction processes. Yeast cells use six MAP-kinases, which respond to diverse conditions such as pheromone signals, osmolarity, cell wall stress and nutritional status. Depending on the external and internal conditions, multiple signals must be integrated leading to a defined output and different signaling kinetics. In this review, we will discuss the molecular architecture of the mating and high-osmolarity-glycerol (HOG) MAPK pathways, and compare their underlying signaling parameters. These two pathways use shared components and by contrasting their activation and regulation we try to deduce common and specific principles for MAPK signaling in yeast. Among those, we will summarize and discuss recent findings, which shed light on positive and negative feedback mechanisms. Finally, we will highlight the importance of quantitative measurements in single cells, which are necessary to unravel stochastic cell-to-cell variations and account for the dynamic signaling kinetics.

1 Introduction to MAPK signaling during mating and high osmolarity conditions in yeast

Activation of the mating pathway by binding of pheromone to its receptor leads to three hallmark events required for the formation of diploid cells: (1) cells synchronize their cell cycle by arresting at "Start", (2) initiate a transcriptional program, and (3) polarize towards the source of pheromone. An impressive body of work identified the components and general wiring of the mating pathway (Dohlman and Thorner 2001). In short, haploid yeast cells exist either as "a" or "α"– mating types. Cells of opposite mating type communicate with each other by pheromones; a-cells secrete a-factor and detect α–factor via the α–factor receptor (Ste2p), while α–cells secrete α–factor and sense a-factor via the a-factor receptor (Ste3p). Both receptors belong to the family of G-protein coupled receptors (GPCR), and are coupled intracellularly to the same heterotrimeric G-protein composed of the α–subunit Gpa1p, the farnesylated ß-subunit Ste4p and the γ–subunit Ste18p (Fig.1A). Receptor-activation triggers dissociation of the α-subunit

Topics in Current Genetics, Vol. 20
F. Posas, A. R. Nebreda (Eds.): Stress-Activated Protein Kinases
DOI 10.1007/4735_2007_0250 / Published online: 1 August 2007

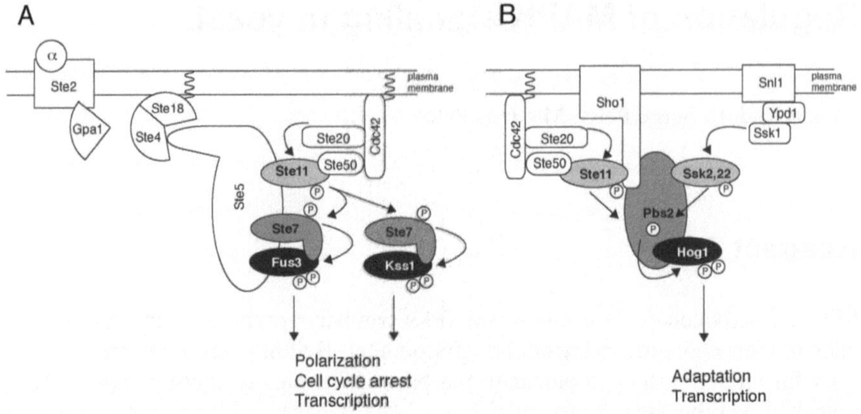

Fig. 1. The mating (A) and HOG (B) MAPK-signaling pathways in budding yeast. The MAPKKKs are shown in light gray, the MAPKKs in dark gray and the MAPKs in black. Arrows indicate activation and phosphorylation events; phosphorylated species are indicated with an encircled "P".

from the obligate β/γ –heterodimer, which binds the scaffold protein Ste5p and thereby recruits the three kinases Ste11p (MAPKKK), Ste7p (MAPKK), and Fus3p (MAPK). Moreover, Ste50p tethers Ste11p to the membrane. Activation of Ste11p requires a phosphorylation signal by the PAK-like kinases Ste20p and Cla4p (Heinrich et al. 2006), which are recruited and activated by the small GTPase Cdc42p in its GTP-bound form (Cdc42p-GTP). Phosphorylated Ste11p activates Ste7p, which in turn doubly phosphorylates Fus3p in its activation loop. Activated Fus3p rapidly dissociates from Ste5p and phosphorylates substrates in the cytoplasm and nucleus, thereby orchestrating the different processes required for mating. In addition to Fus3p, the homologous MAPK Kss1p plays a redundant role, but differs in the way it regulates the pathway. Shared among all MAPK is an association with actively transcribed genes in the nucleus (Pokholok et al. 2006).

In contrast, the high osmolarity glycerol (HOG) pathway transiently adapts cells to a changed environment. The response is triggered by a shift in osmotic pressure of the medium (Fig. 1B) (Hohmann 2002). Although the osmostress sensors remain elusive, two redundant pathways converge on the MAPKK Pbs2p. The Sln1p branch involves a phosphorelay consisting of Sln1p, Ypd1p, Ssk1p, and the two redundant MAPKKK Ssk2p and Ssk22p. In contrast, the Sho1p branch involves the MAPKKK Ste11p, which in turn is activated by the Cdc42p-Ste20p/Cla4p module and recruited to the plasma membrane by Ste50p and Sho1p. Different binding events are needed to activate Pbs2p: Ssk2p and Ssk22p directly dock onto Pbs2p, while the interaction of Ste11p and Pbs2p is bridged by Sho1p. Pbs2p activates the MAPK Hog1p, which induces a transcriptional program to ensure long-term adaptation to the changed environment, while phosphorylated targets of Hog1p located in the cytoplasm and the nucleus aim to rapidly respond by various mechanisms. Crucial among them is a rise of internal osmolarity by retention and increased synthesis of intracellular glycerol to balance osmolarity.

Recently, it has become clear that in addition to the described general wiring of these MAPK pathways, the cells have to deal with the stochastic nature of the reactions. This can lead to a large noise in the response of the system and the cell has to be able to tightly regulate (probably by redundant mechanisms) the overall level of active MAPK produced by the cascade. Quantitative data from single cell measurements have allowed formalizing a methodology to understand which factors contribute to the level and noise of the pathway output. Moreover, single cell analysis provided new insights into positive and negative feedback mechanisms that control signaling kinetics and contribute to signaling specificity. Below, we will first summarize these recent findings starting with the analysis of noise in pathways obtained from single cell data. Then we will focus on the activation, regulation and specificity of the two pathways.

2 Quantifying signaling at the single cell level

Yeast cells exposed to α-factor are forced to take a decision: either to keep proliferating in the haploid state or to mate with their partner and generate a diploid cell. Conceptually, the mating process is an "all or none" response, and thus analogous to a cell fate decision. Since the analysis of whole populations inevitably results in averaging the response, it is important to monitor signaling output in single cells in order to understand cell-to-cell variations. Most popular are transcriptional reporters, which express green fluorescence protein (GFP) under the control of a promoter regulated by the signaling pathway of interest. For example, cell-to-cell variations in the mating pathway were studied by expressing GFP or another color variant from either a mating-specific (e.g. *FUS1*) or a constitutively-active (e.g. *ADH1*) promoter, and fluorescence-levels in single cells were quantified and correlated after exposure of the cells to α-factor (Colman-Lerner et al. 2005). The resulting signal was analyzed under the simplified assumption, that the transcriptional output is composed of two modules (Fig. 2A). (1) The pathway capacity is defined as all the events, which start from signal recognition at the plasma membrane to binding of the transcription factor to the promoter (thus including the MAPK cascade), while (2) the expression capacity is a general measure of the ability to transcribe and translate proteins. Single cell measurements allowed a detailed analysis of the factors influencing these two capacities.

A simple example for stochasticity in chemical reactions is the evolution of activated species as a function of time in different compartments (cells). Conceptually, averaging over all compartments would lead to a "smooth" response vs. time curve, while a single compartment may manifest a more complex behavior (Fig. 2B). In biological systems, stochastic noise can be caused by unequal distribution of the reaction partners. In the mating pathway, the noise in the expression capacity accounts for roughly 80% of the observed cell-to-cell variation, while the pathway capacity is responsible for only 20%. Therefore, it is not surprising that a

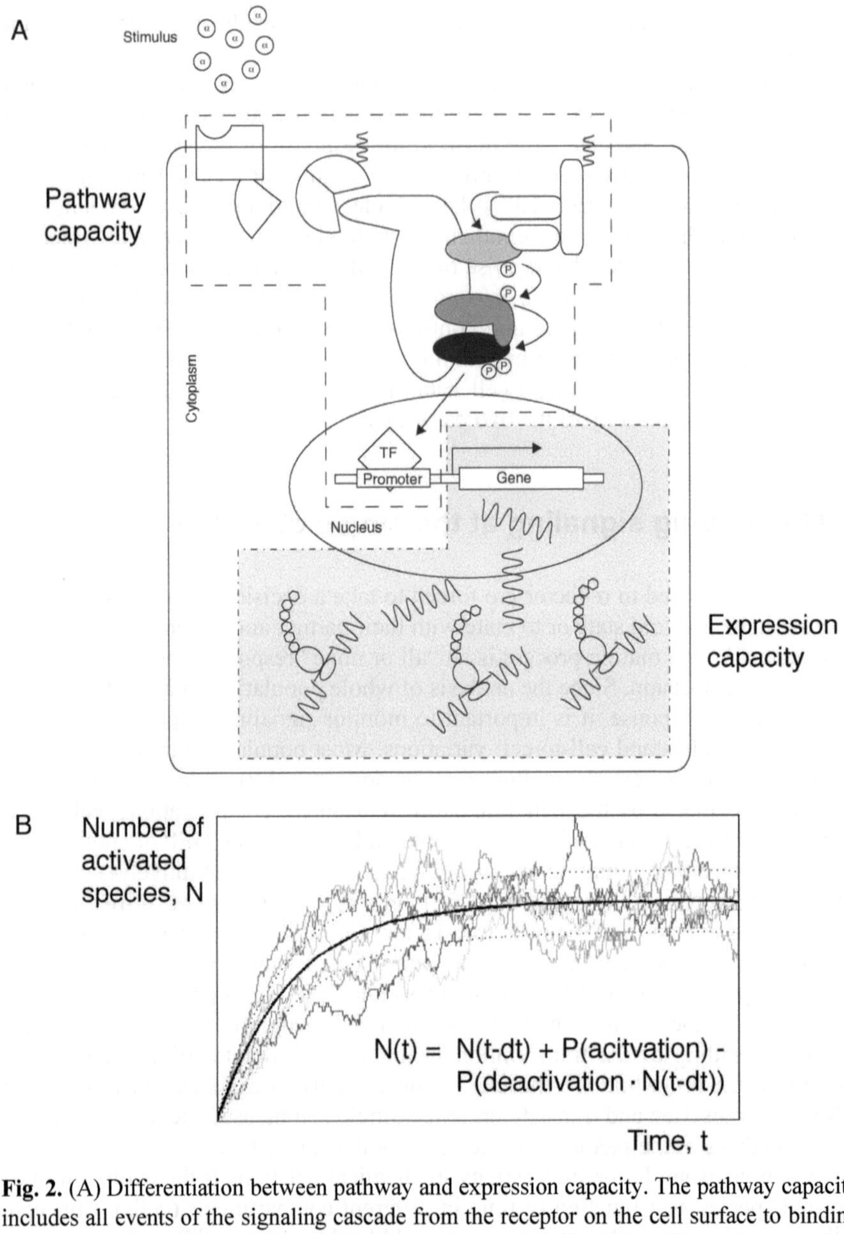

Fig. 2. (A) Differentiation between pathway and expression capacity. The pathway capacity includes all events of the signaling cascade from the receptor on the cell surface to binding of the transcription factor (TF) to DNA. Expression capacity includes all processes required for transcription to protein synthesis. (B) Evolution of the number of active species as function of time representing the stochastic nature of these processes. The solid black line represents the average number of activated species of thousand curves and the dashed line the expected standard deviation.

strong correlation between constitutive and pathway-specific fluorescence emission is observed for a given cell, manifesting the stochastic distribution of the expression machinery. Consistent with this notion, the 5% of cells with the highest and lowest mating specific expression showed a fourfold difference in expression capacity (inferred from the constitutive promoter) at any given α-factor concentration.

In contrast, the variation in pathway capacity is dependent on the α-factor concentration. The noise in pathway capacity will only equal the transcription noise if lower pheromone concentrations are used, while higher concentrations lead to a more uniform reaction. This stimulus-dependant fine-tuning of the signaling seems to rely more on Fus3p (*kss1Δ* cells) while Kss1p (*fus3Δ* cells) leads to larger variation in pathway output (Colman-Lerner et al. 2005). However, the underlying mechanism for this observation needs to be investigated.

3 Regulation of mating signaling

3.1 Pathway activation

The key control responsible for pheromone-induced MAPK signaling involves the activation of the MAPKKK Ste11p. Two interdependent steps are necessary: (1) the MAPKKK needs to be activated (phosphorylated) to be catalytically active and (2) the activated MAPKKK has to physically interact with the downstream MAPKK (Fig. 3A). In some cases, an auxiliary molecule termed scaffold facilitates the second step.

3.1.1 Activation of Ste11p

Ste11p consists of an N-terminal regulatory domain followed by the catalytic kinase domain. Ste11p needs to be activated by Ste20p (a p21-activated like kinase (PAK)), which in turn is activated by Cdc42p-GTP. Ste20p is able to phosphorylate Ste11p on three sites (Drogen et al. 2000), possibly independent of a stimulus. Abolishing these sites by a triple alanine–mutation (Ste11p-Ala3) renders the pathway inactive, while a triple phospho-mimic aspartic acid-mutation leads to a hyperactive allele (Ste11p-Asp3), sufficient to activate the pathway upon overexpression. However, expression of Ste11p-Asp3 or Ste11p-4 (another hyperactive-mutant) from the endogenous promoter only results in a weak constitutive response. While both activated *ste11*-alleles bypass the need for Ste20p (Lamson et al. 2006), they still require α-factor to fully activate the pathway. Thus, a second stimulus-dependent signal is necessary for full activation of the MAPK-cascade *in vivo.*

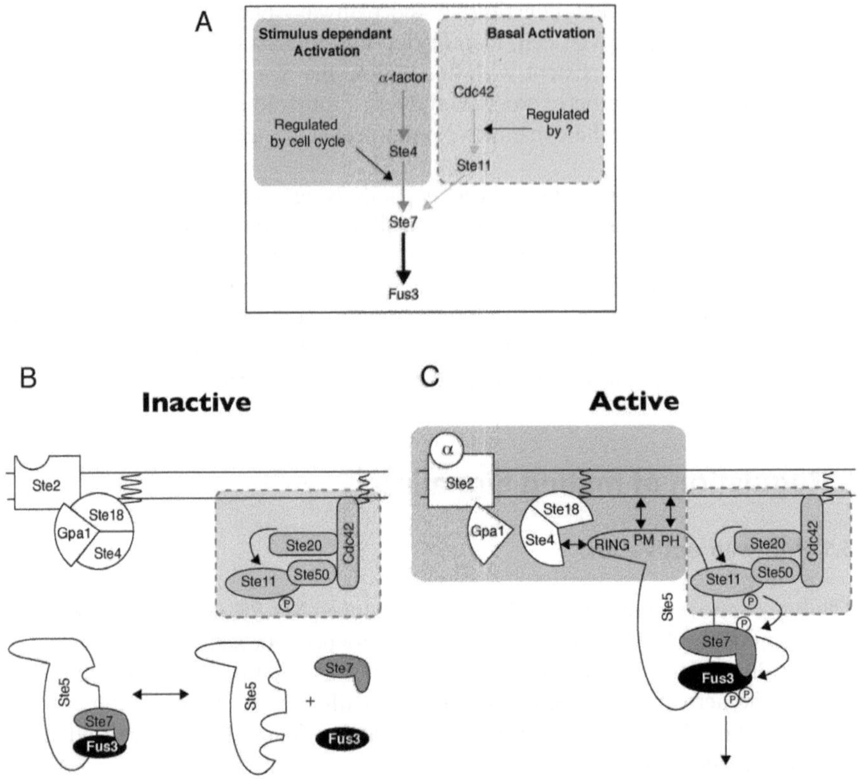

Fig. 3. (A) Schematic drawing of basal (right box) and stimulus-dependent activation (left box) of the mating pathway. (B) Inactive state of the cascade: The receptor is unoccupied, and the heterotrimeric G-protein is assembled. Ste50p and Cdc42p stabilize and activate Ste20p, which phosphorylates and thereby "pre-activates" Ste11p at the plasma membrane, while Ste5p, Ste7p and Fus3p are in the cytoplasm. (C) Activated state of the cascade. Upon α-factor binding, the β/γ subunit of the G-protein dissociates from Gpa1p, allowing Ste4p to interact with the RING domain of Ste5p. This regulated interaction together with two constitutive membrane-binding domains (PM and PH) recruit and stabilize Ste5p at the plasma membrane, and allows its interaction with Ste11p.

3.1.2 Pathway activation upon α-factor treatment

Available evidence suggests that this second activation signal may be membrane recruitment of the scaffold protein Ste5p, which positions phosphorylated Ste11p in close proximity to Ste7p and thereby promotes local activation of Ste7p (Fig. 3B, Fig. 3C). Work on Ste5p demonstrated that membrane recruitment is stimulus regulated and depends on three distinct docking/interaction domains (PM (plasma membrane), RING (really interesting new gene) and PH (pleckstrin homology) domains). The RING domain interacts with the β/γ-heterodimer, which dissociates from the trimeric G-protein upon receptor activation thereby transmitting the signal. However, the RING-domain alone is not sufficient for membrane-localization (Inouye et al. 1997). An amphipathic α-helical stretch in the amino-terminus of Ste5p (PM-domain) leads to an insertion into the cytoplasmic monolayer of the PM (Winters et al. 2005). A Ste5p-mutant protein lacking this stretch fails to localize to the PM and as a consequence such cells are unable to signal. Similarly, the PH-domain specifically binds to phosphatidylinositol 4,5-bisphosphate (PI(4,5)P2) (Garrenton et al. 2006) and a mutant unable to bind this phospholipid is signaling deficient. Therefore, only the additive effect of these three potentially regulatable membrane localization domains allows Ste5p to stably relocate to the plasma membrane, and promote signal transduction. In contrast, the transport of Ste5p to the membrane does not appear to be regulated, because the timescale of translocation is consistent with cytoplasmic diffusion.

The above described two-step activation of mating signaling (Ste11p activation and Ste5p membrane recruitment) ensures a specific signal amplification, prevents spurious activation and signaling during unfavorable times (see below). Additionally, this tightly regulated two-input system can control the main pathway output and is the rate-limiting step of MAPK activation. Mechanistically this could be achieved through regulated membrane localization of a relative small number of Ste5p (compared to the other components of the MAPK signaling pathway), as it is observed in normally growing cells. Ste11p in turn may directly control the pathway output by further increasing its activity in the presence of the mating signal, most likely through phosphorylation of its amino-terminal regulatory domain at the plasma membrane.

3.2 Spatial/temporal regulation

After the regulated assembly of an active signaling complex at the membrane, the pathway output is fine-tuned by several means described below. Most importantly, as the pheromone gradient also contains spatial information, the spatial and temporal regulation of signaling must be modulated by regulated docking interactions between pathway components and kinetic control of the pathway (Remenyi et al. 2006). The latter is also important to alter the signaling strength, which will be discussed in detail below.

3.2.1 Regulation by docking interactions – spatial regulation

For a long time, Ste5p was considered a prototypical scaffold protein with the sole function to position the three kinases of the MAPK module at the right place and time (Elion 2001). Supporting this model, distinct Ste5p sites interact with Ste11p, Ste7p, and Fus3p and mutants lacking individual binding regions were incapable of signal transduction (Choi et al. 1994). Two lines of evidence challenge this passive role: (1) Ste5p needs to homodimerize via its RING domain, and overexpression of an artificially dimerized Ste5p leads to constitutive activation of the pathway (Sette et al. 2000). (2) Careful examination of the Fus3p-binding pocket

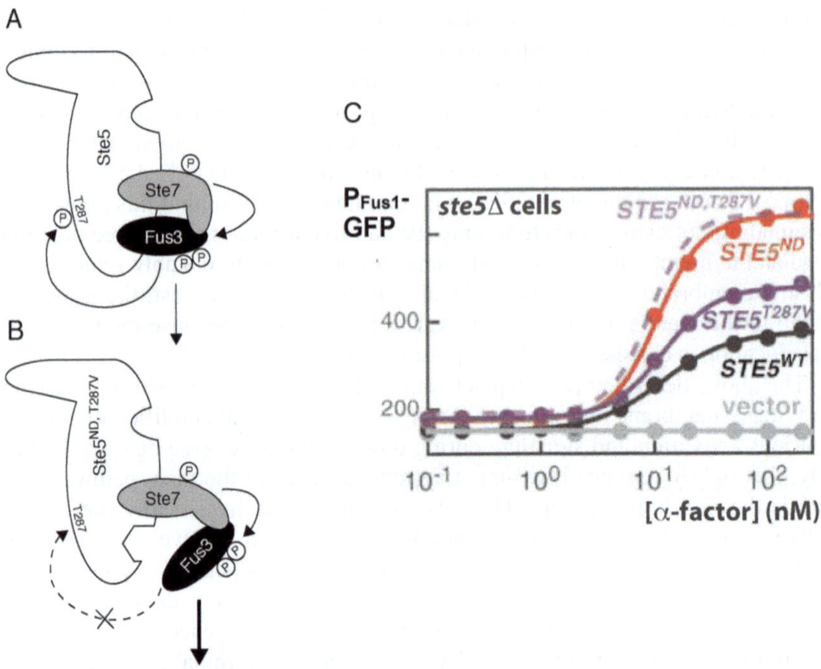

Fig. 4. (A) Negative feedback mechanism of activated Fus3p on the signaling cascade. (B) This feedback can be reduced by using a non-phosphorylatable Ste5p mutant (T287V) or a non-docking mutant (ND) for Fus3p. (C) The amount GFP fluorescence expressed from the FUS1-promotor upon addition of increasing α-factor concentrations provides evidence for a negative feedback effect of Fus3p docking on Ste5p and the resulting phosphorylation of threonine 287. Increased signaling was observed when preventing this negative feedback by a non-docking mutant $STE5^{ND}$ or unphosphorylable mutant $STE5^{T287V}$. Adapted from Bhattacharyya et al. (2006).

demonstrated that a Ste5p-mutant unable to bind Fus3p is not only functional, but the signaling output is even increased (Bhattacharyya et al. 2006).

The latter effect can partially be attributed to an inhibitory phosphorylation of T287, as shown by quantitative analysis of a non-phosphorylatable Ste5p mutant ($STE5^{T287V}$) *in vivo* (Fig. 4C). The non-binding mutation is epistatic to the non-phosphorylatable mutation, suggesting that bound Fus3p phosphorylates T287. Indeed, binding of a Ste5p-peptide to Fus3p slightly distorts the structure such that the activation loop reaches into the kinase domain, giving rise to the autocatalytic production of monophosphorylated Fus3p, which in turn phosphorylates Ste5p on T287 (Bhattacharyya et al. 2006). Taken together, these results demonstrate that Ste5p serves not only as a passive scaffold, but fine tunes pathway capacity via a yet unidentified mechanism. Moreover, as stated above, *fus3*Δ cells also exhibit a higher pathway variation at high α-factor concentrations, but it is unclear whether this is connected to the inhibitory phosphorylation on Ste5p.

3.2.2 Kinetics - temporal regulation

First kinetic parameters were derived using fluorescence recovery after photobleaching (FRAP) measurements. They revealed a highly dynamic membrane association of several signaling components of the mating pathway. Ste5p exhibits a turnover at the plasma membrane of roughly five seconds, while Fus3p exchanges with a half-time of only 0.5 seconds (van Drogen et al. 2001). Moreover, the dissociation rate of activated Fus3p from Ste5p is higher compared to inactive (non-phosphorylated) Fus3p, implying that a single membrane-bound Ste5p scaffold can produce a large number of activated Fus3p molecules, leading to a rapid amplification of the signal by generating doubly-phosphorylated MAPKs. Mechanistically, these phosphorylations can occur in a processive manner, meaning that a single MAPK-MAPKK binding event leads to the doubly-phosphorylated MAPK (Fig. 5A). Alternatively, MAPK-phosphorylation may be distributive and require two separate MAPK-MAPKK binding events, each giving rise to one phosphorylation (Huang and Ferrell 1996). Indeed, distributive phosphorylation is the molecular basis for the sigmoid stimulus/response curve observed with the *Xenopus* p42 MAPK pathway, and therefore follows the kinetics predicted based on theoretical considerations. The mating pathway also shows a sigmoid stimulus/response curve (Fig. 4C), but it remains to be determined whether a distributive mechanism is responsible for this behavior or if the sigmoid stimulus/response kinetic is achieved by other means.

Several mechanisms control the strength and duration of MAPK mating pathway. The signal is dampened by Fus3p-dependent rapid internalization and degradation of pheromone-bound receptors in the vacuole (Hicke and Riezman 1996). Moreover, Fus3p phosphorylates Sst2p, the GTPase-activating protein for the $G\alpha$ subunit, thereby inactivating the heterotrimeric G-protein and therefore downregulates the pathway capacity (Chasse et al. 2006). Additionally, some mechanisms directly inactivate the MAP-kinases. The phosphatase Msg5p physically interacts and dephosphorylates Fus3p (Doi et al. 1994). Cells lacking Msg5p or expressing a mutant form that is unable to bind Fus3p, exhibit an increased signaling output, demonstrating that Msg5p is an integral component of the MAPK-signaling cascade. As the substrate specificity of protein phosphatases is generally rather low, it is possible that Msg5p not only dephosphorylates Fus3p but also contributes to the inactivation of the upstream kinases Ste7p and Ste11p. Active Ste11p and Ste7p are subjected to ubiquitin-dependent degradation and thereby inactivated (Esch and Errede 2002; Wang et al. 2003) (Fig. 5B). Although the importance of these degradation pathways remains to be investigated, available evidence suggests that interfering with the degradation machinery may lead to a prolonged activation of Fus3p.

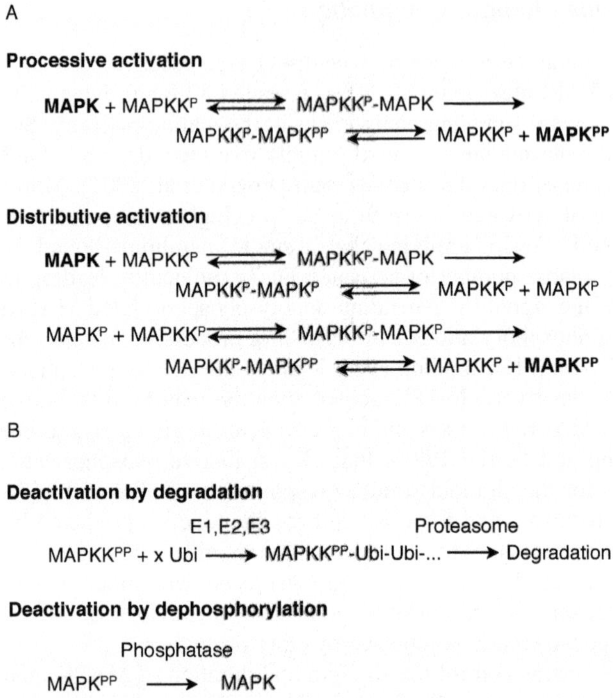

Fig. 5. (A) Scheme for processive or distributive phosphorylation of the MAPK. "P" indicates singly phosphorylated, "PP" doubly phosphorylated MAPK. (B) Scheme for kinase inactivation by degradation (reported for MAPKKK and MAPKK) or dephosphorylation (shown for MAPK) of the doubly-phosphorylated, active species.

4 Regulation of mating signaling by internal and external factors

The potential to respond to pheromone is regulated by internal and external signals including certain cell cycle stages and unfavorable conditions such as stress and poor nutrients. The molecular mechanisms underlying these switches are partially elucidated and probably rely on inhibition of either one of the two activation modes (Ste5p recruitment and Ste11p activation) as this would lead to a block in signal transduction.

Exponentially growing cells are unable to signal during cell cycle stages with high Cdc28p/Cln2p activity, namely between "Start" and the end of "S-phase". Moreover, overexpression of the G1-cyclin Cln2p renders cells insensitive to α-factor (Hartwell 1980; Oehlen and Cross 1994). Indeed, quantitative single cell analysis showed that cell-to-cell variations are cell cycle-dependent (Colman-Lerner et al. 2005) and the overall variability was significantly reduced in synchronized populations and by specific inactivation of the Cdc28p-kinase. Genetic epistasis analysis suggested that the CDK-inhibited step in the mating cascade occurs either on Ste5p, Ste11p, or Ste20p (Wassmann and Ammerer 1997; Oehlen and Cross 1998). Indeed, Ste5p is phosphorylated by Cdc28p/Cln2p at eight sites adjacent to the Ste5p PM-domain (Strickfaden et al. 2007). Mimicking the bulk electrostatic charge mutation to glutamic acid residues inhibits the mating response, while conversely mutating all phosphorylation sites to alanine renders the pathway insensitive to Cln2p overexpression, and abolishes the cell-cycle-dependent regulation of the mating response. These data suggest that the mating pathway senses the cell-cycle state by monitoring Cdc28p-Cln2p activity via multisite phosphorylation of Ste5p, which interferes with its membrane localization and therefore prevents signal transduction (Serber and Ferrell 2007; Strickfaden et al. 2007). It is possible that other signals including general stress or starvation conditions also use Ste5p membrane recruitment as a switch to block signal transduction. Alternatively, these signals may either regulate Ste11p activation by inactivating the Cdc42p/Ste20p branch, or by activating negative regulators such as Msg5p.

5 The Osmotic stress pathway

The HOG pathway allows cells to adapt to a change in the environment. The stimuli triggering the pathway are diverse and can range from a raise in surrounding osmolarity to arsenite poisoning. Adaptation to a changing environment is expected to be a continuous and reversible process and independent of the cell cycle stage. In contrast to the mating pathway, the HOG pathway has different (yet unknown) sensors/receptors and fewer regulatory steps within the cascade are known.

5.1 Pathway activation by stress signals

Unlike the mating pathway, the HOG pathway is activated by multiple stress signals, all leading to activation of the MAPKK Pbs2p. Two distinct branches, named after their trans-membrane proteins Sho1p and Sln1p, have been identified and both appear to function downstream of the unknown osmo-sensor(s) (Fig. 1B). Sho1p and Sln1p activate Ste11p and the two redundant kinases Ssk2p and Ssk22p, respectively. The activated MAPKKK's phosphorylate Pbs2p, which directly interacts and activates the MAPK Hog1p. Abolishing the individual interactions leads to signaling-defective mutants, because Pbs2p serves both as a MAPKK and scaffold for the pathway. The Pbs2p-Hog1p interaction seems to be unregulated, and the critical step for the activation of either osmo-sensing branch is to recruit Pbs2p in physical proximity to the activated, specific MAPKKK's at the plasma membrane.

The Sln1p-branch uses the redundant MAPKKK Ssk2p and Ssk22p, which are activated by a phospho-relay system involving Sln1p, Ypd1p, and Ssk1p. Pathway activation is triggered by a direct physical interaction with Pbs2p, which is mediated by a conserved MAPKKK docking site on Pbs2p (Posas et al. 1996). In contrast, the Sho1p branch is activated by direct, probably simultaneous binding of Ste11p and Pbs2p (Posas and Saito 1997; Zarrinpar et al. 2004). Under normal conditions, Sho1p forms a homooligomer, which dissociates upon osmo-stress in a Hog1p-dependant manner (Hao et al. 2007). The dissociation does not seem to be involved in signal transduction *per se*, but in adaptation to sustained stimulus, as a phospho-mimic mutation diminishes signaling. The effect of this negative feedback loop on HOG signal transduction can be compared to the *SST2*-dependant receptor desensitization in the mating pathway. Activation of Pbs2p is mediated by a direct interaction between a PxxP-binding motif in Pbs2p and the C-terminal SH3 domain in Sho1p (Zarrinpar et al. 2004). *In vivo*, the Sho1p-Pbs2p interaction is transient and regulated by Pbs2p kinase activity, as shown by co-localization of Sho1p with an overexpressed, kinase-inactive Pbs2p-GFP (Reiser et al. 2000). However, more work is needed to elucidate the importance of this regulatory step for Hog1p signaling kinetics. As in the mating pathway, activation of Ste11p requires phosphorylation by the Cdc42p/Ste20p/Cla4p module (Tatebayashi et al. 2006). Consistent with this notion, the non-phosphorylatable Ste11p-Ala3 is unable to activate Fus3p, Kss1p or Hog1p, while Ste11p-Asp3 bypasses the need of the upstream PAK-like kinases and is able to activate all three MAPKs. Interestingly, signaling-specificity is provided by the scaffold Ste5p and Pbs2p, and the two pathways can even be re-wired by engineered fusion proteins of Ste11p with Ste5p and Pbs2p (mating stimulus leads to HOG output and vice versa) (Park et al. 2003).

5.2 Internal regulation of the HOG pathway

Analogous to other MAPKs, Hog1p has to be doubly phosphorylated on T174 and Y176, two residues located in the activation loop, to achieve full catalytic functionality. While mutations in T174 render Hog1p inactive, cells expressing a Hog1p-mutant that cannot be phosphorylated on Y176 (by a mutation of Y176 to phenylalanine) are able to grow on hyperosmotic media and induce a normal transcriptional program upon osmotic shock. Surprisingly, however, these cells are unable to phosphorylate Hog1p substrates (Bell et al. 2001; Bell and Engelberg 2003). The regulatory consequences of this mutation are not entirely clear and await further investigations.

A striking difference between the cellular dynamics of Hog1p and Fus3p is their re-localization to the nucleus. While Fus3p and Kss1p rapidly shuttle between the cytoplasm and the nucleus in the presence or absence of α–factor (van Drogen et al. 2001), Hog1p accumulates in the nucleus upon activation by osmotic stress. However, in both cases physical association with target genes is only detected upon pathway activation (Pokholok et al. 2006; Proft et al. 2006). The relocalization of Hog1p is dependent on double phosphorylation and thus provides a convenient readout to quantify Hog1p-activity in single cell experiments. Phosphorylated Hog1p is retained in the nucleus by tight association with the transcription factors Hot1p, Msn2p and Msn4p (Reiser et al. 1999). The phosphatase Ptp2p counteracts this retention by dephosphorylating Hog1p in the nucleus, and unphosphorylated Hog1p is actively exported into the cytoplasm. In addition, Hog1p activates cytoplasmic phosphatases, thereby, triggering a negative feedback loop to restrict its own activity (Wurgler-Murphy et al. 1997). Other phosphatases may also counteract Pbs2p and Ste11p. In contrast to the mating pathway, degradation of activated components has not been reported to date for the HOG pathway.

6 Specificity/crosstalk between the two pathways

Specificity is used to describe the remarkable transcriptional fidelity of the different MAPK-pathways, despite the fact that they use common components. For example, high salt condition specifically upregulate Hog1p-dependent transcriptional targets, but do not affect targets of the mating pathway. However, when *hog1Δ* or *pbs2Δ* cells are exposed to high osmolarity conditions, these cells ectopically activate the mating pathway (O'Rourke and Herskowitz 1998). This phenomenon has been termed "crosstalk", and generally refers to a situation where a specific input for one pathway also activates the other. The crosstalk of *hog1Δ* cells triggered by high osmolarity conditions depends on an intact mating MAPK-cascade as *ste11Δ*, *ste7Δ*, *fus3Δkss1Δ* or *ste12Δ* double or triple mutant cells do not show transcriptional induction characteristic for the mating response. Further, crosstalk depends on an intact Sho1p-branch of the HOG pathway and on impaired Hog1p-activity as shown by the use of a drug-inhibitable Hog1p-as mutant (Westfall and Thorner 2006). The salt-induced transcriptional program alone is neither required to induce nor sufficient to prevent crosstalk as shown by the afore mentioned Hog1p-Y176F mutation (Bell and Engelberg 2003). Together, these findings imply that Hog1p directly phosphorylates target(s) required to prevent crosstalk. Although several models have been proposed to explain crosstalk, the exact mechanism still remains elusive. An elegant speculation suggests that crosstalk depends on the leakage of activated Ste11p, perhaps by a defect in degrading its activated pool.

7 General principles for yeast MAPK regulation

Conceptually, different modes of MAPK-activation are plausible. The traditional view envisions that inactive MAPKKK and MAPKK are bound to a scaffold molecule and activated upon stimulation of the sensor/receptor. In contrast, we propose that the regulated step is most likely the complex formation between a pre-activated MAPKKK and the MAPKK. This involves two steps directly regulated by the stimuli; membrane-recruitment of the scaffold/MAPKK complex and

MAPKKK activation. Although both parts of this two-step activation scheme can be regulated, membrane-recruitment appears to be rate-limiting.

The signaling complexes are formed and activated at the cell membrane and faithful transduction is achieved by specific docking interactions between the components. The strength of those docking interactions, together with the phosphorylation mode represents a first layer of kinetic pathway regulation. In this respect, docking interactions could be regulated upon kinase activation. As observed for Fus3p, it seems plausible that the turnover of activated MAPK at the membrane represents a mechanism conserved among all MAPK cascades. As discussed, the pathway output is mainly regulated by dephosphorylation and degradation of activated components. It appears that pathway-specific phosphatases play an important role by inactivating MAPK's and possibly other upstream kinases. Most phosphatases involved in MAPK regulation are upregulated at the transcriptional level upon pathway activation (Doi et al. 1994; Jacoby et al. 1997; Gasch et al. 2000; Roberts et al. 2000), which increases the rate of dephosphorylation over time and thereby strengthens negative feedback regulation. On a shorter timescale, degradation of activated components may be more important, however, the physiological role of these mechanisms remains to be investigated. It will be important to determine whether these mechanisms also operate in other MAPK pathways in yeast, and whether they may have been conserved during evolution.

The fundamental physiological difference of the mating and stress MAPK-pathways is reflected by the complex activation and regulation mode of the mating pathway. Mating is a potentially hazardous process, because it alters the cell fate. Therefore, cells must be able to prevent activation of the mating pathway during unfavorable conditions. In contrast, the stress pathway needs to be activated under all conditions to prevent lethal damage. We expect that the sophisticated tasks of the scaffold protein Ste5p (pathway activation, regulation by extracellular signals, regulated docking interactions) may have evolved to solve the complex regulation required to orchestrate the mating process with cellular conditions. Contrary to the long held dogma, we emphasize that the Ste5p-Fus3p interaction is not necessary for signal transduction, but serves to fine-tune/downregulate the pathway output.

8 Single cell measurement – future research

Decades of research on MAPK signaling in yeast lead to the understanding how the different pathways are wired and what general mechanisms are used for their regulation. As a next step it will be important to describe these mechanisms on a molecular level, and quantify their contributions for faithful signal transduction and signaling kinetics. Elucidating these mechanisms will require careful biochemical analysis of the different components, and quantitative single-cell measurements with specific reporters that monitor the individual steps of MAP-kinase signaling. Only quantitative data on the cellular level will provide the accuracy necessary to discriminate between multiple activation patterns. Finally, such data will be crucial to experimentally validate the mathematical models that have been proposed to describe MAP-kinase signaling.

Acknowledgments

We thank the QUASI network for fruitful discussions, and Francesc Posas, Gustav Ammerer, and Reinhard Dechant for critical reading of the manuscript. Work in the laboratory of Matthias Peter is supported by QUASI, SPMD, the SystemsX initiative, the ETHZ, and the Swiss National Science Foundation (SNF).

References

Bell M, Capone R, Pashtan I, Levitzki A, Engelberg D (2001) Isolation of hyperactive mutants of the MAPK p38/Hog1 that are independent of MAPK kinase activation. J Biol Chem 276:25351-25358

Bell M, Engelberg D (2003) Phosphorylation of Tyr-176 of the yeast MAPK Hog1/p38 is not vital for Hog1 biological activity. J Biol Chem 278:14603-14606

Bhattacharyya RP, Remenyi A, Good MC, Bashor CJ, Falick AM, Lim WA (2006) The Ste5 scaffold allosterically modulates signaling output of the yeast mating pathway. Science 311:822-826

Chasse SA, Flanary P, Parnell SC, Hao N, Cha JY, Siderovski DP, Dohlman HG (2006) Genome-scale analysis reveals Sst2 as the principal regulator of mating pheromone signaling in the yeast *Saccharomyces cerevisiae*. Eukaryot Cell 5:330-346

Choi KY, Satterberg B, Lyons DM, Elion EA (1994) Ste5 tethers multiple protein kinases in the MAP kinase cascade required for mating in *S. cerevisiae*. Cell 78:499-512

Colman-Lerner A, Gordon A, Serra E, Chin T, Resnekov O, Endy D, Pesce CG, Brent R (2005) Regulated cell-to-cell variation in a cell-fate decision system. Nature 437:699-706

Dohlman HG, Thorner JW (2001) Regulation of G protein-initiated signal transduction in yeast: paradigms and principles. Annu Rev Biochem 70:703-754

Doi K, Gartner A, Ammerer G, Errede B, Shinkawa H, Sugimoto K, Matsumoto K (1994) MSG5, a novel protein phosphatase promotes adaptation to pheromone response in *S. cerevisiae*. Embo J 13:61-70

Drogen F, O'Rourke SM, Stucke VM, Jaquenoud M, Neiman AM, Peter M (2000) Phosphorylation of the MEKK Ste11p by the PAK-like kinase Ste20p is required for MAP kinase signaling in vivo. Curr Biol 10:630-639

Elion EA (2001) The Ste5p scaffold. J Cell Sci 114:3967-3978

Esch RK, Errede B (2002) Pheromone induction promotes Ste11 degradation through a MAPK feedback and ubiquitin-dependent mechanism. Proc Natl Acad Sci USA 99:9160-9165

Garrenton LS, Young SL, Thorner J (2006) Function of the MAPK scaffold protein, Ste5, requires a cryptic PH domain. Genes Dev 20:1946-1958

Gasch AP, Spellman PT, Kao CM, Carmel-Harel O, Eisen MB, Storz G, Botstein D, Brown PO (2000) Genomic expression programs in the response of yeast cells to environmental changes. Mol Biol Cell 11:4241-4257

Hao N, Behar M, Parnell SC, Torres MP, Borchers CH, Elston TC, Dohlman HG (2007) A systems-biology analysis of feedback inhibition in the Sho1 osmotic-stress-response pathway. Curr Biol 17:659-667

Hartwell LH (1980) Mutants of *Saccharomyces cerevisiae* unresponsive to cell division control by polypeptide mating hormone. J Cell Biol 85:811-822

Heinrich M, Köhler T, Mösch HU (2006) Role of the Cdc42-Cla4 interaction in the, pheromone response of *Saccharomyces cerevisiae*. Eukaryot Cell 6:317-327

Hicke L, Riezman H (1996) Ubiquitination of a yeast plasma membrane receptor signals its ligand-stimulated endocytosis. Cell 84:277-287

Hohmann S (2002) Osmotic stress signaling and osmoadaptation in yeasts. Microbiol Mol Biol Rev 66:300-372

Huang CY, Ferrell JE Jr (1996) Ultrasensitivity in the mitogen-activated protein kinase cascade. Proc Natl Acad Sci USA 93:10078-10083

Inouye C, Dhillon N, Thorner J (1997) Ste5 RING-H2 domain: role in Ste4-promoted oligomerization for yeast pheromone signaling. Science 278:103-106

Jacoby T, Flanagan H, Faykin A, Seto AG, Mattison C, Ota I (1997) Two protein-tyrosine phosphatases inactivate the osmotic stress response pathway in yeast by targeting the mitogen-activated protein kinase, Hog1. J Biol Chem 272:17749-17755

Lamson RE, Takahashi S, Winters MJ, Pryciak PM (2006) Dual role for membrane localization in yeast MAP kinase cascade activation and its contribution to signaling fidelity. Curr Biol 16:618-623

O'Rourke SM, Herskowitz I (1998) The Hog1 MAPK prevents cross talk between the HOG and pheromone response MAPK pathways in *Saccharomyces cerevisiae*. Genes Dev 12:2874-2886

Oehlen LJ, Cross FR (1994) G1 cyclins CLN1 and CLN2 repress the mating factor response pathway at Start in the yeast cell cycle. Genes Dev 8:1058-1070

Oehlen LJ, Cross FR (1998) Potential regulation of Ste20 function by the Cln1-Cdc28 and Cln2-Cdc28 cyclin-dependent protein kinases. J Biol Chem 273:25089-25097

Park SH, Zarrinpar A, Lim WA (2003) Rewiring MAP kinase pathways using alternative scaffold assembly mechanisms. Science 299:1061-1064

Pokholok DK, Zeitlinger J, Hannett NM, Reynolds DB, Young RA (2006) Activated signal transduction kinases frequently occupy target genes. Science 313:533-536

Posas F, Saito H (1997) Osmotic activation of the HOG MAPK pathway via Ste11p MAPKKK: scaffold role of Pbs2p MAPKK. Science 276:1702-1705

Posas F, Wurgler-Murphy SM, Maeda T, Witten EA, Thai TC, Saito H (1996) Yeast HOG1 MAP kinase cascade is regulated by a multistep phosphorelay mechanism in the SLN1-YPD1-SSK1 "two-component" osmosensor. Cell 86:865-875

Proft M, Mas G, de Nadal E, Vendrell A, Noriega N, Struhl K, Posas F (2006) The stress-activated Hog1 kinase is a selective transcriptional elongation factor for genes responding to osmotic stress. Mol Cell 23:241-250

Reiser V, Ruis H, Ammerer G (1999) Kinase activity-dependent nuclear export opposes stress-induced nuclear accumulation and retention of Hog1 mitogen-activated protein kinase in the budding yeast *Saccharomyces cerevisiae*. Mol Biol Cell 10:1147-1161

Reiser V, Salah SM, Ammerer G (2000) Polarized localization of yeast Pbs2 depends on osmostress, the membrane protein Sho1 and Cdc42. Nat Cell Biol 2:620-627

Remenyi A, Good MC, Lim WA (2006) Docking interactions in protein kinase and phosphatase networks. Curr Opin Struct Biol 16:676-685

Roberts CJ, Nelson B, Marton MJ, Stoughton R, Meyer MR, Bennett HA, He YD, Dai H, Walker WL, Hughes TR, Tyers M, Boone C, Friend SH (2000) Signaling and circuitry of multiple MAPK pathways revealed by a matrix of global gene expression profiles. Science 287:873-880

Serber Z, Ferrell JE Jr (2007) Tuning bulk electrostatics to regulate protein function. Cell 128:441-444

Sette C, Inouye CJ, Stroschein SL, Iaquinta PJ, Thorner J (2000) Mutational analysis suggests that activation of the yeast pheromone response mitogen-activated protein kinase pathway involves conformational changes in the Ste5 scaffold protein. Mol Biol Cell 11:4033-4049

Strickfaden SC, Winters MJ, Ben-Ari G, Lamson RE, Tyers M, Pryciak PM (2007) A mechanism for cell-cycle regulation of MAP kinase signaling in a yeast differentiation pathway. Cell 128:519-531

Tatebayashi K, Yamamoto K, Tanaka K, Tomida T, Maruoka T, Kasukawa E, Saito H (2006) Adaptor functions of Cdc42, Ste50, and Sho1 in the yeast osmoregulatory HOG MAPK pathway. Embo J 25:3033-3044

van Drogen F, Stucke VM, Jorritsma G, Peter M (2001) MAP kinase dynamics in response to pheromones in budding yeast. Nat Cell Biol 3:1051-1059

Wang Y, Ge Q, Houston D, Thorner J, Errede B, Dohlman HG (2003) Regulation of Ste7 ubiquitination by Ste11 phosphorylation and the Skp1-Cullin-F-box complex. J Biol Chem 278:22284-22289

Wassmann K, Ammerer G (1997) Overexpression of the G1-cyclin gene CLN2 represses the mating pathway in *Saccharomyces cerevisiae* at the level of the MEKK Ste11. J Biol Chem 272:13180-13188

Westfall PJ, Thorner J (2006) Analysis of mitogen-activated protein kinase signaling specificity in response to hyperosmotic stress: use of an analog-sensitive HOG1 allele. Eukaryot Cell 5:1215-1228

Winters MJ, Lamson RE, Nakanishi H, Neiman AM, Pryciak PM (2005) A membrane binding domain in the ste5 scaffold synergizes with gbetagamma binding to control localization and signaling in pheromone response. Mol Cell 20:21-32

Wurgler-Murphy SM, Maeda T, Witten EA, Saito H (1997) Regulation of the *Saccharomyces cerevisiae* HOG1 mitogen-activated protein kinase by the PTP2 and PTP3 protein tyrosine phosphatases. Mol Cell Biol 17:1289-1297

Zarrinpar A, Bhattacharyya RP, Nittler MP, Lim WA (2004) Sho1 and Pbs2 act as coscaffolds linking components in the yeast high osmolarity MAP kinase pathway. Mol Cell 14:825-832

Pelet, Serge
Institute of Biochemistry, ETH Zürich, CH-8093 Zürich, Switzerland

Peter, Matthias
Institute of Biochemistry, ETH Zürich, CH-8093 Zürich, Switzerland
matthias.peter@bc.biol.ethz.ch

Rudolf, Fabian
Competence Center for Systems Physiology and Metabolic Disease (CC-SPMD)
Institute of Biochemistry, ETH Zürich, CH-8093 Zürich, Switzerland

Modeling the dynamics of stress activated protein kinases (SAPK) in cellular stress response

Edda Klipp and Jörg Schaber

Abstract

Experimental research has revealed components and mechanisms of cellular stress sensing, signaling, and adaptation. In addition, mathematical modeling has proven to foster the understanding of basic principles of signal transduction and signal processing. In the following, we will give an outline of the modeling process and introduce basic concepts for dynamic model analysis. We briefly review basic dynamic properties of stress activated protein (SAP) or mitogen activated protein (MAP) kinase (K) cascades that demonstrate why this type of signaling pathways is so abundantly used in cellular information processing and response.

1 Introduction

In the past few years, a series of modeling approaches have been developed and adopted to support the understanding of the complex behavior of signaling networks in general and of MAPK cascades in particular. The concepts range from very abstract models that elucidate some key properties of signaling pathways (e.g. Heinrich et al. 2002; Papin and Palsson 2004), to very detailed models that precisely monitor the dynamics of specific regulatory events (e.g. Vaseghi et al. 2001; Schoeberl et al. 2002; Swameye et al. 2003; Yi et al. 2003). A systematic overview of structural properties and dynamic features of signaling pathway models are given in Tyson et al. (2003) and Papin et al. (2005). A series of studies conducted by Ferrell and co-workers (Ferrell 1996, 1998, 2002; Ferrell and Bhatt 1997) investigated so-called ultrasensitivity, that is, the ability of the cascade to convert a graded input stimulus into a switch-like all-or-none output signal. Other theoretical studies investigated further striking features of MAPK cascade dynamics (Bhalla and Iyengar 1999).

The complexity of biochemical networks is far from being resolved experimentally. Nevertheless, there is need to understand their behavior in a rational way, which is often hard to achieve by intuition. Establishing models of such networks supports the integration of experimental knowledge into a consistent picture, the formulation of hypotheses, and cognitions in a precise language. It serves to test, support, or falsify hypotheses about the underlying biological mechanism. Model-

Topics in Current Genetics, Vol. 20
F. Posas, A. R. Nebreda (Eds.): Stress-Activated Protein Kinases
DOI 10.1007/4735_2007_0241 / Published online: 1 August 2007

ing may integrate different parts of the whole and, thereby, allow analysis of properties that only emerge upon the interaction of elements in a comprehensive network. A sound model can produce predictions that can be experimentally tested and it can simulate processes that are experimentally hidden.

In the following, we will give an outline of the modeling process and introduce some basic concepts for model analysis that will be exemplified with a typical model of a MAPK cascade.

2 Mathematical modeling in systems biology

2.1 Purpose of modeling

The development of a model serves the abstract and condensed representation of facts or hypotheses to allow for the analysis of their relations and to gain understanding about their internal organization and their communication with the environment.

Although data in biological research is currently exploding in number, such data is useless without sufficient interpretation. A computational model can, on the one hand, serve the data interpretation; on the other hand, it can point to biological aspects that are still not sufficiently experimentally resolved. Within the field of Systems Biology, the view has been established that experimental research and model development should go hand in hand in an iterative manner including formulation of an initial model, hypothesis generation, experimental testing of hypotheses, model-based experimental design, model refinement upon new data, etc.

The iterative modeling and experimentation process is hard to follow in publications (Schuster et al. 2006), since they often only represent the final results. Model improvement with time and with accumulating experimental information is documented, for example, for yeast cell cycle (Novak et al. 1999; Chen et al. 2000, 2004; and others) and for signaling pathways (Bhalla and Iyengar 1999, 2001; Bhalla 2002, 2004).

2.2 Model development in five steps

2.2.1 Defining the scope

Usually, an experimental observation inspires the formulation of a hypothesis as the first step. In the second step, we define what questions the model is supposed to answer; this is referred to as, the *scope* of the model. The scope determines what components and processes the model will take into account or omit and it defines the system's boundaries. The initial model is usually formulated as a word model. The word model itself is also subjected to a process of refinement and sophistication in the course of model development. A graphical representation of the model structure, such as a diagram, is also helpful.

2.2.2 Verification

Subsequently, the word model is translated into a mathematical model (for an overview of mathematical techniques, see below). To assure that our model, in principle, is able to answer our initial question, we must *verify* that our model can achieve this independently of choice of specific parameter values, that is, in a qualitative way. When there is no mathematical theorem available that tells us something about the general properties of our system, verification of the proposed model behavior is generally obtained by playing around with the model structure and its parameters, checking whether it behaves in the way we want or not. Verification of the model structure is an important step in the process of model development because it can save much time and effort later on. When the model is not able to fit observed data, this might be due to a general problem of the model structure. Having checked this in advance, we can avoid validating a model in vain. Generally, it is also desirable to learn more about the general properties of the model; for example, steady states and bifurcation points. When we study metabolic systems, we can apply mathematical tools like metabolic control analysis (see Section 2.4.2) to analyze the system.

2.2.3 Validation

Having verified that the model can principally reproduce our expectations, we can *validate* that the model can also reproduce our observations in a quantitative manner. This is achieved by adjusting the model parameters such that the components of the model match observational data. It is important to gain further support for our model by testing whether it is also able to reproduce independent data without changing the fitted parameters. Independent in this sense means that the data was neither used to fit the parameters nor to develop our model. We need a training data set and test data set. The test data generally describe the same phenomena but under slightly different conditions. It is a prerequisite for a sound model validation, that is, the model should be able to reproduce observed data under different conditions but with the same parameters that were used to reproduce the training data set. This is supposed to reflect the fact that our model accurately describes the intrinsic structure of the studied system and, like nature, is able to adequately adjust its reaction to a changing environment/input without changing internal structure and interactions.

2.2.4 Sensitivity analysis

It is important to know the limits of applicability of a model to determine to what extent possible predictions and conclusions hold. Moreover, it is important to know what parameters are sensitive, and specifically, what changes have a substantial impact on the system's behavior, and thus have to be determined with great accuracy. To achieve this, we must conduct a *sensitivity* analysis. Usually, this is achieved by changing one parameter value at a time and looking at the resulting change of a specific output variable. A classical measure of sensitivity is the relative sensitivity S that is defined as

$$S = \frac{\Delta O}{O} \cdot \frac{p}{\Delta p}, \tag{1}$$

where $\Delta O/O$ is the relative change of some output of interest and $\Delta p/p$ is the relative parameter change, compared to the initial state of parameter, respectively. S is easy to interpret, as $S = 1$ means that a certain percentage change of a parameter yields the same percentage change of the considered output. Usually, when $|S| \leq 1$, p is considered as non-sensitive. When $|S| >> 1$, p is considered as sensitive. The range in which p is changed depends on the uncertainty with which p is determined. This can be the measurement error or some other knowledge about the range in which p can vary.

Sensitivity of parameters contains important information about our system. It shows where small measurement errors can have drastic consequences for the system behavior but also where additional research or measurements might be adequate. Sensitive parameters may indicate interesting drug targets. Sensitivity analysis refers to the robustness and resilience of the system.

It is also important to explore the sensitivity to changes in the input stimuli, since biological systems are subjected to varying environmental conditions. Moreover, a structural sensitivity analysis, that is to say, changing model formulas, provides valuable information about what features of the model that are necessary to exhibit a certain behavior and what parts can be omitted or simplified.

The sensitivity analysis relates to and complements the two preceding steps: verification and validation. Verification analyzes the theoretical properties of our model system, that is to say, the qualitative structure of the state space. Validation determines a concrete state of the system that reflects observed biological phenomena, that is, where our system is quantitatively located in the theoretical state space. Finally, sensitivity analysis provides a quantitative picture of the state space around our system.

2.2.5 Predictions

We can then use the model to explore more systematically regions of the state space that are of particular interest and make predictions. The model ideally should be able to predict future experiments. When the model correctly predicts the experiments we gain confidence in the model and also in the original hypothesis. Moreover, the model can be used to design future experiments. In combination with the sensitivity analysis, we can determine where additional measurements give us most information about the system.

If the model does not correctly predict the experiments, it has to be checked whether the experiments still comply with the original hypothesis. If they do, we have to modify the model; otherwise, we have to modify the hypothesis.

2.3 Mathematical modeling of biochemical reaction networks

2.3.1 Differential equations

In the following, we will consider the concentration change of compounds involved in a signaling process over time. This is frequently done by describing the temporal changes with ordinary differential equations (ODEs). For every individual compound, it holds that its concentration $S_j(t)$ is increased by the reactions producing it and decreased by the reactions consuming it:

$$\frac{dS_j(t)}{dt} = \sum_{producing} v_i - \sum_{consuming} v_k \,. \tag{2}$$

The dynamics of the complete network $\boldsymbol{S}(t) = (S_1(t), S_2(t), \ldots, S_n(t))^T$ is expressed by the balance equations

$$\frac{d\boldsymbol{S}(t)}{dt} = \boldsymbol{N}\boldsymbol{v}(\boldsymbol{S}(t), \boldsymbol{p}), \tag{3}$$

where $\boldsymbol{S}$, $\boldsymbol{v}$, and $\boldsymbol{p}$ denote the vectors of the concentrations, reaction rates, and parameters of the system and t is the time. $\boldsymbol{N}$ is the stoichiometric matrix that contains the information, which compounds are produced and consumed in the individual reactions (see Section 2.4.1). The reaction rates are dependent on the concentrations of their substrates and possible modifiers, but also on thermodynamic conditions and on the specific molecular mechanisms of the reaction. Typical expressions for the reaction rates are the so-called mass action rate law

$$v_j(S_i) = k_j \cdot S_i \tag{4}$$

or the Michaelis-Menten rate law

$$v_j(S_i) = \frac{V_{max} S_i}{K_M + S_i} \tag{5}$$

or the Hill kinetic

$$v_j(S_i) = \frac{V_{max} S_i^n}{K_M^n + S_i^n} \,. \tag{6}$$

In these equations, k_j is a rate constant, V_{max} the maximal rate obtained for infinitive substrate concentrations, K_M is the Michaelis-Menten or half-saturation constant, the concentration at which half of V_{max} is reached, and n is the so-called Hill coefficient that determines the sigmoidal shape of $v_j(S_i)$. The mass action law implies a linear dependence of rate on substrate concentration, while hyperbolic Michaelis-Menten kinetics and sigmoidal Hill kinetics show saturation. Note that more elaborated kinetic mechanisms are described, especially for more substrates and for reversible reactions.

In modeling of metabolic reactions, the rate is proportional to the concentration of the catalyzing enzyme, which is frequently considered as constant and incorporated in the rate constant or maximal velocity, such that $k_j = \tilde{k}_j \cdot E_j$ or $V_{max} = \tilde{k}_j \cdot E_j$. For modeling of signaling pathways, it is more appropriate to consider the catalyzing reagent specifically, since it is another pathway component with temporally changing concentration.

Cellular signaling pathways have to cross several boundaries such as the cell membrane, the nuclear envelope, or the mitochondrial membranes. This may make it necessary to include different compartments into the model. Moving between compartments has different effects in discrete or continuous settings. If one molecule leaves a compartment, then one molecule will arrive in the neighboring compartment. If one μm of a substance leaves a compartment, the concentration change in the neighboring department depends on their relative volumes.

Assume that we consider processes involving two compartments *cx* and *cy*, such as cytosol and nucleus, including transport between both compartments, as well as volume changes of the compartments. Then the temporal evolution of a compound *i* in compartment *cx* reads:

$$\frac{dS_{i,cx}}{dt} = \sum_{j=1}^{r} n_{ij} v_j + \sum_{j=r+1}^{r+t_1} n_{ij} w_j \frac{V_{cy}}{V_{cx}} - \sum_{j=r+t_1+1}^{r+t_1+t_2} n_{ij} w_j - \frac{S_{i,cx}}{V_{cx}} \frac{dV_{cx}}{dt}, \tag{7}$$

(and analogously for compartment *cy*) where r is the number of biochemical reactions with the rates v_j, and t_1 and t_2 are the number of transport steps from nucleus (*cy*) to cytosol (*cx*) and *vice versa* with the rates w_j, respectively. The quantities n_{ij} denote the stoichiometric coefficients of the compounds in the respective reactions or transport steps. The change of the volumes can depend on various processes, which are not in the scope of this study.

2.3.2 Other modeling techniques

Depending on the available experimental information, the purpose of modeling, the experience and preference of the modeler signaling pathways can be described with different techniques. In general, all approaches rely on a description of the network structure with a graph representing as edges for the interaction (activation, inhibition, complex formation) between the nodes, to be exact, the different signal molecules. Boolean networks or Petri nets describe the states of individual nodes in a discrete fashion and these states are updated along a discretized time axis according to the rules assigned to the edges. In their basic version, Boolean networks allow only for two states (1 or 0, i.e., active or not active). Petri nets assign individual tokens to the places (i.e. nodes). More sophisticated approaches tend to consider more different states.

The dynamics on a continuous time scale can also be simulated in a stochastic manner, for example, with one of Gillespie's methods by assuming discrete state values, such as, molecule numbers (Gillespie 1977).

2.3.3 Parameter estimation

The model shall explain observed data. This is usually achieved by adjusting the parameters such that the model simulations match the observed data. This can be done by hand, however, this task should be left to computer programs, because of its enormous complexity and pitfalls. When working with a parameter estimation program, we first have to tell the program what we mean by 'matching the observed data' in mathematical terms, that is, provide it with a *goal* or *objective function*. A common approach is to minimize the sum of squared residuals (SSR) as a function of the parameter vector $\boldsymbol{p}$

$$\arg\min SSR(\boldsymbol{p}) = \arg\min \sum_{i=1}^{n} r_i^2(\boldsymbol{p}), \qquad (8)$$

where the residuals $r_i(\boldsymbol{p}) = y_i - f_i(\boldsymbol{p}, x_i)$ are the distances between the observed data y_i and the simulation results $f_i(\boldsymbol{p}, x_i)$ at the independent variable x_i. This approach is so popular because it comes with an important theoretical result. In case, the residuals $r_i(\boldsymbol{p})$ are identically independently and normally distributed, then, for large number of data points *n*, the parameter vector $\hat{\boldsymbol{p}}$ that minimizes $SSR(\boldsymbol{p})$ is also normally distributed with an approximate covariance matrix $s^2 C^{-1}(\hat{\mathrm{p}})$ where

$$C(\hat{\boldsymbol{p}}) = \left(\frac{\partial f(\hat{\boldsymbol{p}}, x)}{\partial \hat{\boldsymbol{p}}}\right)^T \left(\frac{\partial f(\hat{\boldsymbol{p}}, x)}{\partial \hat{\boldsymbol{p}}}\right) \text{ and } s^2 = SSR(\hat{\boldsymbol{p}})/(n-p). \qquad (9)$$

s^2 is the estimated variance of the residuals, *p* is the number of parameters and superscript T indicates the transposed matrix. From *C* approximate confidence intervals and regions for $\hat{\boldsymbol{p}}$ can be computed (Seber 2003) giving an estimate of the quality of the estimator. There also exist other methods for parameter estimation, like, for example, Bayesian Estimation (Gelman et al. 1995) and Robust Estimation (Rousseuw and Leroy 1987).

It has to be noted that despite the large theoretical body on least squares estimates most of the theorems, like the one indicated in (9), are only valid approximately or asymptotically in the sense that they become correct as *n* approaches infinity. More importantly, all the theory assumes is that the model at hand is the correct one and the estimated parameter is close to the 'true' parameter that actually produced the data. Despite these obvious shortcomings, many concepts of the asymptotic theory are very useful for parameter estimation, even though they should be handled with care.

There is a wealth of algorithms available that help to optimize the goal function. They can be classified into local and global optimization algorithms. Local optimization algorithms usually employ the derivative of the objective function (Newton, Steepest Descent, Conjugate Gradient) to find an optimum near the starting point. Global optimization algorithms (Random Walk, Simulated Annealing, Evolutionary Algorithms) seek to find the global optimum and differ in the way they sample the parameter space and accept new states. In Systems Biology, optimization landscapes are mostly hilly and, therefore, global optimization algorithms are recommendable. A recent study (Moles et al. 2003) compared several global optimization algorithms for fitting a nonlinear biochemical model with 36 parameters. In this study, the algorithm Stochastic Ranking Evolutionary Strategy (SRES) (Runarsson and Yao 2000) performed best. The authors of this chapter have also good experience with the Differential Evolution algorithm (Price and Storn 1997; Storn and Price 1997). Recently, a hybrid approach combining global and local optimization algorithms applied to biochemical systems has been presented that seems promising (Rodriguez-Fernandez et al. 2006).

2.4 Analysis of models

The model can be analyzed in various ways to test whether its behavior really reflects the aspects that we wanted to represent and to deduce predictions based on a presumably appropriate description. We will use the classical form of a MAPK cascade (Fig. 1) to exemplify the following analysis techniques.

2.4.1 Stoichiometric analysis

Purely based on the stoichiometry, that is, on the wiring, is the analysis of the stoichiometric matrix $\boldsymbol{N}$ (equation 3). The linear dependence of rows of the stoichiometric matrix points to moiety conservation in the system, that is to say, it reveals which compounds or moieties are neither produced nor degraded by the network in total, such as the sum of differently modified forms of a protein. In mathematical terms, one has to find a regular matrix $\boldsymbol{G}$ such that $\boldsymbol{G}\cdot\boldsymbol{N}=0$. Then the formula $\boldsymbol{G}\cdot\boldsymbol{S}=const.$ expresses the conservation relations. The linear dependence of columns of $\boldsymbol{N}$ ($\boldsymbol{N}\cdot\boldsymbol{K}=0$ with regular matrix $\boldsymbol{K}$) reveals the dependence of fluxes in steady state; that is to say, steady state fluxes are linear combinations of the columns of matrix $\boldsymbol{K}$. For example, in an unbranched pathway, all fluxes must be the same in case of steady state. Since we focus on transient instead of steady state, we will not consider matrix $\boldsymbol{K}$ further. For the MAPK cascade in Figure 1 the stoichiometric matrix reads:

$$\boldsymbol{N}=\begin{pmatrix}-1 & 1 & 0 & 0 & 0 & 0 & 0 & 0 & 0 & 0\\ 1 & -1 & 0 & 0 & 0 & 0 & 0 & 0 & 0 & 0\\ 0 & 0 & -1 & 1 & 0 & 0 & 0 & 0 & 0 & 0\\ 0 & 0 & 1 & -1 & -1 & 1 & 0 & 0 & 0 & 0\\ 0 & 0 & 0 & 0 & 1 & -1 & 0 & 0 & 0 & 0\\ 0 & 0 & 0 & 0 & 0 & 0 & -1 & 1 & 0 & 0\\ 0 & 0 & 0 & 0 & 0 & 0 & 1 & -1 & -1 & 1\\ 0 & 0 & 0 & 0 & 0 & 0 & 0 & 0 & 1 & -1\end{pmatrix}\begin{matrix}MAPKKK\\ MAPKKK-P\\ MAPKK\\ MAPKK-P\\ MAPKK-PP\\ MAPK\\ MAPK-P\\ MAPK-PP\end{matrix}. \quad (10)$$

$$\begin{matrix}v_1 & v_2 & v_3 & v_4 & v_5 & v_6 & v_7 & v_8 & v_9 & v_{10}\end{matrix}$$

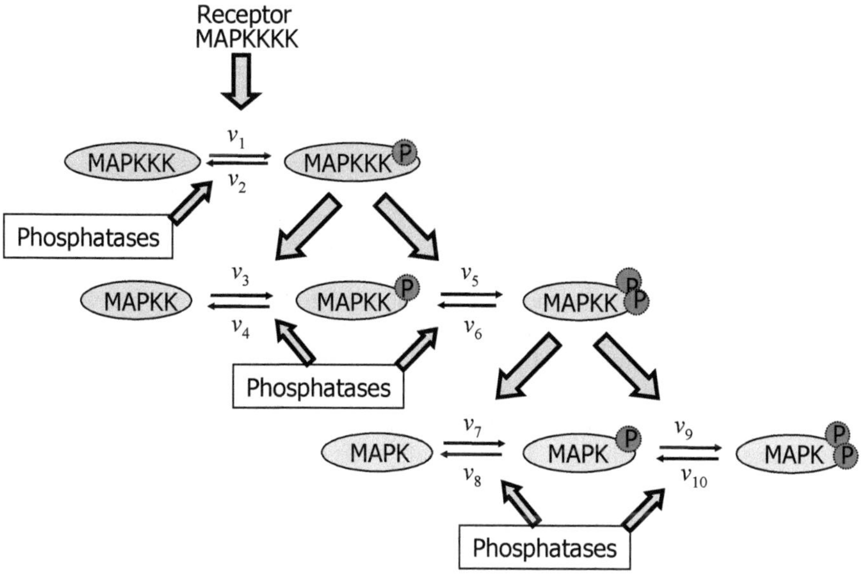

Fig. 1. Schematic representation of the MAPK cascade. Odd numbered reactions are catalyzed by upstream kinases. Even numbered reactions are catalyzed by phosphatases.

The conservation matrix can be calculated as

$$\boldsymbol{G} = \begin{pmatrix} 1 & 1 & 0 & 0 & 0 & 0 & 0 & 0 \\ 0 & 0 & 1 & 1 & 1 & 0 & 0 & 0 \\ 0 & 0 & 0 & 0 & 0 & 1 & 1 & 1 \end{pmatrix}. \tag{11}$$

This can be interpreted as a conservation of the protein concentration at each level within the considered period; that is,

$MAPKKK(t) + MAPKKK\_P(t) = MAPKKK_{tot} = const.$,

$MAPKK(t) + MAPKK\_P(t) + MAPKK\_PP(t) = MAPKK_{tot} = const.$, and

$MAPK(t) + MAPK\_P(t) + MAPK\_PP(t) = MAPK_{tot} = const.$

This finding is in agreement with the assumption that protein synthesis and degradation are slow processes compared to the protein modification upon signal activation.

2.4.2 Metabolic control analysis

Metabolic control analysis (MCA) (Kacser and Burns 1973; Heinrich and Rapoport 1974) seeks to quantify the impact of individual rates or parameters on the steady state values of variables such as fluxes J_j and concentrations S_k. Quantitative measures are the so-called flux control coefficients $C_{v_i}^{J_j} = v_i / J_j \cdot \partial J_j / \partial v_i$ or the concentration control coefficients

$C_{v_i}^{S_k} = v_i / S_k \cdot \partial S_k / \partial v_i$, which are normalized derivatives of the variables (J_j or S_k) with respect to rates v_i. The theorems of MCA (Reder 1988) establish a relation between the control coefficients, which are a property of the whole system, and the local sensitivities of the individual rates v_i with respect to the compound concentrations S_k ($\varepsilon_{S_k}^{v_i} = S_k / v_i \cdot \partial v_i / \partial S_k$) and the network stoichiometry *N*. These relations refer to steady state. The sensitivities $\varepsilon_{S_k}^{v_i}$ can be calculated from the rate laws. Although the more interesting aspect of a signaling pathway is its dynamic behavior, we can analyze the impact of small parameter changes on the initial state before stimulation by using MCA.

In MCA, a parameter sensitivity measure is often applied, the response coefficient *R*. The analysis of time-dependent response coefficients is especially interesting for signaling pathways. They are defined as follows

$$R_p^{S_j}(t) = \frac{p}{S_j(t)} \cdot \frac{\partial S_j(t)}{\partial p}. \qquad (12)$$

They quantify the impact of a parameter value on the dynamics of a compound, not only on its steady state value (Ingalls and Sauro 2003). Figure 2 shows the time-courses for the response coefficient of the concentration of MAPK-PP on the kinetic parameters of the individual reactions. We observe that the value of the response coefficients change over time, such that the impact of individual parameters is different at different time points. Thus, if we would like to increase the concentration of MAPK-PP at its highest peak, we should preferentially increase k_5 and decrease k_6, but if we are interested in the period after return to the original state, we should alter k_1 and k_2.

Please note that control and response coefficients refer only to infinitesimal changes of rates or parameters, which we interpreted in the above discussion as percentage change. Large changes may alter the system strongly and cannot be predicted in a general form but measured by the general sensitivity (equation 1) that has a similar form as (equation 12).

3 Studied phenomena

3.1 Dynamic behavior and parameters

The specific behavior of a biochemical network is determined by (i) its wiring, expressed by the stoichiometric matrix *N*, (ii) by the kinetic laws of the individual reactions including the involvement of modifiers that are not substrate or product of this reaction, (iii) by the values of the kinetic parameters, and (iv) by the compound concentrations such as initial concentrations and conserved moieties.

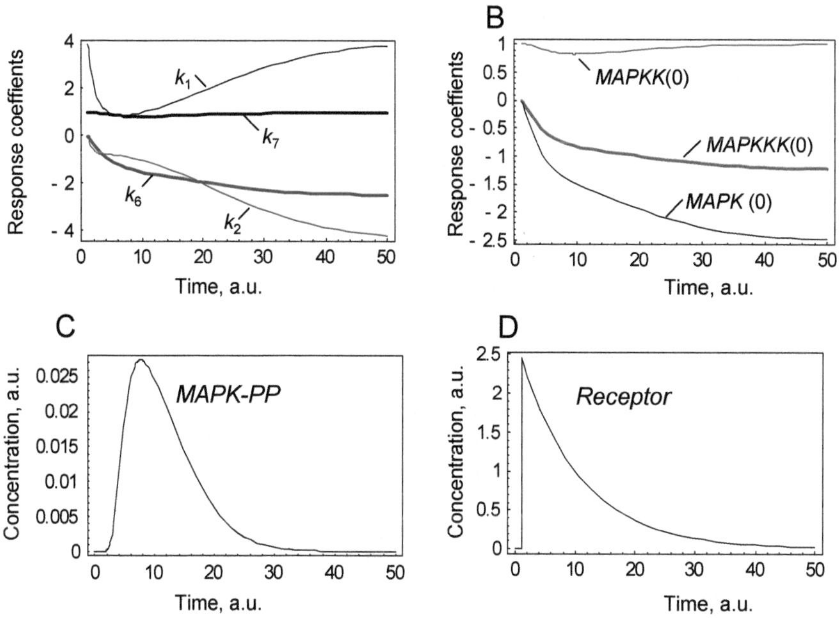

Fig. 2. Time-dependent response analysis with respect to the time course of MAPK-PP (a.u.: arbitrary units). A. Response coefficients for selected parameters. The coefficients with respect to k_1 or k_7 are always positive; their increase would lead to an increase of the time course of *MAPK-PP*. While this is almost constant for k_7, the response coefficient for k_1 assumes a minimum during the rise of *MAPK-PP*. The coefficients with respect to k_2 or k_6 are always negative; their increase would lead to a decrease of the time course of *MAPK-PP*. Both represent kinetic parameters of phosphatase that downregulate the pathway. B. Response coefficients for initial concentrations. The response coefficients for unphosphorylated kinases can be positive or negative with varying magnitude, depending on the initial conditions. C. Time course of *MAPK-PP* for comparison. D. Time course of receptor for comparison. Parameter values: k_i=1, *MAPK*(0)=*MAPKK*(0)=*MAPKKK*(0)=1, all other initial concentrations are zero.

In order to obtain a satisfactory picture of the studied object, all four aspects must be appropriate. The wiring scheme is frequently (but not always!) sufficiently well known from experimental information. For some metabolic reactions, the kinetic mechanism is also determined together with the respective parameters. However, kinetic laws and parameters are often not well defined by experimental information, whereas concentrations or number of molecules are often known to a satisfactory extent. In the following, we will highlight some selected aspects of the dynamic behavior of MAPK cascades.

3.2 Ultrasensitivity, amplification, and robustness

The conserved signaling structure of the MAPK cascade prompts the questions why exactly this architecture is so widely used. It has been hypothesized that stability of the signal off-state, signal amplification, and robustness are important evolutionary targets (Bhalla and Iyengar 1999;Barkai and Leibler 1997; Brown et al. 1997). Stability of the steady state, when there is no signal present at the receptor (signal-off state), is of importance for signaling networks, if they do not want to run the risk of auto activation. Moreover, a signal should be amplified or at least not damped on the way from the cell membrane to its target site. Last but not least, the signaling network should be robust to perturbations like variations in the environment and mutations.

Stability of the off-state is achieved by a property called ultrasensitivity. This means that activation of the cascade is not achieved unless the receptor activity passes a certain threshold beyond which maximal signal strength is quickly achieved. This is also called switch-like behavior. In the MAP kinase cascade of Figure 1, the stimulus-response curves of the second and third layer (MAPKK and MAPK) are sigmoidal due to double phosphorylation (dashed curve in Fig. 3). This can also be seen from the mathematical expression for the concentration of the double phosphorylated form P of the kinases of the second and third level in response to the stimulus s assuming mass-action kinetics (equation 4),

$$P(s)=\frac{s^2 k_1 k_2}{s^2 k_1 k_3 + s k_1 k_4 + k_2 k_4}, \tag{13}$$

where k_1, k_3 and k_2, k_4 are the rate constants of the phosphorylation and dephosphorylation reactions, respectively.

Even stronger ultrasensitivity can be achieved by additional saturation effects of the reactions (Ferrell and Bhatt 1997) (solid curve in Fig. 3). Ultrasensitivity can also be amplified by additional levels in a kinase cascade (Ferrell 1997). Depending on the parameters and molecule numbers of the involved kinases, several cascade steps can also lead to signal amplification (Heinrich et al. 2002). A recent study showed that a cascade of 3 and 4 levels maximizes the sensitivity of the signal amplitude in case of signal amplification, which might be an indication why there are often three levels of kinases (Binder 2005). This can be viewed as an advantageous regulatory property of the transient response, as the signal amplitude can be rapidly regulated for example to avoid crosstalk (see Section 3.6). On an evolutionary time scale, MAP kinase cascades are more robust to parameter perturbation than single molecule systems with similar properties, mainly because effects of parameter changes can compensate each other in networks with more reactions (Bluthgen and Herzel 2003).

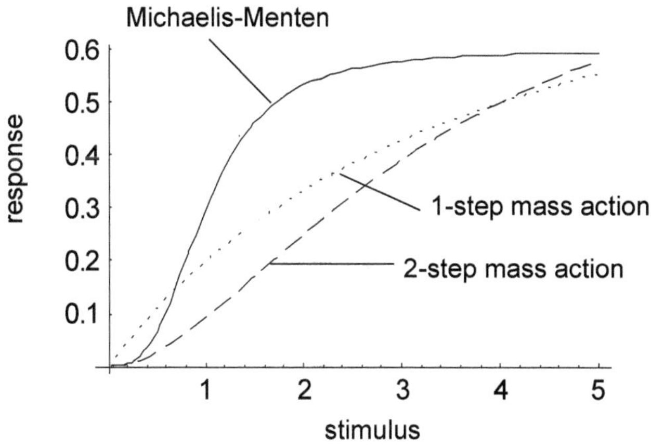

Fig. 3. Stimulus – response curves. On the x-axis shows the strength of the stimulus and the y-axis the strength of the steady state response of the phosphorylated form of a kinase assuming different kinetics and phosphorylated mechanisms. Solid line: 2-step phosphorylation mechanism as in Figure 1 with Michaelis-Menten kinetics (ultrasensitive). Dotted curve: 1-step phosphorylation mechanism with mass action kinetics (not ultrasensitive). Dashed curve: 2-step phosphorylation mechanism with mass action kinetics as in Figure 2 (ultrasensitive).

3.3 Relative importance of kinases and phosphatases

MAPK cascades are regulated by the activity of kinases that phosphorylate the proteins, and by phosphatases that in turn ensure the dephosphorylation. While kinases activate and phosphatases deactivate, both partners are necessary to determine the basic level of activation in absence of external stimuli, but also strength and duration of activation in its presence. It has been discussed that kinases are responsible for the amplitude of the signal, while phosphatases determine its duration (Hornberg et al. 2005). Interestingly, this holds only for weakly activated cascades (Heinrich et al. 2002), while strongly activated cascades show the tendency of prolonged activation upon increase of stimulus. This is based on conservation of MAP kinase proteins on each level, which limits the increase of the active form upon strong activation (Fig. 4). This way, neither kinases nor phosphatases alone determine the properties amplitude or duration. Instead, the critical quantity is their ratio. If the ratio of kinase and phosphatase activity is pushed above a critical value, we observe an increase in the amplitude; if the ratio is below, then we find increase in the duration.

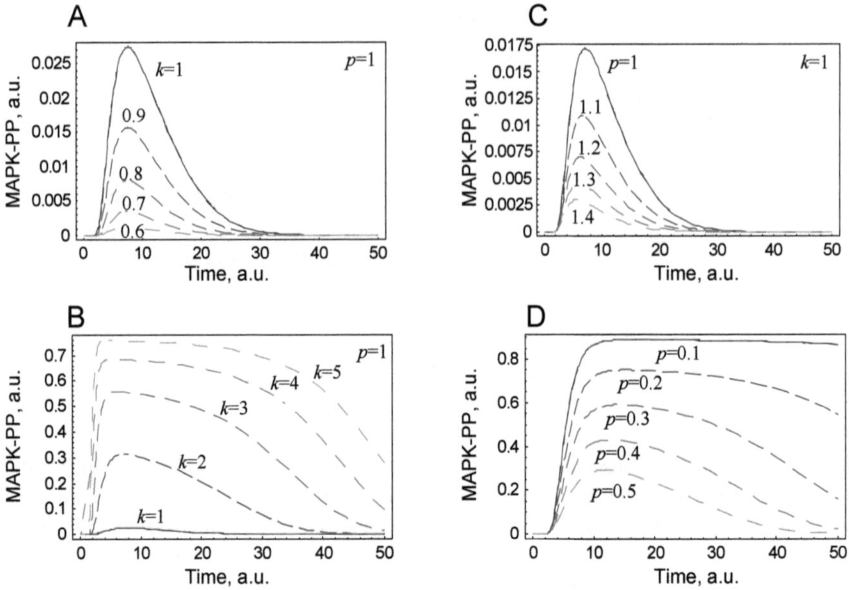

Fig. 4. Impact of kinases and phosphatases. Shown in each case the time course of *MAPK-PP* (Fig. 1) for various combinations of parameter values. It is obvious that changing the values of kinases or phosphatases can in result in changes of amplitude or duration, respectively. A. Equal phosphatases $k_i=p=1$ for even i; varying low values $k_i=k$ of kinases, i.e. odd i (values as indicated). B. Equal phosphatases $k_i=p=1$ for even i; varying high values $k_i=k$ of kinases, i.e. odd i (values as indicated). C. Equal kinases $k_i=k=1$ for odd i; varying high values , $k_i=p$ of phosphatases, i.e. even i (values as indicated). D. Equal kinases $k_i=k=1$ for odd i; varying low values $k_i=p$ of phosphatases, i.e. even i (values as indicated)

3.4 Regulation of MAPK cascade by receptor activity

The function of MAPK cascades is to transfer a signal perceived by a receptor (and possibly transmitted by further protein interactions) to downstream targets. The profile of receptor activation itself (or more precisely, the input function for the activation of MAPKKK) is, therefore, critical for the concentration profile of the cascade output, i.e. MAPK-PP. It is important to consider the duration of receptor activation. If the receptor is only transiently activated, then the MAP kinases will also be transiently activated, since the phosphatases lead to their dephosphorylation after receptor deactivation. If the receptor is permanently activated, then a MAPK cascade without additional regulatory interactions will reach a new steady state with activated kinases. Their level depends, of course, on the specific parameter values. Regulatory interactions such as feedback inhibition (see below) can imply oscillations or pathway downregulation in the presence of active

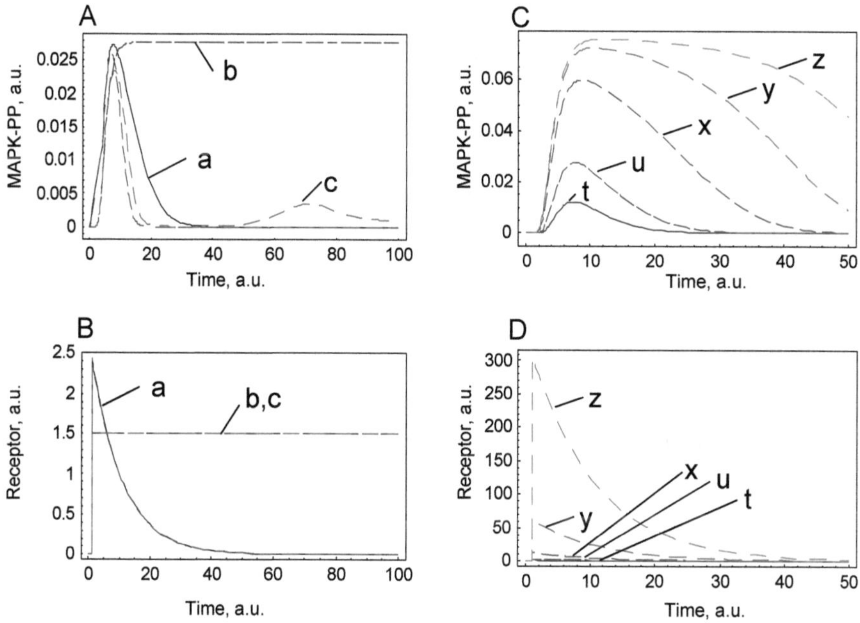

Fig. 5. Relation between receptor profile and profile of MAPK-PP. A and B: A transient rise of *MAPK-PP* (A, curve a) can result from a transient activation of the receptor (B, curve a). Sustained receptor activation (B, curve b) can lead to sustained activation of the kinase (A, curve b). However, if the kinase activation induces processes such as transcription and translation that activate eventually the phosphatases, then sustained receptor activation (B, curve c) may not be reflected in the kinase activation. In the present example, the kinase shows damped oscillations (A, curve c). C and D: transient receptor activation with increasing values of amplitudes causes transient activation of kinases with increasing values of amplitudes and increasing values of duration, respectively. (Curves with same letters correspond). Parameters as in Figure 2.

receptor (Kholodenko 2000b). Figure 5 shows some examples that illustrate the effects.

3.5 Regulation of MAPK cascade by downstream processes - feedback

Figure 1 shows the basic version of MAPK cascade regulation. Various additional regulatory interactions have been suggested, such as the activation of pathway input by MAPK-PP, that is to say, its upstream activators or the receptor (Kholodenko 2000a). An alternative regulatory mechanism is the induction of phosphatases by the downstream kinase. For example, it was proposed that double phosphorylated Hog1 (Hog1-PP) acts as a transcription factor that induces the transcription of genes coding for phosphatases, which in turn are involved in the dephosphorylation of Hog1-PP or its upstream kinases (Jacoby et al. 1997). Since this process involves several time consuming steps, including transcription and translation, it may lead to a delay between the activation of the MAPK and its deactivation. Such delay processes have a high chance of leading to oscillations. A case of damped oscillations is shown in Figure 5 (A, curve c). Furthermore, transcriptional activation of phosphatases is a possible mechanism for pathway downregulation, which on one hand ensures that the pathway target, i.e. the MAPK, was successfully activated; but on the other hand, does not require downregulation of the receptor.

3.6 Crosstalk and dynamics

There are many different ways in which signaling pathways can interact with each other, a phenomenon often called crosstalk. For example, different pathways can be triggered by the same receptor or they can share components that once activated by one pathway leak into another pathway and, thereby, activate it (for an overview of different ways of pathway crosstalk see Schwartz and Baron 1999; Schwartz and Madhani 2004; Cowan and Storey 2003). In modeling crosstalk, there has been the issue of quantifying the amount of crosstalk. Some studies analyzed the topological and structural properties of signaling networks by, for example, classifying modes of interaction (Papin and Palsson 2004) or by counting the theoretically possible interactions between pathways (Binder and Heinrich 2004). However, in order to understand crosstalk mechanisms the dynamic behavior of interacting pathways is also important. One study analyzed the steady state of a simple dynamic model of a three step signaling cascade of two pathways that shared the middle kinase (Fig. 6) (Somsen et al. 2002). In this study, the authors argued that signal specificity, that is to say, activation of the third kinase by one input signal only, could be achieved despite of the fact that there was no physical separation of the cascades (Fig. 6A). However, signal specificity could only be achieved by high inputs. This can be illustrated by the steady state response curve of the output of one pathway (Fig. 6B). It can be seen that component Y is activated at high levels of R, its specific input, when S, its foreign input, is low but also at intermediate levels of S irrespective of R. As a consequence, Y will be activated (YP) by both R and S in the dynamic case; only the response to the foreign signal will be delayed (Fig. 6C). Therefore, considering the dynamic case, no signal specificity is achieved. In signaling, it is rather the transient dynamic behavior than the steady state behavior that is important and in this case it seems unlikely that the proposed model of Somsen et al. (2002) occurs in nature, because there will always be a crossactivation.

Two recent studies address the dynamic features of pathway crosstalk. By analyzing the activation of pathways by a so-called intrinsic and an extrinsic stimulus, respectively, one study defined measures for pathway specificity and fidelity (Komarova et al. 2005). These measures give useful insights how pathways interact with each other. However, it is important to note that these measures refer to responses to one stimulus at a time. However, it can be assumed that cells usually

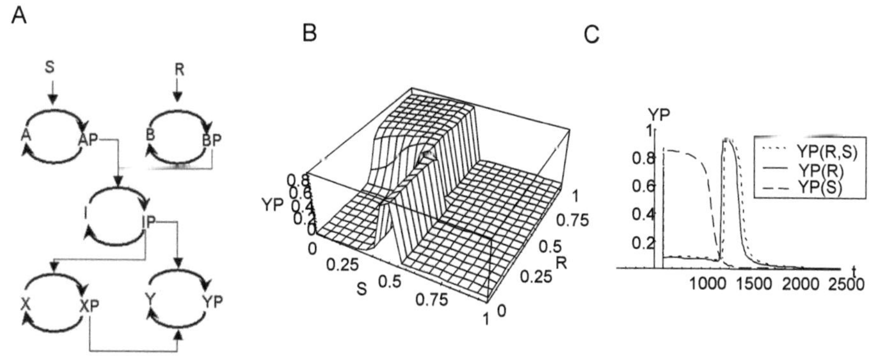

Fig. 6. Crosstalk: comparison of steady state and dynamic behavior exemplified for the Somsen model (Somsen et al. 2002). A: Model scheme, adapted. B: Steady state response of *YP* to input *R* and *S*. C: Dynamics of *YP*, assuming as input a unit step at t=500 with an exponential decay.

process multiple information in parallel and these measures give no clue how signals interact while being transmitted concomitantly. It can be expected that signals amplify or inhibit each other when transmitted at the same time. Thus, it does not suffice to study each signal in isolation but also to study the cell's response to multiple stimuli at the same time. Schaber et al. (2006) proposed crosstalk measures that include parallel multiple pathway activation called the intrinsic and extrinsic specificity that yield a better understanding of how the pathways dynamically interact

4 Discussion/Summary

We have studied different mechanisms that are involved in the temporal regulation of MAPK cascades. We have shown that the dynamics of the activated proteins cannot be predicted solely on the basis of the network structure or on the basis of steady state investigations. MAPK cascades are highly regulated. One can expect that the kinetic properties of the proteins making up such cascades have evolved to ensure the proper behavior, that is, robust response to input signals with variable strength and shape, but also fragility, which appropriately encode the environmental information transmitted by different signal properties.

Acknowledgement

This work was supported by the European Commission via the QUASI project (Grant FP6-2002-LSH-503 230 to E.K.).

References

Barkai N, Leibler S (1997) Robustness in simple biochemical networks. Nature 387:913-917

Bhalla US (2002) Biochemical signaling networks decode temporal patterns of synaptic input. J Comput Neurosci 13:49-62

Bhalla US (2004) Models of cell signaling pathways. Curr Opin Genet Dev 14:375-381

Bhalla US, Iyengar R (1999) Emergent properties of networks of biological signaling pathways. Science 283:381-387

Bhalla US, Iyengar R (2001) Robustness of the bistable behavior of a biological signaling feedback loop. Chaos 11:221-226

Binder B (2005) Design principles and control mechanisms of signal transduction networks. In: Biophysics. Berlin: Humboldt-University, p 131

Bluthgen N, Herzel H (2003) How robust are switches in intracellular signaling cascades? J Theor Biol 225:293-300

Brown GC, Hoek JB, Kholodenko BN (1997) Why do protein kinase cascades have more than one level? Trends Biochem Sci 22:288

Chen KC, Calzone L, Csikasz-Nagy A, Cross FR, Novak B, Tyson JJ (2004) Integrative analysis of cell cycle control in budding yeast. Mol Biol Cell 15:3841-3862

Chen KC, Csikasz-Nagy A, Gyorffy B, Val J, Novak B, Tyson JJ (2000) Kinetic analysis of a molecular model of the budding yeast cell cycle. Mol Biol Cell 11:369-391

Ferrell JE Jr (1996) MAP kinases in mitogenesis and development. Curr Top Dev Biol 33:1-60

Ferrell JE Jr (1997) How responses get more switch-like as you move down a protein kinase cascade. Trends Biochem Sci 22:288-289

Ferrell JE Jr (1998) How regulated protein translocation can produce switch-like responses. Trends Biochem Sci 23:461-465

Ferrell JE Jr (2002) Self-perpetuating states in signal transduction: positive feedback, double-negative feedback and bistability. Curr Opin Cell Biol 14:140-148

Ferrell JE Jr, Bhatt RR (1997) Mechanistic studies of the dual phosphorylation of mitogen-activated protein kinase. J Biol Chem 272:19008-19016

Gelman A, Carlin JB, Stern HS, Rubin DB (1995) Bayesian Data Analysis. Boca Raton: Chapman & Hall

Gillespie DT (1977) Exact Stochastic Simulation of coupled chemical-reactions. J Phys Chem 81:2340-2361

Heinrich R, Neel BG, Rapoport TA (2002) Mathematical models of protein kinase signal transduction. Mol Cell 9:957-970

Heinrich R, Rapoport TA (1974) A linear steady-state treatment of enzymatic chains. General properties, control and effector strength. Eur J Biochem 42:89-95

Hornberg JJ, Bruggeman FJ, Binder B, Geest CR, de Vaate AJ, Lankelma J, Heinrich R, Westerhoff HV (2005) Principles behind the multifarious control of signal transduction. ERK phosphorylation and kinase/phosphatase control. FEBS J 272:244-258

Ingalls BP, Sauro HM (2003) Sensitivity analysis of stoichiometric networks: an extension of metabolic control analysis to non-steady state trajectories. J Theor Biol 222:23-36

Jacoby T, Flanagan H, Faykin A, Seto AG, Mattison C, Ota I (1997) Two protein-tyrosine phosphatases inactivate the osmotic stress response pathway in yeast by targeting the mitogen-activated protein kinase, Hog1. J Biol Chem 272:17749-17755

Kacser H, Burns JA (1973) The control of flux. Symp Soc Exp Biol 27:65-104
Kholodenko BN (2000a) Negative feedback and ultrasensitivity can bring about oscillations in the mitogen-activated protein kinase cascades. Eur J Biochem 267:1583-1588
Kholodenko BN (2000b) Negative feedback and ultrasensitivity can bring about oscillations in the mitogen-activated protein kinase cascades. Eur J Biochem 267:1583-1588
Komarova NL, Zou X, Nie Q, Bardwell L (2005) A theoretical framework for specificity in cell signaling. Mol Sys Biol: Epub
Moles CG, Mendes P, Banga JR (2003) Parameter estimation in biochemical pathways: a comparison of global optimization methods, Genome Res 13:2467-2474
Novak B, Toth A, Csikasz-Nagy A, Gyorffy B, Tyson JJ, Nasmyth K (1999) Finishing the cell cycle. J Theor Biol 199:223-233
Papin JA, Hunter T, Palsson BO, Subramaniam S (2005) Reconstruction of cellular signalling networks and analysis of their properties. Nat Rev Mol Cell Biol 6:99-111
Papin JA, Palsson BO (2004) Topological analysis of mass-balanced signaling networks: a framework to obtain network properties including crosstalk. J Theor Biol 227:283-297
Price K, Storn R (1997) Differential evolution. Dr Dobb's Journal 22:18-24
Reder C (1988) Metabolic control theory: a structural approach. J Theor Biol 135:175-201
Rodriguez-Fernandez M, Mendes P, Banga JR (2006) A hybrid approach for efficient and robust parameter estimation in biochemical pathways. Biosystems 83:248-265
Rousseeuw PJ, Leroy AM (1987) Robust Regression and Outlier Detection. New York: John Wiley & Sons
Runarsson TP, Yao X (2000) Stochastic ranking for constrained evolutionary optimization. IEEE Transactions on Evolutionary Computation 4:284-294
Schaber J, Kofahl B, Kowald A, Klipp E (2006) A modelling approach to quantify dynamic crosstalk between the pheromone and the starvation pathway in baker's yeast. FEBS J 273:3520-3533
Schoeberl B, Eichler-Jonsson C, Gilles ED, Muller G (2002) Computational modeling of the dynamics of the MAP kinase cascade activated by surface and internalized EGF receptors. Nat Biotechnol 20:370-375
Schuster S, Klipp E, Marhl M (2006) The predictive power of molecular network modelling: case studies of predictions with subsequent experimental verification. In: Eisenhaber F (ed) Discovering biomolecular mechanisms with computational biology. Austin, TX: Landes Bioscience
Seber GAF, Wild CJ (2003) Nonlinear Regression. Hoboken, NJ: Wiley-Interscience
Somsen OJ, Siderius M, Bauer FF, Snoep JL, Westerhoff HV (2002) Selectivity in overlapping MAP kinase cascades. J Theor Biol 218:343-354
Storn R, Price K (1997) Differential evolution - A simple and efficient heuristic for global optimization over continuous spaces. Journal of Global Optimization 11:341-359
Swameye I, Muller TG, Timmer J, Sandra O, Klingmuller U (2003) Identification of nucleocytoplasmic cycling as a remote sensor in cellular signaling by databased modeling. Proc Natl Acad Sci USA 100:1028-1033
Tyson JJ, Chen KC, Novak B (2003) Sniffers, buzzers, toggles and blinkers: dynamics of regulatory and signaling pathways in the cell. Curr Opin Cell Biol 15:221-231
Vaseghi S, Macherhammer F, Zibek S, Reuss M (2001) Signal transduction dynamics of the protein kinase-A/phosphofructokinase-2 system in *Saccharomyces cerevisiae*. Metab Eng 3:163-172
Yi TM, Kitano H, Simon MI (2003) A quantitative characterization of the yeast heterotrimeric G protein cycle. Proc Natl Acad Sci USA 100:10764-10769

Klipp, Edda
Max Planck Institute for Molecular Genetics, Ihnestr. 63-73, 14195 Berlin, Germany
klipp@molgen.mpg.de

Schaber, Jörg
Max Planck Institute for Molecular Genetics, Ihnestr. 63-73, 14195 Berlin, Germany

Stress-activated protein kinase signaling in *Drosophila*

Gerasimos P. Sykiotis and Dirk Bohmann

Abstract

Genetic analyses in *Drosophila* have shown that the components of the Stress-Activated Protein Kinase (SAPK) pathways are structurally and functionally conserved in this organism. The genetic tractability of the fruit fly has facilitated *in vivo* analyses of SAPK signaling that would have been exceedingly difficult to carry out in mammalian systems. This approach has yielded important new knowledge about the *in vivo* functions of JNK in development, wound healing, innate immunity, developmental and TNF-mediated apoptosis, oxidative stress resistance and lifespan regulation. Similarly, the *Drosophila* p38 pathway appears to be involved in development, immunity, and stress response, but more research is needed to fully understand its components and functions. It is anticipated that genetic approaches in *Drosophila* will continue to produce novel insights into SAPK signaling that are directly relevant to rodent models and humans.

1 Introduction

The c-Jun N-terminal Kinase (JNK) and p38 are jointly referred to as the Stress-Activated Protein Kinases (SAPKs). SAPK signaling is activated in adverse biological conditions and contributes to the organism's response to various endogenous stress signals, such as Tumor Necrosis Factor (TNF) and interleukins, and to a range of environmental stressors, including UV and ionizing radiation, free oxygen radicals, mechanical stress, and xenobiotics. JNK and p38 signaling has been implicated in diverse physiological and pathological phenomena, including wound healing, inflammation, and cancer. In light of their widespread roles in health and disease, it is not surprising that SAPKs have been the objects of intense scientific scrutiny (>15,000 hits in Pubmed as of October 2006). While the bulk of this research has been done in mammalian cell or animal systems, the SAPKs are structurally and functionally conserved in genetically simpler model organisms like yeast, *Caenorhabditis elegans* and *Drosophila melanogaster* (Kockel et al. 2001; Martin-Blanco 2000; Stronach and Perrimon 1999). The structural and functional conservation of genes and pathways between *Drosophila* and man means that results obtained using this genetically tractable model organism should be relevant also for humans. In addition to the ease and speed of genetic manipulation, experimental advantages of *Drosophila* include the lack of genetic redundancy, the

Topics in Current Genetics, Vol. 20
F. Posas, A. R. Nebreda (Eds.): Stress-Activated Protein Kinases
DOI 10.1007/4735_2007_0251 / Published online: 1 August 2007

use of developmental phenotypes as readouts of functional assays, the generation of clones of homozygous mutant cells within a heterozygous tissue, and the feasibility of gene manipulation via RNA interference *in vivo*. The application of these approaches to the study of JNK and p38 in *Drosophila* has yielded novel insight into the *in vivo* functions of the SAPK pathways. This review will summarize this knowledge, with an emphasis on recent findings on the role of *Drosophila* JNK and p38 in stress defense.

2 Structural conservation of the JNK and p38 pathways in *Drosophila*

The distinguishing feature of Mitogen-Activated Protein Kinase (MAPK) signal transduction pathways is their hierarchical assembly: a MAPK is activated through phosphorylation by an upstream MAPK kinase (MAP2K), which in turn is activated by a MAP3K. The *Drosophila* genome encodes conserved kinases at each level of hierarchy in the JNK and p38 branches of MAPK signaling (and also in the Extracellular Signal-Regulated Kinase, ERK, branch) (Fig. 1). While biochemically these kinases generally conform to the characteristics of their respective groups as defined in mammals, the *Drosophila* SAPK pathways are genetically much less redundant than their mammalian counterparts. In mammals, most nodes in the JNK and the p38 cascades are represented by three or four gene products, which overlap in their function and tissue specificity. Conversely, *Drosophila* typically has but a single gene for each signaling component. For example, mammalian JNKs are encoded by three genes (JNK1, JNK2, and JNK3), which give rise to at least ten splice forms (Gupta et al. 1996); *Drosophila*, on the other hand, has only one JNK gene, named *basket* (*bsk*) (Riesgo-Escovar et al. 1996; Sluss et al. 1996). With regard to the p38 branch, there are only two p38 genes in the fly, D-p38a and D-p38b (Han et al. 1998a, 1998b), compared to four in mammals (Ono and Han 2000). One level higher, the *Drosophila* MAP2Ks are represented by the MKK7 homologue Hemipterous (Hep) in the JNK pathway (Glise et al. 1995), and the MKK3 homologue Licorne (Lic) in the p38 pathway (Suzanne et al. 1999). There is also an MKK4 homologue (*CG9738*), which can specifically activate JNK in cultured *Drosophila* cells (Han et al. 1998b; Boutros et al. 2002); however, this family member is not genetically characterized, and its *in vivo* requirement for JNK responses is unclear. As in mammals, the level of SAP3Ks shows the greatest degree of diversity. The *Drosophila* genome contains six possible SAP3K genes: *slipper* (*slp*), which encodes a homologue of the MLKs (Stronach and Perrimon 2002), *D-Ask1* (also known as *pk92B*) (Kuranaga et al. 2002), *D*-Tak1 (Takatsu et al. 2000), *D-MEKK1* (Inoue et al. 2001), and two others with similarity to Tak and M3K12 (*CG4803* and *CG8789*) that are less well characterized. In contrast to mammals, however, each of these enzymes is encoded by a single gene rather than a gene family (as is also the case for MAPKs and

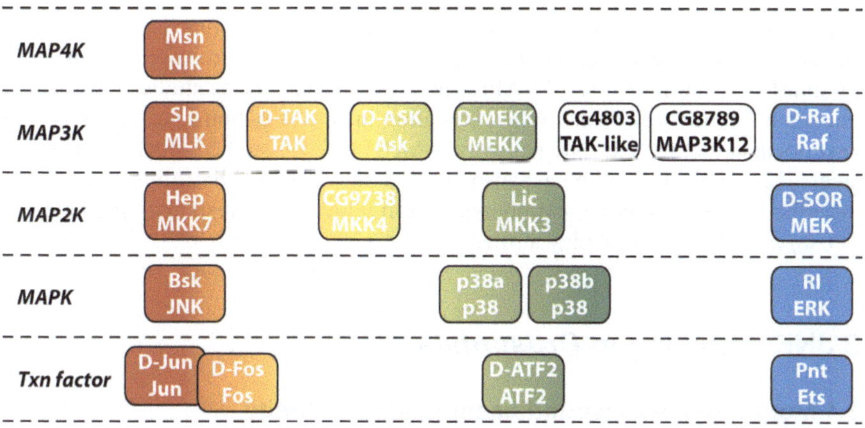

Fig. 1. MAP kinase signaling components in *Drosophila*. The kinases of the different signaling tiers are represented in common rows as indicated. Abbreviated *Drosophila* names are indicated with the name of the cognate family of vertebrate homologues shown below. Elements of SAPK signaling are shown in red/yellow/green. Signaling relationships are not clearly established in all cases and are tentatively assigned by color-coding. Cases where no information is available (CG4803 and CG8789) are left white. The Raf-ERK MAPK pathway is marked in blue.

MAP2Ks). In addition to these three core kinases, some pathways include upstream MAP4Ks and/or downstream MAPK-activated protein kinases (MAPKAPs). For example, the *Drosophila* homolog of NIK (NCK interacting kinase) has been identified as a JN4K that functions as an upstream component of the JNK cascade in development (dorsal closure) and in immune responses (Su et al. 1998). Finally, the MAPK pathways comprise important negative regulators, the MAPK phosphatases (MKPs), which are also conserved from yeast to humans (Keyse 2000). The expression of these MKPs is often upregulated in response to signaling through their respective MAPK branch, thus giving rise to a negative feedback loop. For the *Drosophila* JNK pathway there is only one JNK-specific phosphatase, encoded by the *puckered* (*puc*) gene (Martin-Blanco et al. 1998). A p38-specific phosphatase in *Drosophila* has not yet been reported.

Due to the lack of redundancy, the genetic analysis of SAPK signaling is relatively straightforward in *Drosophila* as compared to mammals. For example, loss-of-function mutations of various members of the same branch cause by and large the same phenotype. This is nicely illustrated by the embryonic development phenotypes of JNK pathway mutants (Kockel et al. 2001). The lack of genetic redundancy simplifies genetic interaction studies, and facilitates the establishment of epistatic relationships. For example, once the developmental phenotype of a mutation in the pathway has been established, this phenotype can serve as the basis for genetic screens to identify additional pathway components. This approach has been very fruitful in the analysis of JNK signaling, and there is a large body of literature describing the developmental functions of JNK. In fact, in their extensive review of JNK in *Drosophila* development, Stronach and Perrimon (1999) remark that relatively little is known about the role of the *Drosophila* SAPKs in actual stress responses. Nevertheless, more recent work has begun to employ *Drosophila* to understand the functions of SAPKs in stress defense; these studies have shown that, as in higher organisms, SAPK signaling in flies is pleiotropic. This has been best documented for the JNK pathway, which has been studied much more extensively, but seems to be the case also for p38 signaling. In the following, we will describe the various biological phenomena in which stress responses mediated by *Drosophila* JNK and p38 play a role.

3 JNK signaling in *Drosophila*

3.1 Morphogenesis, wound healing, and immunity

The study of JNK signaling in *Drosophila* began in 1996, at a time when the role of the mammalian pathway in various stress response paradigms was already well established. Interestingly, the first functions ascribed to JNK pathway components in *Drosophila* had no obvious connection to stress signaling; instead, they illustrated the role of JNK in tissue morphogenesis: it was found that *Drosophila* embryos lacking the JNKK Hep were incapable of completing an epithelial cell sheet movement called dorsal closure (Glise et al. 1995; Harden 2002). Once the failure of dorsal closure was determined to be a phenotype resulting from defective JNK signaling, many other pathway components were rapidly identified. These included Basket (JNK) (Riesgo-Escovar et al. 1996), Slipper (MLK) (Stronach and Perrimon 2002), Kayak (Fos) (Zeitlinger et al. 1997) l(2)IA109 (Jun) (Hou et al. 1997; Kockel et al. 1997; Riesgo-Escovar and Hafen 1997), Misshapen (Msn) (Su et al. 1998), and Puckered (Puc) (Martin-Blanco et al. 1998). Dorsal closure is a developmental function, and has no obvious connections with stress responses. However, it was quickly recognized that the closure of the dorsal opening in the *Drosophila* embryo resembles phenomenologically the epithelial cell movements that mediate the healing of a wound (Goberdhan and Wilson 1998; Riesgo-Escovar and Hafen 1997). It was therefore speculated that JNK might also contribute to injury-induced tissue remodeling and wound repair. This hypothesis was subsequently confirmed, as it was shown that JNK signaling is rapidly and specifically induced in models of injury that use *Drosophila* embryos, larvae, imaginal discs, or adults (Bosch et al. 2005; Galko and Krasnow 2004; Mattila et al. 2005; Ramet et al. 2002); the loss of JNK function results in defects in wound healing. This example demonstrates how the use of *Drosophila* provided new insight into JNK signaling: First, a role for JNK in development was established. Then, the availability of mutants with a defined phenotype facilitated the identification of additional pathway components, and the elucidation of their epistatic relationships. Finally, the developmental mutant phenotype suggested the possible involvement of the pathway in another phenomenon (wound healing), and *Drosophila* was again used as a tool to validate that hypothesis.

An extension of this line of investigation led to the discovery that, in addition to mediating the response to sterile wounding, the *Drosophila* JNK pathway is also activated by microbial infection. Although *Drosophila* does not possess adaptive immunity, it has a conserved innate immune response system. It had been observed that the response of adult flies to sterile injury and infection is remarkably overlapping: the genomic gene expression profiles after sterile wounding and microbial infection are qualitatively very similar (Boutros et al. 2002; Jasper and Bohmann 2002). It was then shown that the JNK pathway, likely activated by D-TAK1, cooperates with the fly version of the NF-κB systems (Dif and Relish pathways) to induce an innate immune response.

3.2 Oxidative stress defense and lifespan regulation

Initially, Goberdhan and Wilson (1998) suggested that, in addition to its role in development, the *Drosophila* JNK might also function as a *bona fide* SAPK. This hypothesis was soon validated: it was shown that Bsk can be activated in cultured *Drosophila* cells treated with LPS, in a way that is reminiscent of the response of the pathway in mammals (Sluss et al. 1996). Later, genomic experiments showed that the transcriptional response of adult flies to oxidative stress was in a significant part dependent on an intact JNK signaling pathway (Jasper et al. 2001; Wang et al. 2003; Zou et al. 2000). Adult flies feeding on the free radical generator Paraquat mount a (partly) JNK-dependent transcriptional anti-oxidant response, which encompasses the induction of defense genes encoding glutathione S-transferases, metallothionines, heat shock proteins, etc. Consistently, flies in which the JNK pathway was modestly activated (for example, by moderate overexpression of JNKK, or by heterozygosity for the JNK phosphatase Puckered), showed increased resistance to oxidative stress (Wang et al. 2003). These and other experiments clearly demonstrated that the *Drosophila* JNK pathway can mediate reactive oxygen species-inducible gene expression and confer oxidative stress resistance to the organism. The oxidative stress theory of aging posits that the ability of an organism to cope with and defend itself against reactive oxygen species contributes to longevity (Stadtman 2001). Consistent with this thesis and with the increased tolerance of JNK gain-of-function flies to Paraquat, it was found that JNK signaling and the transcriptional program it elicits can extend lifespan in *Drosophila* (Wang et al. 2003, 2005). A similar role of JNK in anti-oxidant defense and lifespan regulation was described in *C. elegans* (Oh et al. 2005). These examples demonstrate how the use of simpler model organisms produced knowledge about SAPK signaling that would have been exceedingly difficult to derive from mammalian systems.

3.3 JNK-dependent apoptosis during development

Another example for the utility of *Drosophila* as a genetically accessible model organism for the *in vivo* study of stress signaling mechanisms is the control of apoptosis during organogenesis. Imaginal disc development has been a spectacularly powerful model to study the relationships of patterning, signaling, growth control and morphogenesis in the context of an intact tissue. Much of the signaling circuitry that is now commonly understood to drive metazoan development has been originally deciphered through studies in the *Drosophila* wing, eye, or leg imaginal discs. Such studies have shown that JNK signaling can serve as a system that cleans up developmental mistakes. An illustrative demonstration of such a role came from Adachi-Yamada et al. (1999a), who showed that when cells in the wing imaginal disc receive combinations of morphogen signals (TGF-β and Wnt-type factors in this case) falling outside the physiological range, JNK is activated, leading to apoptotic removal of the potentially miss-programmed cells. Thus, the evident function of JNK in this context is to detect and eliminate mistakes in developmental programming. A different example, with a probably similar mechanistic basis, is the removal of unusually slow growing cells from developing imaginal discs. This phenomenon, which is referred to as cell competition, is observed when neighboring cells in a developing disc proliferate at different rates (Vidal and Cagan 2006). Such a situation can arise, for example, when one cell population has sustained somatic mutations or is otherwise damaged in a way that might compromise the integrity or function of the developing organ. In this scenario, the slower growing cells are again removed via JNK-dependent apoptosis (Moreno et al. 2002a; Ryoo et al. 2004). Interestingly, this process is independent of the absolute rate of cell proliferation: the cells that are removed by cell competition would be able to grow into a full-sized organ if they were not juxtaposed to faster growing cells. Remarkably, even wild type cells will be eliminated if they come under the influence of unusually fast growing cells, such as cells in which Myc is over-activated to oncogenic levels (Moreno and Basler 2004). The mechanism that measures differences in cell growth rates and triggers the activation of JNK signaling and apoptosis is not ultimately clear, but has been connected to uneven distribution of survival signals between the slower and faster growing cells (Adachi-Yamada and O'Connor 2002; Moreno et al. 2002a). The understanding of such interactions among different cell groups within a tissue, and the role that JNK is playing in this context, could be informative for tissue interactions in cancer; for example, the interplay between malignant cells and their neighboring normal or cancer stromal cells. Studies in *Drosophila* have the potential to make further contributions to this important field, which is poorly understood at present.

3.4 JNK in TNF- and irradiation-induced apoptosis

One of the best-studied regulators of apoptosis is TNF. It was recently shown that *Drosophila* possesses a conserved TNF pathway. Although the general functional principles remain conserved between insects and mammals, some of the circuitry wiring that was added during evolution is not present in flies. The *Drosophila* homologue of TNF is encoded by the gene *eiger* (Igaki et al. 2002; Moreno et al. 2002b) and signals via the fly TNF receptor Wengen (Kanda et al. 2002; Kauppila et al. 2003). Reminiscent of TNF signaling in mammals, the activation of Wengen in response to Eiger relays a potent apoptotic stimulus in various cell types. In addition to JNK, Eiger-mediated apoptosis requires the caspase-9 homolog DRONC and the Apaf-1 homolog DARK. However, in contrast to the situation in mammals, the death signal is transduced exclusively via JNK, whereas the Caspase 8-dependent branch of the pathway that leads to cytochrome C release from the mitochondria in mammalian cells is not triggered by Eiger/Wengen signaling in flies (Moreno et al. 2002b). Simplifying the pathway further, the activation of NF-κB, which is considered an integral part of the TNF response in mammals, is not conserved in *Drosophila*. Therefore, it appears that the evolutionarily older TNF response comprises Caspase 9-mediated activation of the JNK pathway; whereas the Caspase 8 and NF-κB branches of TNF signaling emerge as more recent additions. For these reasons, the genetic dissection of TNF signaling and the role of JNK in the control of apoptosis should be more accessible to genetic analyses in *Drosophila* than in mammalian systems.

One of the questions that could be addressed using the fruit fly model is which transcription factors and mediators are involved downstream of JNK in apoptotic responses. Moreover, it has been proposed that the cellular context determines survival or death in response to genotoxic stresses; this tenet has been supported by systems biology approaches in cell culture (Janes et al. 2005), but *in vivo* experimental evidence is scarce. Recent work has begun to investigate these issues by studying the downstream effectors of irradiation-induced JNK-mediated apoptosis. It was shown that both γ-irradiation and UVC irradiation can activate JNK in imaginal discs, presumably by inducing DNA damage (McEwen and Peifer 2005; Luo et al. 2007). In response to γ-irradiation, JNK activity and *puc* expression are increased in a p53-dependent manner; in this case, JNK acts upstream of both the pro-apoptotic gene *reaper* and the effector caspases (McEwen and Peifer 2005). Consistent with other studies (Ryoo et al. 2004), it was found that if cell death is inhibited, JNK signaling actually promotes tissue overgrowth (McEwen and Peifer 2005). These findings demonstrate the importance of cellular context for JNK-mediated life or death decisions. Similarly, JNK is activated in response to UVC irradiation, and promotes apoptosis by transcriptionally upregulating the pro-apoptotic gene *hid* (Luo et al. 2007). The transcription factors Foxo and Fos mediate *hid* induction by JNK. Since Foxo can also mediate cell protective effects in response to JNK signaling (Wang et al. 2005), it appears again that the cellular context determines the outcome on cell survival or apoptosis; in the case of UVC irradiation, the activities of the pro-survival EGFR and insulin pathways and the pro-apoptotic JNK cascade are integrated at the level of Foxo (Luo et al. 2007). Thus, *Drosophila* is providing an *in vivo* system to test and validate systems biology-derived predictions about the interplay of TNF, EGFR, insulin, and JNK signaling in the decision between cell death and survival (Janes et al. 2005).

4 p38 signaling in *Drosophila*

4.1 Identification of the fruit fly p38 homologues

In contrast to the plethora of studies on *Drosophila* JNK signaling, surprisingly little is known about the p38 pathway in this organism; in fact, some of the available data appear controversial and confusing, thus leaving the outstanding questions about the role of this pathway in the fruit fly without definitive answers (Table). It is, therefore, more appropriate to present a historical account of this field of research up to now. p38 signaling in *Drosophila* was first described less than ten years ago (Han et al. 1998a, 1998b). These early studies showed that *Drosophila* possesses two p38 homologues, D-p38a and D-p38b. Thus, there may be some genetic redundancy at the MAPK level in this branch of MAPK signaling in *Drosophila*, which is in contrast to the ERK and JNK cascades. Han et al. (1998a) showed that D-p38a is phosphorylated and activated by osmotic stress, heat shock, serum starvation, and H_2O_2 in cultured *Drosophila* cells. Interestingly, however, LPS (a known activator of mammalian p38) did not activate D-p38a in this study. The functional homology of D-p38a to the yeast p38 homolog HOG1 was shown by its ability to partially rescue the sensitivity of *HOG1* mutants to increased osmolarity. Shortly afterwards, a different group (Han et al. 1998b) independently reported on both D-p38a and D-p38b, and identified their role in the regulation of *Drosophila*'s innate immune response. The two D-p38 proteins are encoded by different genes, and are 75% identical. D-p38a is one amino acid longer than D-p38b (366 *vs.* 365). High amounts of D-p38a mRNA are detectable in the early embryo, suggesting maternal deposition and a possible role in early embryonic development. Both D-p38a and D-p38b are expressed throughout embryonic development, showing higher expression in the gut. Both D-p38s could phosphorylate D-Jun and mammalian ATF-2 in *in vitro* kinase assays. The *Drosophila* MKK3 was identified as an upstream kinase that can activate both D-p38s in cultured *Drosophila* cells. In cultured cells, both D-p38 isoforms were activated by UV light and by LPS, which is in contrast to the report by Han et al. (1998a). Importantly, this study showed that D-p38 downregulates the innate immunity response: the p38 inhibitor SB203580 (an anti-inflammatory drug) increased the induction of anti-microbial peptides by LPS in cultured cells, whereas the overexpression of D-p38a blocked the expression of anti-microbial peptides by bacterial challenge in larvae.

4.2 p38 in *Drosophila* development

The conservation of p38 signaling in *Drosophila* prompted investigations of its potential contribution to developmental processes, a task for which model organisms like the fruit fly are well suited. Three groups provided evidence for roles of the p38 pathway in *Drosophila* development. Adachi-Yamada et al. (1999b) identified D-p38b independently, and used various loss of D-p38b function conditions

Table 1. Outstanding questions about the Drosophila p38 pathway

•	What are the physiological and/or developmental roles of D-p38b? Is this isoform functionally redundant with D-p38a?
•	Is D-p38 a mediator or a suppressor of the immune response?
•	Which MAP3K(s) and MAP2K(s) functions upstream of D-p38 in stress responses *in vivo*? Can D-p38 be activated by stress in a MAP2K-independent manner?
•	Is the developmental function of the MAP2K Licorne mediated by D-p38? Do the D-p38 kinases have roles in development under non-stressful conditions (like JNK does)?
•	Is there a D-p38-specific phosphatase (analogous to Puckered in the JNK branch)?

to address its developmental role (expression of dominant-negative D-p38b, reduction of gene dose using a deficiency uncovering *D-p38b*, expression of D-p38b anti-sense RNA, and treatment with SB203580). Loss of D-p38b function in the wing disc caused a phenotype that mimicked the loss of the morphogen Dpp, and also modified the phenotypes of Dpp loss- and gain-of-function conditions. Moreover, D-p38b was phosphorylated by the activation of Dpp signaling through a constitutively active Dpp receptor, and suppressed the induction of a Dpp target gene. These data showed that p38 signaling can modulate the outcomes of Dpp signaling on wing morphogenesis and vein differentiation. However, the authors acknowledged that it remains unclear whether D-p38 is a downstream component of canonical Dpp signaling in these processes, or whether these effects are only observed under extreme conditions induced by genetic manipulations (which perhaps generate cellular stress).

Suzanne et al. (1999) reported the developmental phenotype caused by the loss of D-MKK3 (which they named Licorne). They generated animals lacking maternally deposited and zygotically expressed Licorne and Hemipterous (interestingly, *lic* and *hep* are neighboring genes) and selectively supplied Lic and/or Hep from transgenes. Their analysis showed that *lic* (but not *hep*) is required during oogenesis in the germline for correct asymmetric development of the egg. Loss of *lic* function affects the dorsal-ventral and anterior-posterior patterning of the eggshell and the embryo. Although this study clearly implicated the D-p38 MAP2K D-MKK3 in development, it did not address whether the developmental effects of Licorne are actually mediated by D-p38. The anatomic location of Licorne function in the egg and its downstream targets in oogenesis remain unknown.

Finally, D-p38 was implicated in planar polarity signaling downstream of the MAP4K *misshapen* (*msn*) and the Frizzled effector *disheveled* (*dsh*) (Paricio et al. 1999). However, the evidence for the role of D-p38 was derived only from the ability of chromosomal deficiencies uncovering the *D-p38a* and *D-p38b* loci to modify strong *msn* and *dsh* gain-of-function phenotypes. Similarly, the suggestion that the MAP3K D-TAK1 might serve as an upstream kinase of D-p38 was also based on the ability of these deficiencies to modify an eye phenotype generated by the overexpression of D-TAK1 (Mihaly et al. 2001). Thus, all of these developmental studies could not provide conclusive evidence for the role of p38 signaling in *Drosophila* development. Ideally, loss-of-function alleles of the two D-p38 isoforms should be employed for this inquiry.

4.3 Genetic analysis of the fruit fly p38 pathway

Before D-p38 mutants became available, however, more insight into the role of p38 signaling *in vivo* was gained from the identification of the *Drosophila* MAP3K D-MEKK1, which is similar to the mammalian MEKK4/MTK1. Through genetic and biochemical studies in *D-MEKK1* mutants, Inoue et al. (2001) provided evidence that D-MEKK1 activates D-p38 in response to heat shock and osmotic stress, and is required for normal resistance to these stresses. However, the phosphorylation of D-p38 in response to heat shock and osmotic stress was reduced but not abolished in D-MEKK1 mutants, suggesting that another MAP3K is also partially responsible for D-p38 activation in the stress response. The downstream effector of D-MEKK1 is also not known. Although, D-MEKK1 was identified as a Licorne-binding protein in a yeast two-hybrid screen, Licorne apparently cannot be phosphorylated by D-MEKK1, and it is not required for the activation of D-p38 by D-MEKK1. Moreover, unlike *lic* mutants, *D-MEKK1* mutants do not show any developmental defects. These observations suggest that one or more MAP2Ks other than Licorne are the direct targets of D-MEKK1 in response to environmental stresses. Also, *D-MEKK1* mutant flies show a normal immune response and normal Dpp signaling. A later study of the *D-MEKK1* mutants found that they are also sensitive to oxidative stress (Paraquat), and confirmed that they have normal survival after septic injury (Brun et al. 2006). Only a small subset of immunity genes (the *turandot* genes) were found to be regulated via D-MEKK1; whether D-p38 mediates their induction downstream of D-MEKK1 is not known. Overall, these results contrast with the original findings of Han et al. (1998b) and Adachi-Yamada et al. (1999b), which suggested a role of D-p38 in the immune response and in wing morphogenesis. These discrepancies could be explained either by genetic redundancy at the level of MAPKs, MAP2Ks, and/or MAP3Ks, by specificity of the signaling response for the various activating stresses or physiological cues, or/and by signaling selectivity from each hierarchical level of MAPK signaling to the lower one (e.g. D-MEKK-1 might activate only one of the two D-p38 isoforms *in vivo*, through one or more MAP2Ks). To address these possibilities, loss-of-function alleles in all identified D-p38 pathway components are clearly required.

The generation of a *D-p38a* null mutation was an important step in this direction (Craig et al. 2004). *D-p38a* mutants are viable and fertile, and they have no observable developmental defects. However, they are susceptible to some environmental stresses, including heat shock, oxidative stress (H_2O_2), and dry starvation. Interestingly, the phenotypes of *D-p38a* mutants only partially overlap with those of *D-MEKK1* mutants: *D-p38a* mutants are sensitive to heat shock but not osmotic stress, whereas *D-MEKK1* mutants are sensitive to osmotic stress but not heat shock (Inoue et al. 2001). These discrepancies suggest that D-p38a might mediate some, but not all of the functions attributed to D-p38 signaling. Moreover,

D-p38a mutants showed a normal response to immune challenge, healed their wounds and survived bacterial infections normally, and had a normal lifespan at 25°C and 29°C (Craig et al. 2004). The observation that *D-p38a* mutants do not show developmental defects suggests that D-p38b might function downstream of Licorne in establishing the polarity of the egg, and/or that it can compensate for D-p38a in this process. However, the overexpression of a D-p38b anti-sense construct in a *D-p38a* mutant background did not result in any developmental phenotypes. Evidently, *D-p38b* null alleles are urgently required (together with null mutations in *D-MKK3/licorne* and *D-MKK4*) to decipher the complexities and the functional roles of p38 signaling *in vivo*.

4.4 RNA interference in cultured *Drosophila* cells

In lieu of such a complete collection of null mutants, a few studies have used RNA interference to dissect p38 signaling in cultured *Drosophila* cells. The approach was to knock down one or multiple MAP2Ks and MAP3Ks by RNAi, and then evaluate the activation of p38 by a range of stimuli. Using this strategy, Zhuang et al. (2006a) showed that the activation of p38 by various stimuli in Schneider cells was not mediated by D-MKK4. This finding is consistent with the fact that D-MKK4 cannot phosphorylate D-p38 in Schneider cells (Han et al. 1998b – please note that this study is often erroneously cited as having shown that D-MKK4 *does* activate D-p38). In contrast, D-MKK3 is the major mediator of D-p38 signaling in Schneider cells: the activation of D-p38 by heat shock and UV radiation was almost abolished by D-MKK3 RNAi, and the activation by NaCl (increased osmolarity) and peptidoglycan (PGN, the active component of LPS) was significantly reduced but not completely blocked. With regard to the MAP3Ks, this study found that UV and PGN activate D-p38 through D-MEKK1, heat shock activates D-p38 through both D-MEKK1 and D-ASK1, and maximal activation of D-p38 by NaCl requires all four *Drosophila* MAP3Ks (D-MEKK1, D-ASK1, D-TAK1, and D-MLK). A similar study showed that the activation of D-p38 by the heavy metals arsenic and cadmium is mediated by D-MEKK1 and does not require D-MKK4 (Ryabinina et al. 2006). In addition to D-MEKK1, the other MAP3Ks were required for full activation of D-p38 by cadmium, but not by arsenic. It was also found that D-ASK1 is required for D-p38 activation by resveratrol. The cell culture approach taken in these two studies is attractive, because of the feasibility of manipulating multiple signaling components in a defined cellular context while applying stressful conditions. However, residual protein activity and off-target effects are general pitfalls of RNAi studies. Perhaps more importantly, the degree to which the results can be extrapolated from the cultured cell line to the organism is unclear. For example, the cell culture experiments identified D-MKK3 as a major mediator of D-p38 activation in response to osmotic stress (Zhuang et al. 2006a). However, it was previously shown that D-MKK3/Licorne is not activated by osmotic stress in larvae (Inoue et al. 2001). Thus, the MAP2K that is relevant to D-p38 activation by stresses *in vivo* is unknown. Importantly, a MAP2K-independent activation mechanism of mammalian p38 has been described, in which auto-phosphorylation of p38 is facilitated by the adaptor protein TAB1 (Ge et al. 2002). The *Drosophila* TAB2 homolog is required for the immune activation of D-JNK and NF-κB (Zhuang et al. 2006b); however, the potential role of the *Drosophila* TAB1 in D-p38 activation has not been addressed.

4.5 Downstream effectors and upstream components of *Drosophila* p38 signaling

Although information about the upstream regulators of D-p38 is accumulating, very little is known about its downstream targets. A recent study showed that the *Drosophila* ATF-2 homolog mediates stress responses downstream of D-p38 (Sano et al. 2005). In *in vitro* kinase assays D-ATF-2 was phosphorylated only by D-p38 and not by D-JNK; in mammals, however, ATF-2 is phosphorylated by both p38 and JNK (Raingeaud et al. 1995; Gupta et al. 1995, 1996). In cultured *Drosophila* cells, the phosphorylation of D-ATF-2 in response to osmotic stress was blocked by the p38 inhibitor SB203580. Overexpressing dominant-negative forms of D-p38 and D-ATF-2 in the developing wing showed a genetic interaction between D-p38 and D-ATF-2 (Sano et al. 2005). Importantly, overexpressing D-ATF-2 in embryos increased their sensitivity to osmotic stress. D-ATF-2 was also found to positively regulate the transcription of several genes in response to osmotic stress. Some of these genes are involved in immunity, suggesting that D-ATF-2 might mediate a positive role of D-p38 in immunity gene expression. However, the precise role of D-p38 in the immune response is still unclear: the early findings of Han et al. (1998b) suggested that D-p38 actually downregulates (rather than activates) immunity gene expression; on the other hand, the *D-p38a* mutant flies showed a normal immune response phenotype (Craig et al. 2004). *D-p38b* and *D-ATF-2* null alleles (if viable) might clarify and reconcile these contradictory findings.

A few additional studies have explored the upstream activators, downstream targets, and physiological roles of D-p38. The overexpression of the *Drosophila* MLK, Slipper, in cultured cells resulted in phosphorylation of D-p38 (Sathyanarayana et al. 2003); however, the activation of D-MLK by ceramide activated the JNK branch of MAPK signaling, but not D-p38 (Sathyanarayana et al. 2002). LK6, the *Drosophila* homolog of the MAPK-interacting kinases 1 and 2 (MNK1 and MNK2), was presumed to be regulated by D-p38 (as the mammalian MNKs are regulated by ERK and p38) (Reiling et al. 2005). However, another study showed that LK6 could bind to D-ERK/Rolled but not to D-p38, and was not activated by arsenic (a p38 activator) (Parra-Pallau et al. 2005). Two reports show that *Drosophila* Schneider cells undergo myogenic differentiation after induction of DNA double strand breaks, or after treatment with drugs that induce replication stress (Hossain et al. 2003, 2005). The pharmacological inhibition of p38 (using SB203580) suppresses this effect (as does the inhibition of the other MAPK branches). These findings are consistent with the role of p38 in vertebrate myogenesis (Keren et al. 2006), and suggest that *Drosophila* might serve as a tractable genetic model to dissect this process.

5 Outlook

The components of the SAPK pathways are structurally and functionally conserved in *Drosophila*, facilitating the *in vivo* analysis of SAPK signaling employing the "awesome power" of this genetic model system. This approach has yielded important new knowledge about the *in vivo* functions of JNK in development, wound healing, innate immunity, apoptosis, oxidative stress resistance, and lifespan regulation. Similarly, the *Drosophila* p38 pathway appears to be involved in development, immunity, and stress response. However, the small degree of genetic redundancy at the p38 kinase level, together with the incomplete repertoire of p38 pathway mutants, still hinders a thorough understanding of the pathway's physiological roles. *D-p38b* mutants, in particular, are absolutely required to address the outstanding questions (Table), before potential breakthroughs similar to those derived from the study of the JNK branch can be achieved. It is anticipated that genetic approaches in *Drosophila* will continue to produce novel insights into SAPK signaling that are directly relevant to rodent models and humans.

References

Adachi-Yamada T, O'Connor MB (2002) Morphogenetic apoptosis: a mechanism for correcting discontinuities in morphogen gradients. Dev Biol 251:74-90

Adachi-Yamada T, Fujimura-Kamada K, Nishida Y, Matsumoto K (1999a) Distortion of proximodistal information causes JNK-dependent apoptosis in *Drosophila* wing. Nature 400:166-169

Adachi-Yamada T, Nakamura M, Irie K, Tomoyasu Y, Sano Y, Mori E, Goto S, Ueno N, Nishida Y, Matsumoto K (1999b) p38 mitogen-activated protein kinase can be involved in transforming growth factor beta superfamily signal transduction in *Drosophila* wing morphogenesis. Mol Cell Biol 19:2322-2329

Bosch M, Serras F, Martin-Blanco E, Baguna J (2005) JNK signaling pathway required for wound healing in regenerating *Drosophila* wing imaginal discs. Dev Biol 280:73-86

Boutros M, Agaisse H, Perrimon N (2002) Sequential activation of signaling pathways during innate immune responses in *Drosophila.* Dev Cell 2:711-722

Brun S, Vidal S, Spellman P, Takahashi K, Tricoire H, Lemaitre B (2006) The MAPKKK Mekk1 regulates the expression of Turandot stress genes in response to septic injury in *Drosophila.* Genes Cells 11:397-407

Craig CR, Fink JL, Yagi Y, Ip YT, Cagan RL (2004) A *Drosophila* p38 orthologue is required for environmental stress responses. EMBO Rep 5:1058-1063

Galko MJ, Krasnow MA (2004) Cellular and genetic analysis of wound healing in *Drosophila* larvae. PLoS Biol 2:E239

Ge B, Gram H, Di Padova F, Huang B, New L, Ulevitch RJ, Luo Y, Han J (2002) MAPKK-independent activation of p38alpha mediated by TAB1-dependent autophosphorylation of p38alpha. Science 295:1291-1294

Glise B, Bourbon H, Noselli S (1995) hemipterous encodes a novel *Drosophila* MAP kinase kinase, required for epithelial cell sheet movement. Cell 83:451-461

Goberdhan DC, Wilson C (1998) JNK, cytoskeletal regulator and stress response kinase? A *Drosophila* perspective. Bioessays 20:1009-1019

Gupta S, Campbell D, Derijard B, Davis RJ (1995) Transcription factor ATF2 regulation by the JNK signal transduction pathway. Science 267:389-393

Gupta S, Barrett T, Whitmarsh AJ, Cavanagh J, Sluss HK, Dérijard B, Davis RJ (1996) Selective interaction of JNK protein kinase isoforms with transcription factors. EMBO J 15:2760-2770

Han S-J, Choi KY, Brey PT, Lee WJ (1998a) Molecular cloning and characterization of a *Drosophila* p38 mitogen-activated protein kinase. J Biol Chem 273:369-374

Han ZS, Enslen H, Hu X, Meng X, Wu IH, Barrett T, Davis RJ, Ip YT (1998b) A conserved p38 mitogen-activated protein kinase pathway regulates *Drosophila* immunity gene expression. Mol Cell Biol 18:3527-3539

Harden N (2002) Signaling pathways directing the movement and fusion of epithelial sheets: lessons from dorsal closure in *Drosophila*. Differentiation 70:181-203

Hossain MS, Akimitsu N, Kurokawa K, Sekimizu K (2003) Myogenic differentiation of *Drosophila* Schneider cells by DNA double-strand break-inducing drugs. Differentiation 71:271-280

Hossain MS, Kurokawa K, Sekimizu K (2005) Induction of fusion-competent myoblast-specific gene expression during myogenic differentiation of *Drosophila* Schneider cells by DNA double-strand breaks or replication inhibition. Biochim Biophys Acta 1743:176-186

Hou XS, Goldstein ES, Perrimon N (1997) *Drosophila* Jun relays the Jun amino-terminal kinase signal transduction pathway to the decapentaplegic signal transduction pathway in regulating epithelial cell sheet movement. Genes Dev 11:1728-1737

Igaki T, Kanda H, Yamamoto-Goto Y, Kanuka H, Kuranaga E, Aigaki T, Miura M (2002) Eiger, a TNF superfamily ligand that triggers the *Drosophila* JNK pathway. EMBO J 21:3009-3018

Inoue H, Tateno M, Fujimura-Kamada K, Takaesu G, Adachi-Yamada T, Ninomiya-Tsuji J, Irie K, Nishida Y, Matsumoto K (2001) A *Drosophila* MAPKKK, D-MEKK1, mediates stress responses through activation of p38 MAPK. EMBO J 20:5421-5430

Janes KA, Albeck JG, Gaudet S, Sorger PK, Lauffenburger DA, Yaffe MB (2005) A systems model of signaling identifies a molecular basis set for cytokine-induced apoptosis. Science 310:1646-1653

Jasper H, Bohmann D (2002) *Drosophila* innate immunity: a genomic view of pathogen defense. Mol Cell 10:967-969

Jasper H, Benes V, Schwager C, Sauer S, Clauder-Münster S, Ansorge W, Bohmann D (2001) The genomic response of the *Drosophila* embryo to JNK signaling. Dev Cell 1:579-586

Kanda H, Igaki T, Kanuka H, Yagi T, Miura M (2002) Wengen, a member of the *Drosophila* tumor necrosis factor receptor superfamily, is required for Eiger signaling. J Biol Chem 277:28372-28375

Kauppila S, Maaty WS, Chen P, Tomar RS, Eby MT, Chapo J, Chew S, Rathore N, Zachariah S, Sinha SK, Abrams JM, Chaundary PM (2003) Eiger and its receptor, Wengen, comprise a TNF-like system in *Drosophila*. Oncogene 22:4860-4867

Keren A, Tamir Y, Bengal E (2006) The p38 MAPK signaling pathway: a major regulator of skeletal muscle development. Mol Cell Endocrinol 252:224-230

Keyse SM (2000) Protein phosphatases and the regulation of mitogen-activated protein kinase signalling. Curr Opin Cell Biol 12:186-192

Kockel L, Homsy JG, Bohmann D (2001) *Drosophila* AP-1: lessons from an invertebrate. Oncogene 20:2347-2364

Kockel L, Zeitlinger J, Staszewski LM, Mlodzik M, Bohmann D (1997) Jun in *Drosophila* development: redundant and non-redundant functions, and regulation by two MAPK signal transduction pathways. Genes Dev 11:1748-1758

Luo X, Puig O, Hyun J, Bohmann D, Jasper H (2007) Foxo and Fos regulate the decision between cell death and survival in response to UV irradiation. EMBO J 26:380-390

Kuranaga E, Kanuka H, Igaki T, Sawamoto K, Ichijo H, Okano H, Miura M (2002) Reaper-mediated inhibition of DIAP1-induced DTRAF1 degradation results in activation of JNK in *Drosophila.* Nat Cell Biol 4:705-710

Martin-Blanco E (2000) p38 MAPK signalling cascades: ancient roles and new functions. Bioessays 22:637-645

Martin-Blanco E, Gampel A, Ring J, Virdee K, Kirov N, Tolkovsky AM, Martinez-Arias A (1998) puckered encodes a phosphatase that mediates a feedback loop regulating JNK activity during dorsal closure in *Drosophila.* Genes Dev 12:557-570

Mattila J, Omelyanchuk L, Kyttala S, Turunen H, Nokkala S (2005) Role of Jun N-terminal Kinase (JNK) signaling in the wound healing and regeneration of a *Drosophila melanogaster* wing imaginal disc. Int J Dev Biol 49:391-399

McEwen DG, Peifer M (2005) Puckered, a *Drosophila* MAPK phosphatase, ensures cell viability by antagonizing JNK-induced apoptosis. Development 132:3935-1946

Mihaly J, Kockel L, Gaengel K, Weber U, Bohmann D, Mlodzik M (2001) The role of the *Drosophila* TAK homologue dTAK during development. Mech Dev 102:67-79

Moreno E, Basler K (2004) dMyc transforms cells into super-competitors. Cell 117:117-129

Moreno E, Basler K, Morata G (2002a) Cells compete for decapentaplegic survival factor to prevent apoptosis in *Drosophila* wing development. Nature 416:755-759

Moreno E, Yan M, Basler K (2002b) Evolution of TNF signaling mechanisms: JNK-dependent apoptosis triggered by Eiger, the *Drosophila* homolog of the TNF super-family. Curr Biol 12:1263-1268

Oh SW, Mukhopadhyay A, Svrzikapa N, Jiang F, Davis RJ, Tissenbaum HA (2005) JNK regulates lifespan in *Caenorhabditis elegans* by modulating nuclear translocation of forkhead transcription factor/DAF-16. Proc Natl Acad Sci USA 102:4494-4499

Ono K, Han J (2000) The p38 signal transduction pathway: activation and function. Cell Signal 12:1-13

Paricio N, Feiguin F, Boutros M, Eaton S, Mlodzik M (1999) The *Drosophila* STE20-like kinase misshapen is required downstream of the Frizzled receptor in planar polarity signaling. EMBO J 18:4669-4678

Parra-Palau JL, Scheper GC, Harper DE, Proud CG (2005) The *Drosophila* protein kinase LK6 is regulated by ERK and phosphorylates the eukaryotic initiation factor eIF4E *in vivo.* Biochem J 385:695-702

Ramet M, Lanot R, Zachary D, Manfruelli P (2002) JNK signaling pathway is required for efficient wound healing in *Drosophila.* Dev Biol 241:145-156

Raingeaud J, Gupta S, Rogers JS, Dickens M, Han J, Ulevitch RJ, Davis RJ (1995) Pro-inflammatory cytokines and environmental stress cause p38 mitogen-activated protein kinase activation by dual phosphorylation on tyrosine and threonine. J Biol Chem 270:7420-7426

Reiling JH, Doepfner KT, Hafen E, Stocker H (2005) Diet-dependent effects of the *Drosophila* Mnk1/Mnk2 homolog Lk6 on growth via eIF4E. Curr Biol 15:24-30

Riesgo-Escovar JR, Hafen E (1997) *Drosophila* Jun kinase regulates expression of decapentaplegic via the ETS-domain protein Aop and the AP-1 transcription factor DJun during dorsal closure. Genes Dev 11:1717-1727

Riesgo-Escovar JR, Jenni M, Fritz A, Hafen E (1996) The *Drosophila* Jun-N-terminal kinase is required for cell morphogenesis but not for DJun-dependent cell fate specification in the eye. Genes Dev 10:2759-2768

Ryabinina OP, Subbian E, Iordanov MS (2006) D-MEKK1, the *Drosophila* orthologue of mammalian MEKK4/MTK1, and Hemipterous/D-MKK7 mediate the activation of D-JNK by cadmium and arsenite in Schneider cells. BMC Cell Biol 7:7

Ryoo HD, Gorenc T, Steller H (2004) Apoptotic cells can induce compensatory cell proliferation through the JNK and the Wingless signaling pathways. Dev Cell 7:491-501

Sano Y, Akimaru H, Okamura T, Nagao T, Okada M, Ishii S (2005) *Drosophila* activating transcription factor-2 is involved in stress response via activation by p38, but not c-Jun NH(2)-terminal kinase. Mol Biol Cell 16:2934-2946

Sathyanarayana P, Barthwal MK, Kundu CN, Lane ME, Bergmann A, Tzivion G, Rana A (2002) Activation of the *Drosophila* MLK by ceramide reveals TNF-alpha and ceramide as agonists of mammalian MLK3. Mol Cell 10:1527-1533

Sathyanarayana P, Barthwal MK, Lane ME, Acevedo SF, Skoulakis EM, Bergmann A, Rana A (2003) *Drosophila* mixed lineage kinase/slipper, a missing biochemical link in *Drosophila* JNK signaling. Biochim Biophys Acta 1640:77-84

Sluss HK, Han Z, Barrett T, Davis RJ, Ip YT (1996) A JNK signal transduction pathway that mediates morphogenesis and an immune response in *Drosophila*. Genes Dev 10:2745-2758

Stadtman ER (2001) Protein oxidation in aging and age-related diseases. Ann N Y Acad Sci 928:22-38

Stronach BE, Perrimon N (1999) Stress signaling in *Drosophila*. Oncogene 18:6172-6182

Stronach BE, Perrimon N (2002) Activation of the JNK pathway during dorsal closure in *Drosophila* requires the mixed lineage kinase, slipper. Genes Dev 16:377-387

Su YC, Treisman JE, Skolnik EY (1998) The *Drosophila* Ste20-related kinase misshapen is required for embryonic dorsal closure and acts through a JNK MAPK module on an evolutionarily conserved signaling pathway. Genes Dev 12:2371-2380

Suzanne M, Irie K, Glise B, Agnes F, Mori E, Matsumoto K, Noselli S (1999) The *Drosophila* p38 MAPK pathway is required during oogenesis for egg asymmetric development. Genes Dev 13:1464-1474

Takatsu Y, Nakamura M, Stapleton M, Danos MC, Matsumoto K, O'Connor MB, Shibuya H, Ueno N (2000) TAK1 participates in c-Jun N-terminal kinase signaling during *Drosophila* development. Mol Cell Biol 20:3015-3026

Vidal M, Cagan RL (2006) *Drosophila* models for cancer research. Curr Opin Genet Dev 16:10-16

Wang MC, Bohmann D, Jasper H (2003) JNK signaling confers tolerance to oxidative stress and extends lifespan in *Drosophila*. Dev Cell 5:811-816

Wang MC, Bohmann D, Jasper H (2005) JNK extends life span and limits growth by antagonizing cellular and organism-wide responses to insulin signaling. Cell 121:115-125

Zeitlinger J, Kockel L, Peverali FA, Jackson DB, Mlodzik M, Bohmann D (1997) Defective dorsal closure and loss of epidermal decapentaplegic expression in *Drosophila* fos mutants. EMBO J 16:7393-7401

Zhuang ZH, Zhou Y, Yu MC, Silverman N, Ge BX (2006a) Regulation of *Drosophila* p38 activation by specific MAP2 kinase and MAP3 kinase in response to different stimuli. Cell Signal 18:441-448

Zhuang ZH, Sun L, Kong L, Hu JH, Yu MC, Reinach P, Zang JW, Ge BX (2006b) *Drosophila* TAB2 is required for the immune activation of JNK and NF-kappaB. Cell Signal 18:964-970

Zou S, Meadows S, Sharp L, Jan LY, Jan YN (2000) Genome-wide study of aging and oxidative stress response in *Drosophila melanogaster*. Proc Natl Acad Sci USA 97:13726-13731

Bohmann, Dirk

Department of Biomedical Genetics, University of Rochester Medical Center, Rochester, New York 14642, USA

Dirk_Bohmann@urmc.rochester.edu

Sykiotis, Gerasimos P.

Department of Biomedical Genetics, University of Rochester Medical Center, Rochester, New York 14642, USA

Protein kinases as substrates for SAPKs

Alexey Kotlyarov and Matthias Gaestel

Abstract

The stress-activated protein kinases (SAPKs) consist of the c-Jun N-terminal kinases (JNKs or SAPK1α, β, γ) and of the four p38 MAPK-isoforms SAPK2a, b, SAPK3 and SAPK4 and phosphorylate a wide variety of different substrate proteins such as transcription factors, enzymes, and structural proteins. Of these kinases, only SAPK2s are known to phosphorylate and regulate downstream kinases; thereby, appending an additional level to this stress activated kinase cascade. Historically, these downstream kinases were designated as mitogen- and stress-activated protein kinases – MSKs, as MAPK-interacting protein kinases – MNKs, and as MAPK-activated protein kinases – MAPKAPKs or MKs. Although their kinase domains are phylogenetically related and show some similarities in regulation, the downstream kinases regulate gene expression at different levels, which contribute to orchestration of the stress response.

1 Definition of kinases downstream to SAPKs

Excellent reviews which summarize and discuss specific groups of MAPK- and SAPK-activated protein kinases already exist (Roux and Blenis 2004; Gaestel 2006; Hauge and Frodin 2006) and, therefore, we will mainly focus on providing recent new information on the group of SAPK-regulated protein kinases. According to the definition of SAPK-regulated protein kinases, the number of these enzymes is limited and contains the different subgroups of Mitogen- and stress-activated kinases (MSKs), MAPK-interacting kinases (MNKs) and MAPK-activated protein kinases (MAPKAPKs or MKs). Remarkably, there are no protein kinases known to be downstream to SAPK1/JNKs, instead all known SAPK-regulated kinases are substrates for SAPK2/p38. Since activation by SAPK2s *in vivo* could not be detected for some enzymes of the MNK- and MK-subfamilies, these enzymes have to be excluded from this review. As shown in Figure 1A, protein kinases which are activated by the SAPK2s *in vivo* are MK2, MK3, MNK1, MSK1, and MSK2. MK5 (also known as PRAK) neither binds to SAPK2a/p38a nor is activated by SAPK2-specific stimuli (Shi et al. 2003). Recently, it has been shown that the atypical protein kinases ERK3 (Schumacher et al. 2004; Seternes et al. 2004) and ERK4 (Aberg et al. 2006; Kant et al. 2006) are responsible for its activation by cytoplasmic anchoring. However, in oncogenic ras-induced senescence

Topics in Current Genetics, Vol. 20
F. Posas, A. R. Nebreda (Eds.): Stress-Activated Protein Kinases
DOI 10.1007/4735_2007_0252 / Published online: 1 August 2007

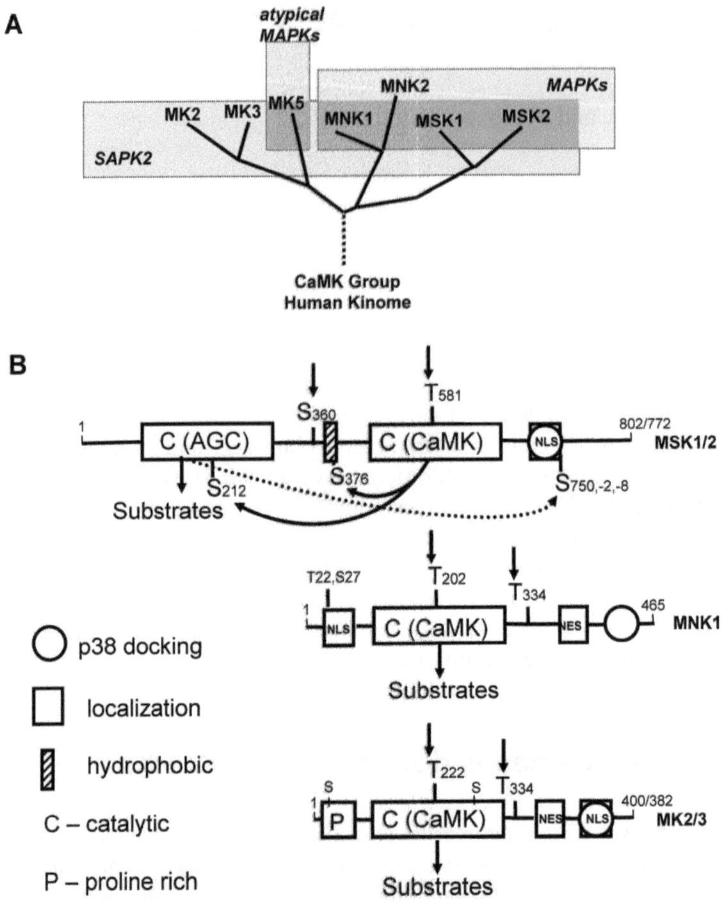

Fig. 1. A) Phylogenetic tree of the group of kinases activated by MAPKs and SAPK2s. By homology of the catalytic domain all of these kinases belong to the calcium/calmodulin-regulated family of protein kinases (CaMK)(Manning et al. 2002). Protein kinases activated by the SAPK2s *in vivo* are MAPK-activated protein kinase (MAPKAPK or MK) 2 and MK3, MAPK-interacting kinase (MNK)1, Mitogen- and stress-activated kinase (MSK)1 and MSK2. B) Schematic presentation of the primary structure, domains and sequence motifs of MSK1/2, MNK1, and MK2/3 from human. Incoming arrows represent regulation phosphorylations by SAPKs. Phosphorylation sites are indicated. NES – nuclear export signal, NLS – nuclear localisation signal.

activation of MK5 apparently proceeds in an SB203580-dependent manner (Sun et al. 2007). MNK2 lacks the SAPK2/p38-specific residues in its interaction domain leading to exclusive activation by classical MAPKs such as ERK2 but not by SAPK2 *in vivo* (Scheper et al. 2003; Parra et al. 2005).

2 Primary structure and overview

The primary structure of the five remaining SAPK-regulated protein kinases is shown in Figure 1B. While MK2, MK3, and MNK1 contain a single calcium/calmodulin-regulated kinase (CaMK)-type catalytic domain, MSK1 and MSK2 display two different catalytic domains, a N-terminal kinase domain (NTK) of the PKA/G/C (AGC)-type and a C-terminal kinase domain (CTK) of the CaMK-type. Accordingly, regulation of MSKs is more complex and involves intramolecular regulatory phosphorylations between the two catalytic domains (McCoy et al. 2005). All five protein kinases are activated by SAPK2-phosphorylation (incoming arrows in Fig. 1B) within the catalytic domain at the kinase activation loop as well as in regulatory regions outside the catalytic domain (Ben-Levy et al. 1995; Engel et al. 1995). Furthermore, all five kinases contain SAPK2-docking sequence motifs and signal sequences which may determine subcellular localization (Ben-Levy et al. 1998; Deak et al. 1998; Engel et al. 1998; Parra-Palau et al. 2003). Since these signals partially overlap (MSK1/2, MK2/3) and are in the vicinity of regulatory phosphorylation sites as well, steric or allosteric co-regulation of these signals seems possible and probable. In MK2/3 and MNK1 the CaMK-like catalytic domain is responsible for substrate phosphorylations while in MSK1/2 the AGC-like catalytic domain phosphorylates the known substrates (See Table 1).

3 SAPK-regulated kinases in detail

3.1 MSKs

MSK1/2 are structurally very similar to another group of kinases, the p90 ribosomal S6 kinases (RSKs), which also contain two catalytic domains (NTK, CTK) separated by a linker region containing a hydrophobic motif (Hauge and Frodin 2006). In contrast to RSKs, where a MAPK docking site for ERK1/2 is present in the C-terminus, this docking site in MSKs contains two more basic amino acid residues, which now also enable p38 to bind to the C-terminus. Furthermore, MSK1/2 display bipartite nuclear localisation signals overlapping with this basic MAPK docking site. Hence, in contrast to RSKs, which are localised in cytoplasm and nucleus and are activated exclusively by the classical MAPKs, MSK1/2 are restricted to the nuclear compartment and are activated by both ERKs and p38.

3.1.1 Regulation by SAPKs

There are also differences in the mechanisms of activation of RSKs and MSK1/2. While for full activation of RSKs the cooperation between ERKs phosphorylating the CTK and the linker domain including the hydrophobic motif and

Table 1. Three subgroups of SAPK-regulated kinases

Subfamily	**MSK1/2**	**MNK1**	**MK2/3**
Phosphorylation site motifs	-RRXS(P)- -ARXS(P)- -KRKXS(P)- -RRXXS(P)- -QRXXS(P)	-KXXS(P)- -RXS(P)- -HXRXXS(P)- -ASASSS(P)- -YGGSSS(P)-	-L/F/GXRXXS(P)- -RXXS(P)- -LRXXXS(P)- -KXXS(P)-
Sub-cellular localisation	Nucleus (Deak et al. 1998)	Cytoplasm (and nucleus) – constitutively exported from nucleus (Parra-Palau et al. 2003)	Nucleus and cytoplasm – activation-dependent shuttling (Ben-Levy et al. 1998; Engel et al. 1998)
Substrate proteins	CREB S133/ATF1 S63 (Deak et al. 1998; Arthur and Cohen 2000; Wiggin et al. 2002) Histone H3 S10, 28, HMG-14 S6 (Soloaga et al. 2003) NF-kB p65 S276 (Vermeulen et al. 2003) ER81 S191,S216 (Janknecht 2003)	eIF4E S209 (Waskiewicz et al. 1997, 1999; Pyronnet et al. 1999) hnRNP A1 S192, S310 (Buxade et al. 2005) hSpry2 S112, S121 (DaSilva et al. 2006)	Hsp25 S15, S86; Hsp27 S15,S78,S82 (Gaestel et al. 1991; Landry et al. 1992; Stokoe et al. 1992b) TTP (Chrestensen et al. 2004; Stoecklin et al. 2004) Others see (Gaestel 2006)
Target genes	Nur77, Nurr1, Nor1(Darragh et al. 2005) c-fos (Schiller et al. 2006)	TNF (Buxade et al. 2005)	TNF, IL-6, IL-1, IFNγ (Kotlyarov et al. 1999)
Level of control of gene expression	transcription	translational control	mRNA stability and translation
Small molecule inhibitors	H89 Ro 318220 (Deak et al. 1998) Imidazo(4,5-c)pyridine 1,2,5-oxadiazol-3-yl (Bamford et al. 2005)	CGP57380 (Knauf et al. 2001)	MK2-I (2,4-diamino-5H-chromeno-(2,3-b)pyridine-3-carbonitrile) (Vertii et al. 2006)
Physiological functions	Chromatin remodelling, Transcriptional regulation of the inflammatory and stress response, Induction of immediate early genes	Regulation of protein translation	Inflammation, Regulation of phosphorylation of sHsps and stress resistance, Generation of 14-3-3 binding sites
Knockout mice	MSK1/2 DKO is viable and shows reduction in c-fos and junB gene transcription in response to anisomycin or UV (Wiggin et al. 2002)	MNK1/2 DKO is viable and shows lack of phosphorylation of eIF4E (Ueda et al. 2004)	MK2/3 DKO is viable (Ronkina et al., 2006), MK2 knockout is resistant against endotoxic shock (Kotlyarov et al. 1999) and more susceptible to Listeria infection (Lehner et al. 2002)

subsequent docking of PDK1 and phosphorylation of the NTK by PDK1 is necessary, MSK1/2 can be fully activated by p38 or ERKs in a complex scenario without the help of further kinases (McCoy et al. 2005). First, p38 (or ERKs) phosphorylates the activation loop of the CTK of MSK1/2 (T581, cf. Fig. 1B) and a serine residue within the linker region (S360). The activated CTK than phosphorylates two other serines, one in the hydrophibic motif (S376) and the other at the activation loop of the NTK (S212). In the non-phosphorylated state the NTK displays a sterically blocked ATP-binding site due to formation of a three-stranded beta sheet unique for MSKs (Smith et al. 2004). Especially phosphorylation of the hydrophobic motif leads to large conformational changes (Frodin et al. 2002) probably reforming the alphaB helix from the three-stranded beta sheet which is necessary for NTK activity (Smith et al. 2004). The activated NTK is responsible for substrate phosphorylation and for auto-phosphorylation of the C-terminus of MSK1/2 (S750, S752, S758). It is assumed that this auto-phosphorylation is the first step of inactivation of MSKs (Hauge and Frodin 2006).

3.1.2 Substrates and function of MSK1/2

According to the constitutive nuclear localisation of MSK1/2, the targets of MSKs are exclusively nuclear and comprise transcription factors and components of the chromatin. Phosphorylation of the cAMP-responsive element binding protein (CREB) and the closely related transcription factor ATF1 at S133 and S63, respectively, by MSKs (Deak et al. 1998; Arthur and Cohen 2000) explains the stress- and cytokine-dependent activation of cAMP-responsive genes and was verified in MSK1/2 double knockout mice (Wiggin et al. 2002). Accordingly, MSK1 has been recently described as the protein kinase responsible for interleukin-1-stimulated CREB-mediated expression of the immediate early gene *c-fos* in keratinocytes (Schiller et al. 2006). In addition, the transcription factor ER81, which is involved in ontogenesis and breast tumour formation, is phosphorylated by MSK1 at two different sites, serine 191 and serine 216, and activated in its transcription capability (Janknecht 2003). Unexpectedly, mutation of both MSK1 phosphorylation sites of ER81 did not abrogate the ability of MSK1 to stimulate ER81 function. Since MSK1 also associates with CBP/p300, it is supposed that MSK1 might target ER81 cofactors as well (Janknecht 2003). Interestingly, MSK1 is also involved in an essential crosstalk to NF-kB signalling and, hence, contributes directly to the inflammatory response. It associates with the p65 NF-kB subunit in a stimulus-dependent manner and phosphorylates it at serine 276 (Vermeulen et al. 2003). This mechanism is responsible for the stimulation of the trans-activating capacity of p65 NF-kB. Another cross talk of SAPKs with TGFβ-signalling exists via MSK1-dependent binding of Smad3 to the transcriptional co-activator CBP/p300 (Abecassis et al. 2004). Whether TGFβ-signalling is significantly impaired in MSK knockout mice is not known so far.

Apart from regulation of transcription factors, MSKs also contribute to transcriptional regulation by modification of chromatin proteins as shown by the reduced or abolished phosphorylation of histone H3 at serines 10 and 28 and high mobility group protein HMG-14 at serine 6 in mice lacking MSK1 and MSK2 (Soloaga et al. 2003). Since a distinct sub-fraction of H3 has been shown to be phosphorylated both on S10 and S28 within the same molecule during nuclear interface (Dyson et al. 2005), specific immediate early gene activation by MSKs seems probable at this level as well. MSK-mediated H3 phosphorylation might also influence other H3 modifications such as methylation, ubiquitylation, or acetylation in a highly dynamic manner (Hazzalin and Mahadevan 2002; Clayton et al. 2006). This idea is supported by the finding that changes in histone H3 phospho-acetylation during early embryonic stem cell differentiation are directly mediated by mitogen- and stress-activated protein kinase 1 (Lee et al. 2006). In contrast to activation of immediate early genes by H3 phosphorylation, MSK1 has also been described to phosphorylate serine 1 in histone H2A and to inhibit transcription in a chromatin-dependent manner (Zhang et al. 2004). In oncogene-transformed fibroblasts with upregulation of Ras-activity an increased H3 serine 10 phosphorylation by MSK1 is observed. However, this phosphorylation seems to be exclusively due to ERK-mediated activation of MSK1 independent of SAPKs (Drobic et al. 2004). An increased level of phospho-MSK1 has also been recently reported for vulnerable neurons from patients with Alzheimer's disease compared to age matched controls (Webber et al. 2005).

Besides interaction of MSKs with SAPKs and MAPKs, in endothelial cells the CTD of MSK1 seems also to interact with group V secretory phospholipase A_2 (Marchand et al. 2006). The facts that vascular endothelia growth factor (VEGF) stimulates MSK1 activation and platelet-activating factor (PAF) synthesis by sPLA2 in parallel and that mutants of MSK1 interfere with PAF synthesis support the notion that MSK1 plays an essential role in PAF synthesis (Marchand et al. 2006). This idea should be proofed using MSK-deficient mice.

3.2 MNKs

There are two human MNK genes, mnk1 and mnk2, each gene coding for the long (a) and short (b) alternatively spliced isoforms (MNK1a, MNK1b, MNK2a, and MNK2b) (Scheper et al. 2003; O'Loghlen et al. 2004). Alternative splicing affects neither the N-terminus, bearing a stretch of basic residues (KR-box) that functions as a nuclear localisation signal and as a binding region for the eukaryotic initiation factor 4G, nor the kinase domain, but leads to a changed C-terminus. MNK1a and MNK2a contain putative nuclear export motifs and MAP kinase-binding sites C-terminal of the central kinase domains, both of which are removed by alternative splicing in MNK1b and MNK2b. These shorter MNK-b isoforms were not yet identified in mice.

MNKs kinase domains are architecturally distinct from other protein kinases. MNKs contain three insertions in the conserved kinase domain, one of which with a cluster of four cysteines, arranged in a manner suggesting a metal ion binding site. This zinc finger-like module within the kinase domain could provide a binding platform for proteins and/or nucleic acids (Jauch et al. 2005). Additionally, MNKs contain an atypical magnesium binding motif and this results in unusual arrangement of the activation segment and an ATP-competitive conformation of the magnesium-binding DFD-motif that differs from the conserved DFG sequence of most other kinases. Recent studies revealed that in non-phosphorylated MNK1 the activation segment is converted into an auto-inhibitory module which exhibits a conformation that obstructs ATP binding (Jauch et al. 2006). Non-canonical arrangement of the kinase domain and structural transitions accompanying MNK activation can be implied for the design of MNK-specific inhibitors.

3.2.1 Regulation by MAPKs

Both MNK1 and MNK2 are directly activated *in vitro* by the MAPKs/ERK1,2 and SAPK2/p38, which phosphorylate one or two threonine residues (e.g. T197 and T202 of the mouse MNK1), located in the activation loop. Studies with ectopically expressed MNKs shown that MNK1 is strongly activated by growth factors, cellular stresses, and inflammatory cytokines via ERK and/or the p38 MAPK, depending on the signalling context, whereas MNK2 has a relatively high basal activity that is hardly affected by changes in MAPK activity. Studies with MNK1/2 chimeric proteins (Parra et al. 2005) demonstrated that specific features of the catalytic domain and C-terminus cause constant MNK2 binding to phospho-ERK, which can continue to phosphorylate MNK2 and keep it in an activated state, but not to SAPK2.

Examination of *in vivo* phosphorylation state of MNKs specific substrates in available SAPK2 and MAPK knockout models could be used to further define the impact of these kinases in MNK activation.

3.2.2 Substrates and function

The first well characterized substrate for the MNKs is eukaryotic initiation factor 4E (eIF4E) (Waskiewicz et al. 1997). eIF4E is regulated by eIF4E binding proteins (4E-BPs), which in their hypophosphorylated form inhibit cap-dependent translation, and by direct phosphorylation by MNKs. MNKs binding to the C-terminal region of eIF4G containing two atypical HEAT repeats (Bellsolell et al. 2006) is required to recruits the kinase to the eIF4F and subsequent eIF4E phosphorylation at S209 (in human eIF4E). Stimulation of cap-dependent mRNA translation by mitogens, growth factors, serum, and other nutrients is correlated with phosphorylation of eIF4E at this site. Phosphorylation of eIF4E reduces its affinity for the cap structure and may also limit cap-dependent mRNA translation (Knauf et al. 2001; Zuberek et al. 2004).

Whether phosphorylation of eIF4E at Ser209 exerts a positive or inhibitory effect on translation efficiency has remained controversial (reviewed in Scheper and Proud 2002). Genetic characterization of the *Drosophila* homologue of Mnk1/2, Lk6 was simultaneously reported by two groups. In both studies Lsk6 loss of function mutation leads to dramatically reduced eIF4E phosphorylation, but comparison of phenotypes reveals surprising differences. Arguier et al. (2005) observed a significant growth reduction of Lk6 mutants in contrast to Reiling et al. (2005) who stated dispensable Lk6 function, at least under high protein diet. The effects of Lk6 activity could be context-dependent and lead either to growth stimulation or inhibition, depending on the dietary conditions.

Targeted disruption of the mouse MNK1 and MNK2 genes was described by Ueda et al (2004). In this study exons 5 and 6 of the MNK1 gene or exons 5 to 9 of the MNK2 gene were replaced with a Neo^r cassette. The single KO mice, as well as $MNK1^{-/-}$ $MNK2^{-/-}$ DKO mice were viable, fertile, and did not show developmental abnormalities. LPS or insulin-induced upregulation of eIF4E phosphorylation in the spleen, liver, or skeletal muscle was abolished in $MNK1^{-/-}$ mice, whereas the basal eIF4E phosphorylation levels were decreased in $MNK2^{-/-}$ mice. In MNK1-MNK2 DKO mice, no phosphorylated eIF4E was detected in any tissue studied, even after LPS or insulin injection. However, neither general protein synthesis nor cap-dependent translation, as assayed by a bicistronic reporter assay system, was affected in MNK-deficient embryonic fibroblasts, despite the absence of phosphorylated eIF4E. Thus, MNK1 and MNK2 are exclusive eIF4E kinases both in cultured fibroblasts and adult tissues, and they regulate inducible and constitutive eIF4E phosphorylation, respectively. These results strongly suggest that eIF4E phosphorylation at Ser209 is not essential for cell growth during development. Instead, the phosphorylation of eIF4E may play a role in the fine control of limited genes through regulation of mRNA translation, stabilization, or processing.

Recently, evidence was provided that the MNKs play roles in the posttranscriptional regulation of TNFα synthesis in T cells by mechanisms that involve its 3`UTR (Buxade et al. 2005). Inhibition of TNF production is paralleled by inhibition of the phosphorylation of eukaryotic initiation factor 4E (eIF4E). Overexpression of MNK1 enhances protein expression from a reporter containing the TNFα 3`UTR, whereas inhibition or knockdown of MNKs blocks synthesis of endogenous TNFα. T-cell activation results in phosphorylation of known AU-rich binding protein hnRNP A1 at two MNK sites, and decreases its ability to bind the TNFα mRNA *in vivo*, in a MNK-dependent manner. Binding of MNK to the eIF4G, which interact with eIF4E and the poly(A) binding protein can potentially recruit the MNKs to the 3`UTR and associated ARE-binding proteins as hnRNP A1. Participation of MNKs in mRNA processing (van der Houven van Oordt et al. 2000) could also explain the altered nuclear-cytoplasmic transport of TNF mRNA in TPL2 knockout mice (Dumitru et al. 2000).

In residential macrophages, the combination of p38 and ERK1/2 regulate translation of TNFα, presumably through MNK1 - an addition to the well established p38-MK2/3 axis. TNFα production is inhibited in a concentration-dependent manner by MNK1 inhibitor CGP57380 (Andersson and Sundler 2006; Cherla et al. 2006). Recently, a MNK-dependent relocalisation of hnRNP A1 to the stress granules (SGs) under the stress conditions was demonstrated (Guil et al. 2006). SGs and processing bodies are distinct RNA-containing cytoplasmic domains playing important role in RNA metabolism and posttranslational regulation of gene expression, recently reviewed in (Anderson and Kedersha 2006). These RNA granules contain various ribosomal subunits, translation factors, decay enzymes, helicases, scaffold proteins, and RNA-binding proteins, including MK2/3 substrates such as TTP and Hsp25. Dynamic nature and extreme complexity of RNA bodies hinder our understanding of this crucial regulatory mechanism. Detailed biochemical examination of interactions, combined with immuno-localisation and comparative phenotypic characterisation of knockout animals could help to unravel their spatial organization and mode of action in future.

Lately, Sprouty (hSpry2) was identified as a new MNK substrate (DaSilva et al. 2006). Sprouty proteins are negative feedback modulators of receptor tyrosine kinase (RTK) pathways in *Drosophila melanogaster* and mammals. Phosphorylation of hSpry2 on S112 and S121 by MNK1 antagonizes the phosphorylation of hSpry2 on Y55. Consequently, serine phosphorylation of hSpry2 stabilizes the protein by interfering with tyrosine-dependent degradation via the ubiquitin-proteasome pathway. Involvement of MNK1 in the regulation of RTK pathways through Sprouty provide the explanation for suppression of Ras/ERK signalling in *Drosophila* eye development by Lk6 (Huang and Rubin 2000).

3.3 MK2/3

MK2 was first purified in 1992 (Stokoe et al. 1992a) and characterised as a small heat shock protein Hsp25/Hsp27 kinase (Stokoe et al. 1992b) activated by SAPK2s (Freshney et al. 1994; Rouse et al. 1994). MK3 was identified as a gene product possibly involved in small cell lung cancer (Sithanandam et al. 1996) and by homology cloning (McLaughlin et al. 1996). Both kinases are structurally very closely related (cf. Fig. 1) and activated exclusively by SAPK2s.

3.3.1 Regulation by SAPKs

MK2 and 3 form a relatively stable complex with SAPK2 in which both proteins are mutually stabilised (Kotlyarov et al. 2002; Sudo et al. 2005). After activation of SAPK2, for full activation of MK2/3 phosphorylation of two or three sites is necessary (Ben-Levy et al. 1995; Engel et al. 1995). One regulatory phosphorylated threonine residue is located at the activation loop of the kinase implicating direct activation and, for MK2 there is also another serine residue within the catalytic domain, which is phosphorylated and might contribute to direct regulation of catalytic activity as well. Interestingly, in both MK2 and MK3 there is a second phosphorylated threonine residue, which has been described as the major phosphorylation site of MK2 located in a hinge region between catalytic domain and C-terminal extension (Stokoe et al. 1992a). The C-terminal extension contains an auto-inhibitory A-helix which might bind to the catalytic domain (Engel et al. 1995; Zu et al. 1995) as well as signals for sub-cellular localisation (NLS, NES). Hence, the phosphorylation site in the hinge region might regulate kinase activity and localisation in parallel. Indeed, it could be demonstrated that phosphorylation of this threonine residue leads to parallel activation and nuclear-cytoplasmic translocation of MK2 (Ben-Levy et al. 1998; Engel et al. 1998) due to a switch from an inactive closed conformation where the A-helix binds to the catalytic core and the NES is masked to an active open conformation where the NES can bind to the export receptor (Neininger et al. 2001; Meng et al. 2002). A similar mechanism is supposed for MK3 which also shows activation-dependent nuclear export (Tanoue et al. 2001; Zakowski et al. 2004). Both MK2 and MK3 show the same SAPK2-dependent characteristics of activation by cytokines and stress (Clifton et al. 1996; Ronkina et al. 2007).

3.3.2 Substrates and function of MK2/3

A wide variety of substrates have been described for MK2/3 ranging from proteins involved in regulation of actin polymerisation, such as small Hsps (Stokoe et al. 1992b), via transcription factors, such as SRF (Heidenreich et al. 1999), E47 (Neufeld et al. 2000), ER81 (Janknecht 2001) and HSF1 (Wang et al. 2006), to RNA-binding proteins, such as hnRNP A0 (Rousseau et al. 2002), poly A-binding protein (Bollig et al. 2003), or TTP (Chrestensen et al. 2004; Stoecklin et al. 2004) (see Gaestel 2006, and supplementary Table thereof). Recently, interesting targets of MK2/3 involved in chromatin remodelling and cell cycle regulation, such as the component of the polycomb gene repressive complex Bmi1 (Voncken et al. 2005), the phosphatases Cdc25B,C (Manke et al. 2005), and the ubiquitin-ligase HDM2 (Weber et al. 2005) have also been identified. Although some of the substrates have been demonstrated only for one of both kinases, there is no obvious difference in substrate specificity between MK2 and MK3 (Clifton et al. 1996). The MK2/3 substrate recognition motif overlaps with the motif of 14-3-3 binding sites, opening the intriguing possibility that MK2/3 may regulate 14-3-3 binding of their substrate proteins (Manke et al. 2005; Gaestel 2006).

An idea about MK2 and MK3 function came from their mouse knockouts. MK2-deficient mice are viable and fertile, but show a compromised expression of TNF and other inflammatory cytokines in response to LPS-challenge which results in increased resistance against endotoxic shock (Kotlyarov et al. 1999) and collagen-induced arthritis (Hegen et al. 2006) as well as increased susceptibility to infection (Lehner et al. 2002). Furthermore, MK2 deficient cells show defects in migration and chemotaxis (Hannigan et al. 2001) and altered susceptibility to stresses such as heat shock or arsenite treatment (Vertii et al. 2006). Interestingly, the defect in production of TNF and other cytokines in MK2-deficient macrophages is due to changes in post-transcriptional regulation of cytokine-mRNA stability and translation and depends on the presence of AU-rich elements in the 3'-region of the mRNA (Han and Ulevitch 1999; Kotlyarov et al. 1999; Winzen et al. 1999; Neininger et al. 2002). Hence, RNA-binding substrates of MK2 are of special interest to understand the molecular mechanism of this regulation. The LPS-induced phosphorylation-dependent binding of the MK2-substrate hnRNP A0 to specific cytokine mRNAs (Rousseau et al. 2002) and the MK2-dependent stabilisation and inactivation of the TNF-mRNA destabilising protein TTP (Stoecklin et al. 2004; Brook et al. 2006; Hitti et al. 2006) provide first examples for the complex mechanisms of this regulation of mRNA stability and translation. However, especially the mechanism of regulation of translation of specific cytokine mRNAs by phosphorylation is not well understood so far.

As mentioned above MK3 displays extensive structural similarities to MK2 and identical functional properties *in vitro,* but the clear phenotype of MK2-deficient mice indicates that MK3 can not compensate for MK2 due to the significantly lower level of expression and activity of MK3 (Ronkina et al. 2007). Adequately, MK3-deficient mice do not show a significant phenotype and an effect of deletion of MK3 on TNF production and TTP stability can only be detected in a MK2-free background by comparing MK2/3 DKO with MK2 KO (Ronkina et al. 2007).

Hence, it seems that both kinases share the same physiological function *in vivo* but are expressed to different levels. However, a specific unknown function for MK2 or MK3 or compensation between both kinases for another function, such as chromatin remodelling or cell cycle regulation cannot be excluded at the moment.

4 Summary and Perspectives

Protein kinases downstream to SAPK2/p38 contribute to the regulation of gene expression at different levels. While MSKs act at the level of chromatin remodelling and transcription, MKs can determine specific transcript stability and translation and MNK regulates general translation initiation. However, the borders between these functions are leaky, since MKs are also involved in chromatin remodelling (Voncken et al. 2005) and transcriptional regulation. MNK also regulates translation of specific AU-rich element carrying mRNAs (Buxade et al. 2005). In addition to regulation of gene expression, MKs contribute to cell migration and stress resistance. The future understanding of concerted action of protein kinases downstream to SAPK2/p38 will significantly contribute to the physiology of stress response and inflammation.

Acknowledgement

The authors' work is supported by grants from the Deutsche Forschungsgemeinschaft and from the European Community

References

Abecassis L, Rogier E, Vazquez A, Atfi A, Bourgeade MF (2004) Evidence for a role of MSK1 in transforming growth factor-beta-mediated responses through p38alpha and Smad signaling pathways. J Biol Chem 279:30474-30479

Aberg E, Perander M, Johansen B, Julien C, Meloche S, Keyse SM, Seternes OM (2006) Regulation of MAPK-activated protein kinase 5 activity and subcellular localization by the atypical MAP kinase ERK4/MAPK4. J Biol Chem 281:35499-35510

Anderson P, Kedersha N (2006) RNA granules. J Cell Biol 172:803-808

Andersson K, Sundler R (2006) Posttranscriptional regulation of TNFalpha expression via eukaryotic initiation factor 4E (eIF4E) phosphorylation in mouse macrophages. Cytokine 33:52-57

Arquier N, Bourouis M, Colombani J, Leopold P (2005) *Drosophila* Lk6 kinase controls phosphorylation of eukaryotic translation initiation factor 4E and promotes normal growth and development. Curr Biol 15:19-23

Arthur JS, Cohen P (2000) MSK1 is required for CREB phosphorylation in response to mitogens in mouse embryonic stem cells. FEBS Lett 482:44-48

Bamford MJ, Alberti MJ, Bailey N, Davies S, Dean DK, Gaiba A, Garland S, Harling JD, Jung DK, Panchal TA, Parr CA, Steadman JG, Takle AK, Townsend JT, Wilson DM, Witherington J (2005) (1H-imidazo[4,5-c]pyridin-2-yl)-1,2,5-oxadiazol-3-ylamine derivatives: a novel class of potent MSK-1-inhibitors. Bioorg Med Chem Lett 15:3402-3406

Bellsolell L, Cho-Park PF, Poulin F, Sonenberg N, Burley SK (2006) Two structurally atypical HEAT domains in the C-terminal portion of human eIF4G support binding to eIF4A and Mnk1. Structure 14:913-923

Ben-Levy R, Hooper S, Wilson R, Paterson HF, Marshall CJ (1998) Nuclear export of the stress-activated protein kinase p38 mediated by its substrate MAPKAP kinase-2. Curr Biol 8:1049-1057

Ben-Levy R, Leighton IA, Doza YN, Attwood P, Morrice N, Marshall CJ, Cohen P (1995) Identification of novel phosphorylation sites required for activation of MAPKAP kinase-2. Embo J 14:5920-5930

Bollig F, Winzen R, Gaestel M, Kostka S, Resch K, Holtmann H (2003) Affinity purification of ARE-binding proteins identifies poly(A)-binding protein 1 as a potential substrate in MK2-induced mRNA stabilization. Biochem Biophys Res Commun 301:665-670

Brook M, Tchen CR, Santalucia T, McIlrath J, Arthur JS, Saklatvala J, Clark AR (2006) Posttranslational regulation of tristetraprolin subcellular localization and protein stability by p38 mitogen-activated protein kinase and extracellular signal-regulated kinase pathways. Mol Cell Biol 26:2408-2418

Buxade M, Parra JL, Rousseau S, Shpiro N, Marquez R, Morrice N, Bain J, Espel E, Proud CG (2005) The Mnks are novel components in the control of TNF alpha biosynthesis and phosphorylate and regulate hnRNP A1. Immunity 23:177-189

Cherla RP, Lee SY, Mees PL, Tesh VL (2006) Shiga toxin 1-induced cytokine production is mediated by MAP kinase pathways and translation initiation factor eIF4E in the macrophage-like THP-1 cell line. J Leukoc Biol 79:397-407

Chrestensen CA, Schroeder MJ, Shabanowitz J, Hunt DF, Pelo JW, Worthington MT, Sturgill TW (2004) MAPKAP kinase 2 phosphorylates tristetraprolin on *in vivo* sites including Ser178, a site required for 14-3-3 binding. J Biol Chem 279:10176-10184

Clayton AL, Hazzalin CA, Mahadevan LC (2006) Enhanced histone acetylation and transcription: a dynamic perspective. Mol Cell 23:289-296

Clifton AD, Young PR, Cohen P (1996) A comparison of the substrate specificity of MAPKAP kinase-2 and MAPKAP kinase-3 and their activation by cytokines and cellular stress. FEBS Lett 392:209-214

Darragh J, Soloaga A, Beardmore VA, Wingate AD, Wiggin GR, Peggie M, Arthur JS (2005) MSKs are required for the transcription of the nuclear orphan receptors Nur77, Nurr1 and Nor1 downstream of MAPK signalling. Biochem J 390:749-759

DaSilva J, Xu L, Kim HJ, Miller WT, Bar-Sagi D (2006) Regulation of sprouty stability by Mnk1-dependent phosphorylation. Mol Cell Biol 26:1898-1907

Deak M, Clifton AD, Lucocq LM, Alessi DR (1998) Mitogen- and stress-activated protein kinase-1 (MSK1) is directly activated by MAPK and SAPK2/p38, and may mediate activation of CREB. Embo J 17:4426-4441

Drobic B, Espino PS, Davie JR (2004) Mitogen- and stress-activated protein kinase 1 activity and histone h3 phosphorylation in oncogene-transformed mouse fibroblasts. Cancer Res 64:9076-9079

Dumitru CD, Ceci JD, Tsatsanis C, Kontoyiannis D, Stamatakis K, Lin JH, Patriotis C, Jenkins NA, Copeland NG, Kollias G, Tsichlis PN (2000) TNF-alpha induction by LPS is regulated posttranscriptionally via a Tpl2/ERK-dependent pathway. Cell 103:1071-1083

Dyson MH, Thomson S, Inagaki M, Goto H, Arthur SJ, Nightingale K, Iborra FJ, Mahadevan LC (2005) MAP kinase-mediated phosphorylation of distinct pools of histone H3 at S10 or S28 via mitogen- and stress-activated kinase 1/2. J Cell Sci 118:2247-2259

Engel K, Kotlyarov A, Gaestel M (1998) Leptomycin B-sensitive nuclear export of MAPKAP kinase 2 is regulated by phosphorylation. Embo J 17:3363-3371

Engel K, Schultz H, Martin F, Kotlyarov A, Plath K, Hahn M, Heinemann U, Gaestel M (1995) Constitutive activation of mitogen-activated protein kinase-activated protein kinase 2 by mutation of phosphorylation sites and an A-helix motif. J Biol Chem 270:27213-27221

Engelman JA, Lisanti MP, Scherer PE (1998) Specific inhibitors of p38 mitogen-activated protein kinase block 3T3-L1 adipogenesis. J Biol Chem 273:32111-32120

Freshney NW, Rawlinson L, Guesdon F, Jones E, Cowley S, Hsuan J, Saklatvala J (1994) Interleukin-1 activates a novel protein kinase cascade that results in the phosphorylation of Hsp27. Cell 78:1039-1049

Frodin M, Antal TL, Dummler BA, Jensen CJ, Deak M, Gammeltoft S, Biondi RM (2002) A phosphoserine/threonine-binding pocket in AGC kinases and PDK1 mediates activation by hydrophobic motif phosphorylation. Embo J 21:5396-5407

Gaestel M (2006) MAPKAP kinases - MKs - two's company, three's a crowd. Nat Rev Mol Cell Biol 7:120-130

Gaestel M, Schroder W, Benndorf R, Lippmann C, Buchner K, Hucho F, Erdmann VA, Bielka H (1991) Identification of the phosphorylation sites of the murine small heat shock protein hsp25. J Biol Chem 266:14721-14724

Guil S, Long JC, Caceres JF (2006) hnRNP A1 relocalization to the stress granules reflects a role in the stress response. Mol Cell Biol 26:5744-5758

Han J, Ulevitch RJ (1999) Emerging targets for anti-inflammatory therapy. Nat Cell Biol 1:E39-E40

Hannigan MO, Zhan L, Ai Y, Kotlyarov A, Gaestel M, Huang CK (2001) Abnormal migration phenotype of mitogen-activated protein kinase-activated protein kinase 2(-/-) neutrophils in zigmond chambers containing formyl-methionyl-leucyl-phenylalanine gradients. J Immunol 167:3953-3961

Hauge C, Frodin M (2006) RSK and MSK in MAP kinase signalling. J Cell Sci 119:3021-3023

Hazzalin CA, Mahadevan LC (2002) MAPK-regulated transcription: a continuously variable gene switch? Nat Rev Mol Cell Biol 3:30-40

Hegen M, Gaestel M, Nickerson-Nutter CL, Lin LL, Telliez JB (2006) MAPKAP kinase 2-deficient mice are resistant to collagen-induced arthritis. J Immunol 177:1913-1917

Heidenreich O, Neininger A, Schratt G, Zinck R, Cahill MA, Engel K, Kotlyarov A, Kraft R, Kostka S, Gaestel M, Nordheim A (1999) MAPKAP kinase 2 phosphorylates serum response factor *in vitro* and *in vivo*. J Biol Chem 274:14434-14443

Hitti E, Iakovleva T, Brook M, Deppenmeier S, Gruber AD, Radzioch D, Clark AR, Blackshear PJ, Kotlyarov A, Gaestel M (2006) Mitogen-activated protein kinase-activated protein kinase 2 regulates tumor necrosis factor mRNA stability and translation mainly by altering tristetraprolin expression, stability, and binding to adenine/uridine-rich element. Mol Cell Biol 26:2399-2407

Huang AM, Rubin GM (2000) A misexpression screen identifies genes that can modulate RAS1 pathway signaling in *Drosophila melanogaster*. Genetics 156:1219-1230

Janknecht R (2001) Cell type-specific inhibition of the ETS transcription factor ER81 by mitogen-activated protein kinase-activated protein kinase 2. J Biol Chem 276:41856-41861

Janknecht R (2003) Regulation of the ER81 transcription factor and its coactivators by mitogen- and stress-activated protein kinase 1 (MSK1). Oncogene 22:746-755

Jauch R, Cho MK, Jakel S, Netter C, Schreiter K, Aicher B, Zweckstetter M, Jackle H, Wahl MC (2006) Mitogen-activated protein kinases interacting kinases are autoinhibited by a reprogrammed activation segment. Embo J 25:4020-4032

Jauch R, Jakel S, Netter C, Schreiter K, Aicher B, Jackle H, Wahl MC (2005) Crystal structures of the Mnk2 kinase domain reveal an inhibitory conformation and a zinc binding site. Structure 13:1559-1568

Kant S, Schumacher S, Singh MK, Kispert A, Kotlyarov A, Gaestel M (2006) Characterization of the atypical map kinase ERK4 and its activation of the MAPK-activated protein kinase MK5. J Biol Chem 281:35511-35519

Knauf U, Tschopp C, Gram H (2001) Negative regulation of protein translation by mitogen-activated protein kinase-interacting kinases 1 and 2. Mol Cell Biol 21:5500-5511

Kotlyarov A, Neininger A, Schubert C, Eckert R, Birchmeier C, Volk HD, Gaestel M (1999) MAPKAP kinase 2 is essential for LPS-induced TNF-alpha biosynthesis. Nat Cell Biol 1:94-97

Kotlyarov A, Yannoni Y, Fritz S, Laass K, Telliez JB, Pitman D, Lin LL, Gaestel M (2002) Distinct cellular functions of MK2. Mol Cell Biol 22:4827-4835

Landry J, Lambert H, Zhou M, Lavoie JN, Hickey E, Weber LA, Anderson CW (1992) Human HSP27 is phosphorylated at serines 78 and 82 by heat shock and mitogen-activated kinases that recognize the same amino acid motif as S6 kinase II. J Biol Chem 267:794-803

Lee ER, McCool KW, Murdoch FE, Fritsch MK (2006) Dynamic changes in histone H3 phosphoacetylation during early embryonic stem cell differentiation are directly mediated by mitogen- and stress-activated protein kinase 1 via activation of MAPK pathways. J Biol Chem 281:21162-21172

Lehner MD, Schwoebel F, Kotlyarov A, Leist M, Gaestel M, Hartung T (2002) Mitogen-activated protein kinase-activated protein kinase 2-deficient mice show increased susceptibility to listeria monocytogenes infection. J Immunol 168:4667-4673

Manke IA, Nguyen A, Lim D, Stewart MQ, Elia AE, Yaffe MB (2005) MAPKAP kinase-2 is a cell cycle checkpoint kinase that regulates the G2/M transition and S phase progression in response to UV irradiation. Mol Cell 17:37-48

Manning G, Whyte DB, Martinez R, Hunter T, Sudarsanam S (2002) The protein kinase complement of the human genome. Science 298:1912-1934

Marchand C, Favier J, Sirois MG (2006) Role of MSK1 in the signaling pathway leading to VEGF-mediated PAF synthesis in endothelial cells. J Cell Biochem 98:1095-1105

McCoy CE, Campbell DG, Deak M, Bloomberg GB, Arthur JS (2005) MSK1 activity is controlled by multiple phosphorylation sites. Biochem J 387:507-517

McLaughlin MM, Kumar S, McDonnell PC, Van Horn S, Lee JC, Livi GP, Young PR (1996) Identification of mitogen-activated protein (MAP) kinase-activated protein kinase-3, a novel substrate of CSBP p38 MAP kinase. J Biol Chem 271:8488-8492

Meng W, Swenson LL, Fitzgibbon MJ, Hayakawa K, Ter Haar E, Behrens AE, Fulghum JR, Lippke JA (2002) Structure of mitogen-activated protein kinase-activated protein (MAPKAP) kinase 2 suggests a bifunctional switch that couples kinase activation with nuclear export. J Biol Chem 277:37401-37405

Neininger A, Kontoyiannis D, Kotlyarov A, Winzen R, Eckert R, Volk HD, Holtmann H, Kollias G, Gaestel M (2002) MK2 targets AU-rich elements and regulates biosynthesis of tumor necrosis factor and interleukin-6 independently at different post-transcriptional levels. J Biol Chem 277:3065-3068

Neininger A, Thielemann H, Gaestel M (2001) FRET-based detection of different conformations of MK2. EMBO Rep 2:703-708

Neufeld B, Grosse-Wilde A, Hoffmeyer A, Jordan BW, Chen P, Dinev D, Ludwig S, Rapp UR (2000) Serine/Threonine kinases 3pK and MAPK-activated protein kinase 2 interact with the basic helix-loop-helix transcription factor E47 and repress its transcriptional activity. J Biol Chem 275:20239-20242

O'Loghlen A, Gonzalez VM, Pineiro D, Perez-Morgado MI, Salinas M, Martin ME (2004) Identification and molecular characterization of Mnk1b, a splice variant of human MAP kinase-interacting kinase Mnk1. Exp Cell Res 299:343-355

Parra-Palau JL, Scheper GC, Wilson ML, Proud CG (2003) Features in the N and C termini of the MAPK-interacting kinase Mnk1 mediate its nucleocytoplasmic shuttling. J Biol Chem 278:44197-44204

Parra JL, Buxade M, Proud CG (2005) Features of the catalytic domains and C termini of the MAPK signal-integrating kinases Mnk1 and Mnk2 determine their differing activities and regulatory properties. J Biol Chem 280:37623-37633

Pyronnet S, Imataka H, Gingras AC, Fukunaga R, Hunter T, Sonenberg N (1999) Human eukaryotic translation initiation factor 4G (eIF4G) recruits mnk1 to phosphorylate eIF4E. Embo J 18:270-279

Reiling JH, Doepfner KT, Hafen E, Stocker H (2005) Diet-dependent effects of the *Drosophila* Mnk1/Mnk2 homolog Lk6 on growth via eIF4E. Curr Biol 15:24-30

Ronkina N, Kotlyarov A, Dittrich-Breiholz O, Kracht M, Hitti E, Milarski K, Askew R, Marusic S, Lin L-L, Gaestel M, Telliez J-B (2006) The MAPKAP kinases MK2 and MK3 cooperate in stimulation of TNF biosynthesis and stabilization of p38 MAPK. Mol Cell Biol 27:170-181

Rouse J, Cohen P, Trigon S, Morange M, Alonso-Llamazares A, Zamanillo D, Hunt T, Nebreda AR (1994) A novel kinase cascade triggered by stress and heat shock that stimulates MAPKAP kinase-2 and phosphorylation of the small heat shock proteins. Cell 78:1027-1037

Rousseau S, Morrice N, Peggie M, Campbell DG, Gaestel M, Cohen P (2002) Inhibition of SAPK2a/p38 prevents hnRNP A0 phosphorylation by MAPKAP-K2 and its interaction with cytokine mRNAs. Embo J 21:6505-6514

Roux PP, Blenis J (2004) ERK and p38 MAPK-activated protein kinases: a family of protein kinases with diverse biological functions. Microbiol Mol Biol Rev 68:320-344

Scheper GC, Parra JL, Wilson M, Van Kollenburg B, Vertegaal AC, Han ZG, Proud CG (2003) The N and C termini of the splice variants of the human mitogen-activated protein kinase-interacting kinase Mnk2 determine activity and localization. Mol Cell Biol 23:5692-5705

Scheper GC, Proud CG (2002) Does phosphorylation of the cap-binding protein eIF4E play a role in translation initiation? Eur J Biochem 269:5350-5359

Schiller M, Bohm M, Dennler S, Ehrchen JM, Mauviel A (2006) Mitogen- and stress-activated protein kinase 1 is critical for interleukin-1-induced, CREB-mediated, c-fos gene expression in keratinocytes. Oncogene 25:4449-4457

Schumacher S, Laass K, Kant S, Shi Y, Visel A, Gruber AD, Kotlyarov A, Gaestel M (2004) Scaffolding by ERK3 regulates MK5 in development. Embo J 23:4770-4779

Seternes OM, Mikalsen T, Johansen B, Michaelsen E, Armstrong CG, Morrice NA, Turgeon B, Meloche S, Moens U, Keyse SM (2004) Activation of MK5/PRAK by the atypical MAP kinase ERK3 defines a novel signal transduction pathway. Embo J 23:4780-4791

Shi Y, Kotlyarov A, Laaß K, Gruber AD, Butt E, Marcus K, Meyer HE, Friedrich A, Volk HD, Gaestel M (2003) Elimination of protein kinase MK5/PRAK activity by targeted homologous recombination. Mol Cell Biol 23:7732-7741

Sithanandam G, Latif F, Duh FM, Bernal R, Smola U, Li H, Kuzmin I, Wixler V, Geil L, Shrestha S (1996) 3pK, a new mitogen-activated protein kinase-activated protein kinase located in the small cell lung cancer tumor suppressor gene region. Mol Cell Biol 16:868-876

Smith KJ, Carter PS, Bridges A, Horrocks P, Lewis C, Pettman G, Clarke A, Brown M, Hughes J, Wilkinson M, Bax B, Reith A (2004) The structure of MSK1 reveals a novel autoinhibitory conformation for a dual kinase protein. Structure 12:1067-1077

Soloaga A, Thomson S, Wiggin GR, Rampersaud N, Dyson MH, Hazzalin CA, Mahadevan LC, Arthur JS (2003) MSK2 and MSK1 mediate the mitogen- and stress-induced phosphorylation of histone H3 and HMG-14. Embo J 22:2788-2797

Stoecklin G, Stubbs T, Kedersha N, Wax S, Rigby WF, Blackwell TK, Anderson P (2004) MK2-induced tristetraprolin:14-3-3 complexes prevent stress granule association and ARE-mRNA decay. Embo J 23:1313-1324

Stokoe D, Campbell DG, Nakielny S, Hidaka H, Leevers SJ, Marshall C, Cohen P (1992a) MAPKAP kinase-2; a novel protein kinase activated by mitogen-activated protein kinase. Embo J 11:3985-3994

Stokoe D, Engel K, Campbell DG, Cohen P, Gaestel M (1992b) Identification of MAPKAP kinase 2 as a major enzyme responsible for the phosphorylation of the small mammalian heat shock proteins. FEBS Lett 313:307-313

Sudo T, Kawai K, Matsuzaki H, Osada H (2005) p38 mitogen-activated protein kinase plays a key role in regulating MAPKAPK2 expression. Biochem Biophys Res Commun 337:415-421

Sun P, Yoshizuka N, New L, Moser BA, Li Y, Liao R, Xie C, Chen J, Deng Q, Yamout M, Dong MQ, Frangou CG, Yates JR 3rd, Wright PE, Han J (2007) PRAK is essential for ras-induced senescence and tumor suppression. Cell 128:295-308

Tanoue T, Maeda R, Adachi M, Nishida E (2001) Identification of a docking groove on ERK and p38 MAP kinases that regulates the specificity of docking interactions. Embo J 20:466-479

Ueda T, Watanabe-Fukunaga R, Fukuyama H, Nagata S, Fukunaga R (2004) Mnk2 and Mnk1 are essential for constitutive and inducible phosphorylation of eukaryotic initiation factor 4E but not for cell growth or development. Mol Cell Biol 24:6539-6549

van der Houven van Oordt W, Diaz-Meco MT, Lozano J, Krainer AR, Moscat J, Caceres JF (2000) The MKK(3/6)-p38-signaling cascade alters the subcellular distribution of hnRNP A1 and modulates alternative splicing regulation. J Cell Biol 149:307-316

Vermeulen L, De Wilde G, Van Damme P, Vanden Berghe W, Haegeman G (2003) Transcriptional activation of the NF-kappaB p65 subunit by mitogen- and stress-activated protein kinase-1 (MSK1). Embo J 22:1313-1324

Vertii A, Hakim C, Kotlyarov A, Gaestel M (2006) Analysis of properties of small heat shock protein Hsp25 in MAPK-activated protein kinase 2 (MK2)-deficient cells: MK2-dependent insolubilization of Hsp25 oligomers correlates with susceptibility to stress. J Biol Chem 281:26966-26975

Voncken JW, Niessen H, Neufeld B, Rennefahrt U, Dahlmans V, Kubben N, Holzer B, Ludwig S, Rapp UR (2005) MAPKAP kinase 3pK phosphorylates and regulates chromatin association of the polycomb group protein Bmi1. J Biol Chem 280:5178-5187 Epub 2004 Nov 5124

Wang X, Khaleque MA, Zhao MJ, Zhong R, Gaestel M, Calderwood SK (2006) Phosphorylation of HSF1 by MAPK-activated protein kinase 2 on serine 121, inhibits transcriptional activity and promotes HSP90 binding. J Biol Chem 281:782-791

Waskiewicz AJ, Flynn A, Proud CG, Cooper JA (1997) Mitogen-activated protein kinases activate the serine/threonine kinases Mnk1 and Mnk2. Embo J 16:1909-1920

Waskiewicz AJ, Johnson JC, Penn B, Mahalingam M, Kimball SR, Cooper JA (1999) Phosphorylation of the cap-binding protein eukaryotic translation initiation factor 4E by protein kinase Mnk1 *in vivo*. Mol Cell Biol 19:1871-1880

Webber KM, Smith MA, Lee HG, Harris PL, Moreira P, Perry G, Zhu X (2005) Mitogen-and stress-activated protein kinase 1: convergence of the ERK and p38 pathways in Alzheimer's disease. J Neurosci Res 79:554-560

Weber HO, Ludwig RL, Morrison D, Kotlyarov A, Gaestel M, Vousden KH (2005) HDM2 phosphorylation by MAPKAP kinase 2. Oncogene 24:1965-1972

Wiggin GR, Soloaga A, Foster JM, Murray-Tait V, Cohen P, Arthur JS (2002) MSK1 and MSK2 are required for the mitogen- and stress-induced phosphorylation of CREB and ATF1 in fibroblasts. Mol Cell Biol 22:2871-2881

Winzen R, Kracht M, Ritter B, Wilhelm A, Chen CY, Shyu AB, Muller M, Gaestel M, Resch K, Holtmann H (1999) The p38 MAP kinase pathway signals for cytokine-induced mRNA stabilization via MAP kinase-activated protein kinase 2 and an AU-rich region-targeted mechanism. Embo J 18:4969-4980

Zakowski V, Keramas G, Kilian K, Rapp UR, Ludwig S (2004) Mitogen-activated 3p kinase is active in the nucleus. Exp Cell Res 299:101-109

Zhang Y, Griffin K, Mondal N, Parvin JD (2004) Phosphorylation of histone H2A inhibits transcription on chromatin templates. J Biol Chem 279:21866-21872

Zu YL, Ai Y, Huang CK (1995) Characterization of an autoinhibitory domain in human mitogen-activated protein kinase-activated protein kinase 2. J Biol Chem 270:202-206

Zuberek J, Jemielity J, Jablonowska A, Stepinski J, Dadlez M, Stolarski R, Darzynkiewicz E (2004) Influence of electric charge variation at residues 209 and 159 on the interaction of eIF4E with the mRNA 5' terminus. Biochemistry 43:5370-5379

Gaestel, Matthias

Department of Biochemistry, Medical School Hannover, Carl-Neuberg-Str. 1, 30625 Hannover, Germany
gaestel.matthias@mh-hannover.de

Kotlyarov, Alexey
Department of Biochemistry, Medical School Hannover, Carl-Neuberg-Str. 1, 30625 Hannover, Germany

List of abbreviations

CBP: CREB-binding protein
CREB: cAMP-responsive element binding protein
DKO: double knockout
hnRNP: heterogeneous nuclear ribonucleoprotein
KO: knockout
MAPK: Mitogen-activated protein kinase
MK: MAPK-activated protein kinase
MNK: MAPK-interacting protein kinases
MSK: Mitogen- and stress-activated protein kinase
SAPK: Stress-activated protein kinase
SG: Stress granules
TTP: Tristetraprolin

Functions of stress-activated MAP kinases in the immune response

Mercedes Rincón and Roger J. Davis

Abstract

The stress-activated protein kinases (SAPK) represent one group of mitogen-activated protein kinases (MAPKs) that are activated by antigen receptors, Toll-like receptors, cytokine receptors, and physical-chemical changes in the environment. The SAPK are established to be important mediators of intracellular signaling during both adaptive and innate immune responses. Here, we summarize recent findings concerning the role of two sub-groups of SAPK – cJun NH_2-terminal kinase (JNK) and p38 MAPK.

1 Introduction

The stress-activated protein kinases JNK and p38 MAPK are activated during both innate and adaptive immune responses. The trigger for SAPK activation can be mediated by multiple cell surface receptors, including antigen receptors, Toll-like receptors, and cytokine receptors. Detailed studies have implicated both JNK and p38 in the development of normal immune responses. Nevertheless, the identification of the specific roles of SAPK in immune responses has been difficult because SAPK signaling pathways appear to function in many of the highly specialized cell types that mediate the immune response.

The purpose of this review is to provide a summary of the known roles of the JNK and p38 sub-groups of SAPK in different immune cells. We will also discuss how these specific responses in individual cell types contribute to the generation of an *in vivo* immune response.

2 SAPK Functions in macrophages and dendritic cells

2.1 Role of JNK

Macrophages and dendritic ells are key components of the innate immune system. Macrophages (together with neutrophils) represent the first defense against pathogen infections in part due to their active phagocytic function. Macrophages are

Topics in Current Genetics, Vol. 20
F. Posas, A. R. Nebreda (Eds.): Stress-Activated Protein Kinases
DOI 10.1007/4735_2007_0253 / Published online: 1 August 2007

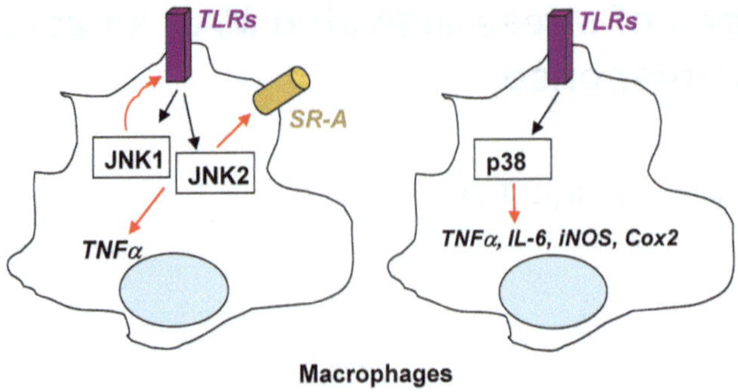

Fig. 1. Regulation of macrophage function by JNK and p38 MAPK.

also one of the first source of inflammatory cytokines such TNFα, IL-1β, and IL-6. A variety of Toll-like receptors (TLR) in these cells mediates the response to different components of pathogens. JNK is one of the pathways stimulated by these receptors. While a large number of studies have described the activation of JNK in macrophages in response to different stimuli (e.g. cytokines, LPS, oxidative stress) less is known about the specific contribution of the isoforms JNK1 and JNK2 to macrophage functions. Overexpression of a dominant negative SAPKb mutant in a macrophage cell line inhibits TNFα translation, but not transcription, in response to LPS (Swantek et al. 1997). Pharmacological inhibitors and peptide inhibitors of JNK have been shown to induce apoptosis of macrophages and inhibit CFS-1-dependent differentiation of bone marrow cells into macrophages (Himes et al. 2006). Activation of macrophages by *Plasmodium falciparum* GPI anchors (pfGPI) and secretion of cytokines such as TNFα play a role in the pathology of malaria. pfGPI has been shown to activate JNK in macrophages, and the production of TNFα induced by pfGPI is decreased in $Jnk2^{-/-}$ macrophages, but not in $Jnk1^{-/-}$ macrophages (Lu et al. 2006), showing the specific role of JNK2 in response to pfGPI signaling (Fig. 1).

Although there have not been any reports describing a requirement of JNK for macrophage phagocytic activity, a study by Ricci et al. has shown a defective uptake and degradation of modified lipoproteins and foam cell formation in $Jnk2^{-/-}$ macrophages (Ricci et al. 2004). No significant defect was found in $Jnk1^{-/-}$ macrophages. The apparent mechanism for the impaired uptake in $Jnk2^{-/-}$ macrophages is the requirement of JNK2-mediated phosphorylation of the scavenger receptor A (SR-A) for internalization of lipids (Fig. 1). This defect in $Jnk2^{-/-}$ macrophages has been associated with the resistance of these mice to develop atherosclerosis (see below).

Despite the number of studies indicating that the JNK pathway is triggered by TLRs, no reports have shown a requirement of JNK for TLR function. A recent study using JNK1 and JNK2 deficient mice has shown that JNK1, but not JNK2, is required for expression of TLR1 in macrophages (Anguita et al., personal communication).

The role of JNK in dendritic cell (DC) functions (e.g. antigen processing and presentation, cytokine production, costimulation) remains unclear. Studies using pharmacological inhibitors have suggested a possible role for JNK in chemokine-induced DC migration (Hu and Ivashkiv 2006; Iijima et al. 2005) and DC apoptosis (Kriehuber et al. 2005; Handley et al. 2005). Pharmacological inhibition of JNK also appears to interfere with the upregulation of CD80, CD83, and CD86 in human DCs by LPS (Nakahara et al. 2004). However, the conclusions of these studies are exclusively based on the use of drugs that have a poorly established specificity. To date, no studies using JNK-deficient mice have described a defect in dendritic cell function.

2.2 Role of p38 MAPK

The function of the p38 MAP kinase pathway in macrophages and DC has been investigated in detail, although most studies are based on the use of a pharmacological inhibitor of this kinase and only limited data are available from genetic analysis. The p38 MAP kinase pathway is activated by TLRs. Activation of p38 MAP kinase by different TLR ligands in macrophages has been shown to regulate the expression a number of pro-inflammatory factors including cytokines (e.g. TNFα, IL-6), Cox2, iNOS. This effect is mediated by p38 MAP kinase acting through at both transcriptional and posttranscriptional mechanisms (e.g. mRNA stability and translation) (Fig. 1).

Several studies have proposed that p38 MAP kinase is required for maturation of monocyte-derived DC, since drug-induced inhibition of this pathway prevents DC maturation induced by LPS, TNF or UV-B (Nakagawa et al. 2004; Puig-Kroger et al. 2001; Arrighi et al. 2001; Xie et al. 2003). Pharmacological inhibition of p38 MAP kinase also interferes with LPS-stimulated endocytosis of antigens by mature DC, an effect that appears to be mediated by blocking podosome (F-actin reach structure) disassembly (West et al. 2004), suggesting that p38 MAP kinase can also be involved in antigen uptake. Upon internalization, the phagosome fuses with lysosomes. This event has been shown to depend on TLR-mediated p38 MAP kinase activity (Blander and Medzhitov 2004), although other reports have disputed this finding (Yates and Russell 2005). Thus, p38 MAP kinase may control more than one aspect of the internalization and destruction of infectious agents. The role of p38 MAP kinase on the phagocytic uptake of microorganisms has not been established. A recent study has suggested that p38 MAP kinase negatively regulates the expression of CTIIA, a master regulator for MHC class II, in dendritic cells and macrophages by decreasing histone acetylation of the CTIIA promoter (Yao et al. 2006), suggesting that activation of p38 MAP kinase can interfere with MHC class II antigen presentation.

3 SAPK functions in B cells

3.1 Role of JNK

The regulation of JNK in B cells has been extensively examined. JNK is activated in these cells in response to B cell receptor (BCR) ligation, CD40-CD40 ligand interaction, LMP1 (Epstain Barr oncogenic latent membrane protein 1) ligation, LPS, CD22 crosslinking, CD53 ligation, and CpG motifs among others. However, little is known about the function of JNK in B cell development, B cell proliferation/survival, or antibody production. No apparent defect in proliferation, survival, and function of B cells were found in JNK1 or JNK2 -deficient mice (Yang et al. 1998; Dong et al. 1998; Sabapathy et al. 1999, 2001) or in Mkk4$^{-/-}$Rag2$^{-/-}$ chimeric mice (Swat et al. 1998). *Mkk7*$^{-/-}$ *Rag1*$^{-/-}$ chimeras also have normal numbers of B cells in bone marrow and peripheral tissues, and B cell differentiation was also normal. However, peripheral B cells in these mice are hyper-proliferative in response to BCR and CD40 ligation as well as in response to LPS (Sasaki et al. 2001). Since MKK7 does not seem to activate other MAPK groups except the JNK subfamily, it is possible that the hyperproliferation observed in these MKK7-deficient B cells is due to insufficient activation of JNK. A recent study in Map3k1$^{\Delta KD}$ mice lacking activity of the MAPKK kinase MEKK1, an upstream activator of both the JNK and p38 MAP kinase signaling pathways, have defective germinal center formation and reduced production of antibodies associated with a hypo-proliferation of B cells upon CD40 stimulation (Collins et al. 1996). It is, however, unclear whether this phenotype is mediated by JNK since the activation of both JNK and p38 MAP kinase was impaired in these cells.

Other studies have also supported a role of JNK in survival in B cells. JNK1 can provide survival signals for BCR-ABL-transformed B lymphoblasts since defective transformation of pre-B cells by BCR-ABL has been observed *in vivo* and *in vitro* in JNK1-deficient mice that could be rescued by expression of Bcl2 (Hess et al. 2002). Interestingly, a recent study also shows that inactivation of JNK in *Theileria*-transformed B lymphocytes leads to lymphocyte apoptosis, further supporting an anti-apoptotic role for JNK activation (Lizundia et al. 2006). Thus, while JNK may mediate cell death or inhibit proliferation in non-transformed B lymphocytes, it provides survival signals in transformed B lymphoblasts.

3.2 Role of p38 MAPK

The p38 MAP kinase pathway has been shown to be activated in B cells in response to BCR and CD40 signals. However, to date, no role for p38 MAP kinase in B cell development, B cell activation, antibody production and generation of memory cells has yet been described. p38α MAPK –deficient mice are not viable because of a placental development defect and Epo deficiency (Adams et al. 2000; Allen et al. 2000; Tamura et al. 2000). p38$\alpha^{-/-}$ Rag$^{-/-}$ chimeras have normal B cell

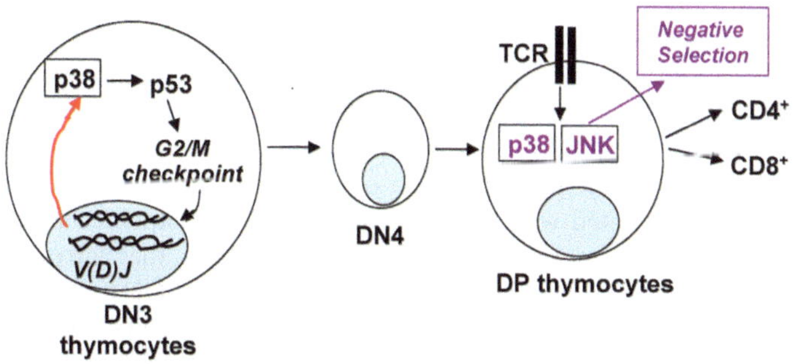

Fig. 2. Regulation of thymocyte development by JNK and p38 MAPK.

development and B cell proliferation in response to antigen receptor (Kim et al. 2005), but this could be due to a compensation effect by the presence of other members of the p38 family.

4 SAPK Functions in T cell development

4.1 Role of JNK

Several processes are critical for the generation of $CD4^+$ and $CD8^+$ T cells in the thymus: a) proliferation and expansion of T cell precursors (early thymocyte development), b) V(D)J recombination of TCRβ and allelic exclusion for clonal specificity (early T cell development), c) negative selection of autoreactive T cells, d) positive selection in the context of appropriate MHC, e) lineage commitment to CD4 or CD8 T cells. JNK does not seem to be involved in early thymocyte development. However, several studies using JNK deficient mice or transgenic mice expressing a dominant negative JNK mutant have implicated JNK in the negative selection of thymocytes (Behrens et al. 2001; Rincón et al. 1998; Sabapathy et al. 1999; Sabapathy et al. 2001) (Fig. 2). The potential contribution of JNK to regulate the presence of autoreactive cells in the periphery and the development of autoimmune disease, due to impaired negative selection, is not yet established.

4.2 Role of p38 MAPK

p38 MAPK, like JNK, has also been implicated with the deletion of $CD4^+$ $CD8^+$ double positive (DP) thymocytes (Sugawara et al. 1998; Tanaka et al. 2002) (Fig. 2), but the contribution of p38 MAPK to the development of autoimmune diseases, due to the presence of autoreactive T cells in the periphery, remains to be established. However, it has been demonstrated that p38 MAPK plays an important role in early thymocyte development. p38 MAPK is activated in immature double negative (DN)3 thymocytes, but it is inactivated once these cells differentiate into DN4 thymocytes (Diehl et al. 2000; Sen et al. 1996). Constitutive activation of p38 MAP kinase *in vivo* has been shown to block thymocyte development at the DN3 stage (Diehl et al. 2000). p38 MAP kinase induces a G2/M cell cycle checkpoint in DN3 thymocytes by phosphorylating and activating p53 (Pedraza-Alva et al. 2006). V(D)J recombination of TCRβ occurs specifically at the DN3 stage and it is associated with the generation of double stranded breaks (DSBs). The presence of DSBs represents the DNA damage signal that activates p38 MAP kinase to trigger a G2/M cell cycle checkpoint that allows DNA repair prior to the cell division and differentiation into DN4 thymocytes (Pedraza-Alva et al. 2006) (Fig. 2). Thus, p38 MAP kinase is activated not only by cytokines, but also by other stress signals in the immune system, and it could be involved in maintenance of genomic stability in thymocytes and T cells.

It has been shown that p38α deficiency in $p38\alpha^{-/-}$ $Rag^{-/-}$ chimeras does not affect T cell development (Kim et al. 2005). No defects in T cell development have been reported in p38β, p38δ, and p38γ (Beardmore et al. 2005; Sabio et al. 2005). This could be due, in part, to compensation by other p38 MAP kinase members since the four members (p38α, p38β, p38δ, and p38γ) are expressed in thymocytes (Beardmore et al. 2005), and studies using double or triple knockout mice may be needed.

5 SAPK functions in $CD4^+$ T cells

5.1 Role of JNK

While JNK3 expression is restricted to brain, heart, and testis, JNK1 and JNK2 are ubiquitously expressed in most analyzed tissues. However, JNK1 and JNK2 are expressed at very low levels in the spleen. Furthermore, JNK1 and JNK2 gene and protein expression is extremely low naïve $CD4^+$ and $CD8^+$ T cells (and also B cells) prior to stimulation (Weiss et al. 2000). Activation through the T cell receptor (TCR) and, most likely a combination of cytokines, induces the expression of both JNK1 and JNK2 in T cells (Weiss et al. 2000) (Fig. 3). In addition, the expression level of the upstream JNK activators, MKK4 and MKK7, is also low in $CD4^+$ T cells, but is upregulated following activation. This increase in JNK pathway gene expression has also been identified using microarray analysis of non-activated and activated T cells (Teague et al. 1999). It remains unknown why the expression of these otherwise ubiquitous protein kinases is turned off specifically in these cells. Whether the reduced expression of the JNK pathway in naïve T cells increases cell survival or whether this represents an additional mechanism to prevent unwanted activation of naïve T cells remains to be examined. Since the upregulation of JNK expression during activation of $CD4^+$ T cells parallels that of IL-2 it is unlikely that JNK could play a key role in the induction of this cytokine. Indeed, IL-2 production in $CD4^+$ T cells from $Jnk2^{-/-}$ mice (Yang et al. 1998) and

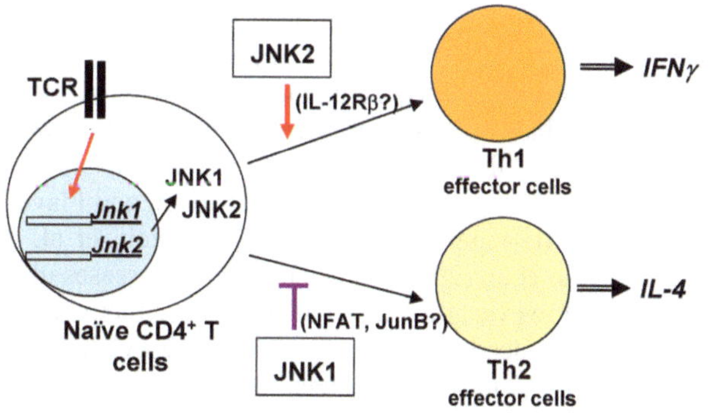

Fig. 3. Regulation of $CD4^+$ Th1 cell differentiation by JNK.

Jnk1$^{-/-}$ mice (Dong et al. 1998) is normal. Furthermore, $CD4^+$ T cells from *Jnk1*$^{-/-}$ *Jnk2*$^{-/-}$ *Rag1*$^{-/-}$ chimeras and *Mkk7* $^{-/-}$ *Rag*$^{-/-}$ chimeras produce increased levels of IL-2 (Dong et al. 2000, 1998). Reduced IL-2 production by total T cells from *Jnk2*$^{-/-}$ mice has been described (Sabapathy et al. 1999); however, this could be due to the presence of both $CD4^+$ and $CD8^+$ T cells in the same cell culture.

Although JNK does not appear to play a role in the early activation of naïve $CD4^+$ T cells, both JNK1 and JNK2 are involved in the differentiation of these cells into effector Th1 and Th2 cells. The expression JNK1 and JNK2 in Th1 and Th2 cells are similar and are greatly increased compared with naïve $CD4^+$ T cells. However, while JNK activity can be rapidly induced in effector Th1 cells, minimal JNK activity could be detected in effector Th2 cells upon restimulation (Yang et al. 1998). This selective activation of the JNK pathway in Th1 cells is likely due to the selective expression in these cells, but not in Th2 cells, of GADD45γ, a protein that mediates activation of MEKK4, an upstream activator of JNK and p38 MAP kinases (Chi et al. 2004; Lu et al. 2001). These observations suggest a non-redundant role of JNK in Th1 differentiation and/or secretion of cytokines upon antigen stimulation. Accordingly, Th2 differentiation and IL-4 production by $CD4^+$ T cells from *Jnk2*$^{-/-}$ mice is normal, but Th1 differentiation and IFNγ production are impaired (Yang et al. 1998). A more recent study has shown that analog peptides of an immunodominant epitope of myelin basic protein that activate JNK (and p38), but not ERK, differentiate a Th0 reactive clone into Th1 cells. In contrast, analog peptides that activate ERK, but not JNK or p38, differentiate Th0 cells into Th2 cells (Singh and Zhang 2004). Although the mechanism by which JNK regulates Th1 differentiation remains unclear, the expression of IL-12Rβ2 on differentiating $CD4^+$ T cells may represent one example of a relevant gene that is regulated by JNK (Yang et al. 1998) (Fig. 3).

Differentiation of $CD4^+$ T cells from *Jnk1*$^{-/-}$ mice into effector Th1 cells is apparently normal, but Th2 differentiation is enhanced. This observation suggests that JNK1 is a negative regulator of Th2 cytokine expression (Dong et al. 1998). A Th2 bias has also been observed in $CD4^+$ T cells from *dnJNK*$^+$ /*Jnk2*$^{-/-}$ mice (*Jnk2*$^{-/-}$ crossed with transgenic mice expressing a dominant negative JNK1 mutant) and *Mkk7*$^{-/-}$ *Rag1*$^{-/-}$ chimeric mice (Dong et al. 2000) (Fig. 3). The negative effect of

JNK1 on IL-4 and other Th2 cytokines has been proposed to be mediated, in part, by an inhibitory effect of JNK on NFATc1 since JNK has been shown to phosphorylate and inhibit some of the NFAT members (Chow et al. 1997, 2000). However, an alternative mechanism has been recently proposed by Gao et al. (2004). JunB, a member of the Jun family of transcription factor, has been shown to specifically accumulate in Th2 relative to Th1 cells (Rincon and Flavell 1999) and be a positive regulator of IL-4 gene expression (Hartenstein et al. 2002; Li et al. 1999). The study by Gao et al. (2004) shows that JNK increases ubiquitination and degradation of JunB. However, they propose that this effect is not due to a direct effect of JNK on JunB. Instead, JNK appears to phosphorylate and activate the E3 ubiquitin ligase Itch which, in turn, mediates ubiquitination of JunB, and thereby its degradation. JunB accumulation has been observed in T cells from both Itch deficient mice and in *Jnk1* $^{-/-}$ mice (and slight reduction in *Jnk2*$^{-/-}$ mice). Thus, it is possible that increased IL-4 production and Th2 differentiation in *Jnk1*$^{-/-}$ mice is mediated by the accumulation of JunB in these cells.

5.2 Role of p38 MAPK

The role and regulation of p38 MAP kinase in $CD4^+$ T cells has been recently reviewed by (reviewed by Dodeller and Schulze-Koops 2006). Ligation of TCR causes activation of p38 MAPK in CD4+ T cells. Full activation of p38 MAPK appears to require ligation of other receptors (e.g. TCR CD28, 4-1BB, ICOS, and Fas) and also cytokines (e.g. IL-12, IL-18, TNFα) The relative contribution of all these stimuli and the co-integration of these different signals to the regulation of p38 MAP kinase during the activation and differentiation of T cells remains to be established.

p38 MAPK activation is not required for IL-2 production and cell proliferation, but it is implicated in the differentiation of naïve T cells into effector cells (Dodeller and Schulze-Koops 2006). Indeed, several studies using both pharmacological inhibitors and genetically manipulated mice have demonstrated the involvement of the p38 MAPK pathway in the production of IFNγ by $CD4^+$ T cells (Rincon and Pedraza-Alva 2003). However, it is unclear whether TCR ligation provides the major signal that activates p38 MAP kinase during the differentiation of Th1 cells, or whether cytokines (e.g. IL-12 or IL-18) are key p38 MAP kinase stimuli in this process (Berenson et al. 2006) (Fig. 4). Although less explored, other biological components present during the activation of T cells such histamine) can also contribute to the regulation of this signaling pathway (Noubade and Teuscher, personal communication). Similar caveat needs to be considered for the activation of the JNK pathway during $CD4^+$ and $CD8^+$ T cell activation and differentiation.

p38 MAP kinase is normally activated by phosphorylation at Thr^{180} and Tyr^{182} by upstream the MAPKK MKK3, MKK6 and, to a lesser extent, MKK4 (Brancho et al. 2003). An alternative pathway of activation has been recently identified in T cells. In this pathway, p38 MAP kinase is phorphorylated at Tyr^{323} by Zap70 upon TCR ligation (Ashwell 2006; Salvador et al. 2005). A recent study on p38 MAP

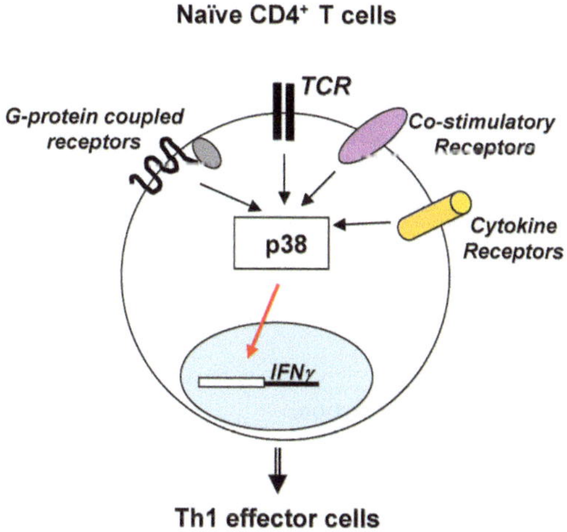

Fig. 4. Regulation of Th1 cell development by p38 MAPK.

kinase structure has proposed that phosphorylation of p38 MAP kinase in this residue can induce a conformational change in the L16 loop causing p38 MAP kinase dimerization (Diskin et al. 2007). The proximity of p38 MAP kinase in the dimers could lead to autophosphorylation of p38 MAP kinase at Thr^{180} and Tyr^{182}. Dlgh1 has been proposed to be the adapter molecule that connects TCR, Lck, and Zap-70 with p38 MAP kinase (Round et al. 2007). It has also been suggested that activation of p38 MAP kinase through Dlgh1 positively regulates NFAT (Round et al. 2007). However, other studies have shown that Dlgh1 and p38 MAP kinase are negative regulators of this transcription factor (Gomez del Arco et al. 2000; Xavier et al. 2004). In addition, while NFATc1 and NFATc2 factors are highly expressed in cell lines and activated T cells, the levels of these factors in naïve $CD4^+$ T cells are minimal (Dienz et al. 2007), suggesting that NFAT is probably not the major target of p38 MAP kinase in these cells. While the phosphorylation of p38 MAP kinase at Ty^{323} contributes to the activation of this kinase in a T cell line, there is no evidence that this phosphorylation is important for activation of p38 MAP kinase in $CD4^+$ and $CD8^+$ T cells.

6 SAPK functions in $CD8^+$ T cells

6.1 Role of JNK

Although both $CD4^+$ T cells and $CD8^+$ T cells are both activated by antigen via the TCR, $CD8^+$ T cells have specific functions and requirements that clearly distinguished from $CD4^+$ T cells. Thus, $CD8^+$ T cells produce substantially lower levels of IL-2 and they require need $CD4^+$ T cell help for optimal proliferation. $CD8^+$ T cells produced IFNγ more rapidly than $CD4^+$ T cells. In addition, $CD8^+$ T cells

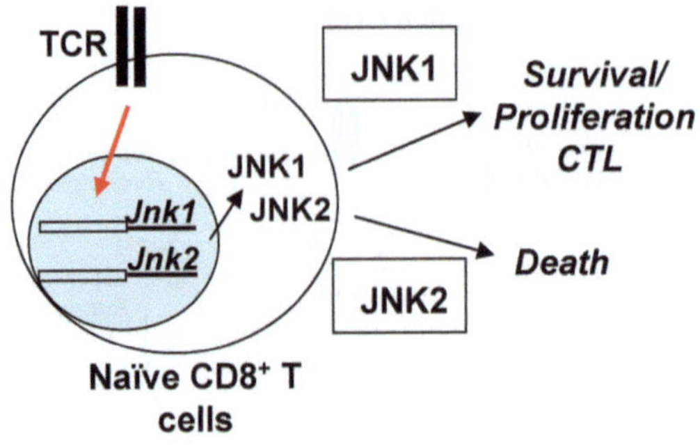

Fig. 5. Role of JNK in CD4$^+$ T cell function.

also become cytotoxic effector cells and express and secrete molecules that are not present in activated CD4$^+$ T cells or effector CD4$^+$ Th1 and Th2 cells. The role of JNK in CD8$^+$ T cells is also different than in CD4$^+$ T cells. The loss of *Jnk2* causes hyperproliferation of CD8$^+$ T cells *in vitro* that does not appear to be due to a decreased activation induced cell death (Conze et al. 2002). Instead, it seems to be caused by increased production of IL-2, a major growth factor of CD8$^+$ T cells (Conze et al. 2002). Thus, not only is JNK2 dispensable for IL-2 production by CD8$^+$ T cells, it is also a negative factor for the expression of this cytokine. Increased expansion of antigen (virus) specific CD8$^+$ T cells in *Jnk2*$^{-/-}$ mice has also been observed *in vivo* upon infection with lymphocytic choriomeningitis virus (LCMV) (Arbour et al. 2002) (Fig. 5).

CD8$^+$ T cells that lack JNK1 are hypoproliferative *in vitro* and they produce lower levels of IL-2. However, the hypoproliferation is not caused by an impaired IL-2 production since the addition of exogenous IL-2 does not rescue the defect. Instead, it appears that the defect is caused by impaired expression of IL-2Rα CD25) in these cells (Conze et al. 2002). In this case, decreased JNK1 may cause a decrease in c-Jun levels that may be responsible for impaired IL-2Rα gene expression. The hypoproliferation of JNK1-deficient CD8$^+$ T cells *in vitro* also correlates with decreased expansion of virus-specific CD8$^+$ T cells in Jnk1$^{-/-}$ mice upon infection with LCMV (Arbour et al. 2002). However, although no effect of JNK1-deficiency was observed in cell death of CD8$^+$ T cells *in vitro*, increased apoptosis of virus-specific CD8$^+$ T cells were observed *in vivo* (Arbour et al. 2002) (Fig. 5). A recent study has shown that the pharmacological inhibitor of JNK can prevent activation induced cell death of a CTL clone (Mehrotra et al. 2006). Thus, JNK1 is required for expansion of CD8$^+$ T cells by regulating the levels of growth factor receptors as well as the sensitivity of these to cells to activation-induced death. This role of JNK1 is consistent with the finding that *Jnk1-/-* mice exhibit major defects in tumor immunosurveillance mediated by CD8$^+$ T cells that is associated with decreased CTL function and reduced expression of T-bet, eomesodermin, and perforin (Gao et al. 2005).

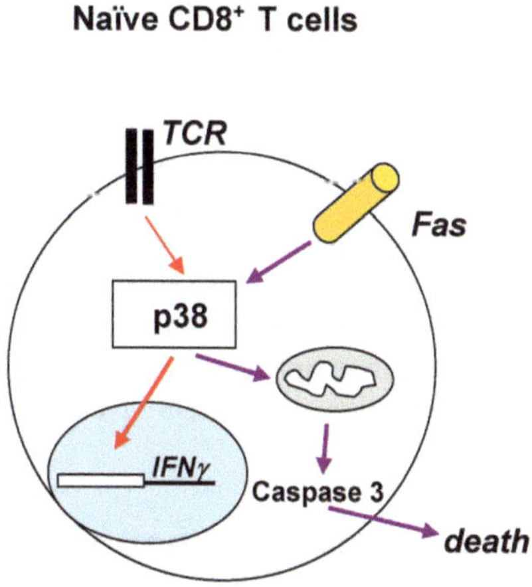

Fig. 6. Role of p38 MAPK in CD4$^+$ T cell function.

6.2 Role of p38 MAPK

The p38 MAPK is activated during stimulation of CD8$^+$ T cells. Studies using transgenic mice expressing dominant-negative p38 MAPK in T cells have shown that p38 MAP kinase is needed for production of IFNγ by CD8$^+$ T cells (Merritt et al. 2000). However, activation of p38 MAP kinase *in vivo* caused by transgenic expression of a constitutive active MKK6, an upstream activator of p38 MAP kinase, induces selective death of CD8$^+$ T cells, but not CD4$^+$ T cells (Merritt et al. 2000) (Fig. 6). Thus, activation of the same signaling pathway causes different outcomes in CD8$^+$ T cells and CD4$^+$ T cells. The molecular mechanism that accounts for this difference between p38 MAPK function in these T cell sub-sets remains unclear. There are four members of this family of kinases (p38α, p38β, p38γ and p38δ); differential expression of these isoforms may contribute to the different responses to p38 MAPK activation. However, no obvious differences in the expression of the four p38 MAPK isoforms has been detected (N. Farley and M. Rincon, unpublished observations). Alternatively, differential expression of pro- or anti-apoptotic targets of p38 MAPK may confer resistance to CD4$^+$ T cells to p38 MAPK-induced cell death.

Pharmacological inhibition of p38 MAPK reduces IL-2 production by CD8$^+$ T cells in response to peptide and APCs (Tham and Mescher 2001) and inhibition of p38 MAPK by expression of dominant-negative p38 MAPK results in increased proliferation (Merritt et al. 2000). p38 MAPK may therefore contribute to the proliferation of CD8$^+$ T cells. No studies have implicated p38 MAP kinase in the regulation of cytotoxic activity of CD8$^+$ T cells.

CD8$^+$ T cells appear to be more sensitive to Fas-induced death than CD4$^+$ T cells (Fortner and Budd 2005). Indeed, a recent study has been shown that p38 MAPK plays a role in Fas-induced death in CD8$^+$ T cells through activation of the mitochondrial apoptosis pathway and subsequent caspase activation (Farley et al. 2006) (Fig. 6). Fas ligation activates p38 MAP kinase, decreases the amount of mitochondrial Bcl2 and BclXL, and activates caspase 9 and caspase 3 in $CD8^+$ T cells. Inhibition of p38 MAP kinase by expression of a dominant-negative p38 MAPK mutant, or treatment with a pharmacological inhibitor, prevents caspase 3 activation and apoptosis of $CD8^+$ T cells in response to Fas ligation. This effect appears to be mediated by translocation of BclXL and Bcl2 from mitochondria to cytosol, most likely because of phosphoryation by p38 MAP kinase. Pharmacological inhibition of p38 MAPK also blocks Fas-mediated apoptosis of human $CD8^+$ T cells by HIV-infected macrophages (Muthumani et al. 2005). Thus, p38 MAP kinase appears to play a major role in $CD8^+$ T cell death.

7 SAPK functions in other T cell populations

Conventional $CD4^+$ T helper and $CD8^+$ T cytotoxic cells represent the majority of T cells in the peripheral immune system. However, recently, studies have focused on the important roles of other subset of T cells. The studies described above for $CD4^+$ and $CD8^+$ T cells were based primarily on T cells bearing variable αβ TCRs. A minor subgroup of these cells defined as invariant NKT (iNKT) cells are mostly $CD4^+$ T cells bearing an invariant TCR (Va14 Ja18 associated with Vβ8, Vβ7 and Vβ2) that is restricted by the CD1d MHC I-like molecule and recognizing glycolipids instead of protein (Bendelac et al. 2006). While iNKT cells represent a minor subpopulation of T cells (less than 5% in spleen), they can play a major role in the initiation and direction of the immune response because they are capable of secreting large amounts of cytokines (e.g. TNFα, IL-6, IL-4 and IFNγ) very rapidly (within 90-120 minutes) upon activation *in vivo*. Interestingly, no studies have addressed the potential role of JNK in the development of these cells or cytokine secretion by these cells. The production of IL-4 by iNKT cells can have profound effects on Th2 differentiation of $CD4^+$ T cells. It is possible that some of the effects of JNK that have been attributed to $CD4^+$ T cells may have been due to the role of JNK in iNKT cells if these cells were present in the preparation of $CD4^+$ T cells. It remains unknown if p38 MAP kinase is important for development and/or activation of iNKT cells.

T regulatory (Treg) cells have been the major focus of research in immunology because of the important role of these cells in the control of *in vivo* immune responses. Treg cells differ from conventional $CD4^+$ T cells in their impaired proliferation capacity as well as the pattern of cytokines that they secrete. Potential roles of JNK or p38 MAPK in cytokine expression and/or proliferation of these cells have not yet been explored. Interestingly, fibroblasts from $Jnk1^{-/-}Jnk2^{-/-}$ mice constitutively express TGFβ1, indicating that JNK is a repressor of TGFβ1 gene expression (Ventura et al. 2004). Since TGFβ has been implicated as one of the mechanisms by which Treg suppress $CD4^+$ T cell activation, it is possible that JNK may play a role in these cells.

In addition to αβ T cells, γδ T cells play a major role in tissues including skin. Little is known about signal transduction in these cells probably due to their limited number, and no studies have examined the role of JNK or p38 MAPK in these cells.

Finally, unlike naïve $CD4^+$ and $CD8^+$ T cells, memory $CD4^+$ and $CD8^+$ T cells are capable of secreting large amounts of effector cytokines very rapidly. The regulation and the role of the JNK and p38 MAPK pathways in these cells remains unexplored. Interestingly, while the levels of JNK1 are practically undetectable in naïve $CD4^+$ T cells, JNK1 expression is increased in memory-like phenotype $CD4^+$ T cells prior to activation (M. Rincon, unpublished studies).

In conclusion, studies of the function of JNK and p38 MAPK pathways in these minor T cells subsets will be a fruitful area for future research. It is likely that these protein kinase signaling pathways play a major role in these cells during an *in vivo* immune response. Indeed, impaired function of these minor T cell sub-sets may contribute to some of the observed phenotypes of *Jnk1-/-* and *Jnk2-/-* mice.

8 SAPK functions during an *in vivo* immune response

Several studies have examined the role of SAPK using pharmacological inhibitors and genetically modified mice that lack expression of specific SAPK isoforms in different models of infectious diseases, autoimmune disease, and allergic diseases. These studies indicate that SAPK contribute to *in vivo* immune responses. Indeed, studies of MAPK phosphatase-deficient mice (*Mkp1*$^{-/-}$ and *Mkp5*$^{-/-}$ mice) indicate that the resolution of immune responses involves the downregulation of JNK and p38 MAPK activation (Chi et al. 2006; Zhang et al. 2004; Zhao et al. 2006). Nevertheless, the final outcome of diseases responses reflects the combined functions of SAPK in different cells of the immune system.

In a mouse model of malaria, *Jnk2*$^{-/-}$ mice exhibit higher survival rates when challenged with *Plasmodium berghei*. In contrast, Jnk1$^{-/-}$ mice respond like wild type mice (Lu et al. 2006). The protective effect in *Jnk2*$^{-/-}$ mice has been attributed an impaired production of TNFα by macrophages.

Infection of *Jnk1*$^{-/-}$ mice with LCMV results in a lower number of virus-specific $CD8^+$ T cells, and lower activated $CD8^+$ T cells than wild type mice. However, *Jnk1*$^{-/-}$ mice clear LCMV infection with the same kinetics than wild type mice (Arbour et al. 2002). *Jnk2*$^{-/-}$ mice have increased virus-specific $CD8^+$ T cells, but there is no difference in virus clearance either. Thus, in addition to $CD8^+$ T cells, other cells may account for virus clearance and those may not be affected by the JNK-deficiency. *Jnk1*$^{-/-}$ mice appear to be more susceptible to *Leishmania* infection than wild type mice, and this phenotype has been associated to the *in vitro* Th2 bias of $CD4^+$ T cells from these mice (Constant et al. 2000). Finally, studies of *Influenza* virus demonstrate a role of p38 MAPK in the immune response to infection (Conze et al. 2000).

SAPK have been associated with the development of arthritis and increased JNK and p38 MAPK activation have been found in joint synoviocytes. Overexpression of human TNFα in transgenic mice causes the development of inflammatory arthritis. The disruption of JNK1 expression (*Jnk1*$^{-/-}$ mice) in these mice did not affect synovial inflammation, bone destruction and cartilage damage characteristic of this disease (Koller et al. 2005), indicating that JNK1 is not required for the development of inflammatory arthritis. The absence of JNK2 (*Jnk2*$^{-/-}$ mice) has little effect on inflammation, but decreases joint damage probably by interfering with MMP expression in collagen-induced model of arthritis (Han et al. 2001). In contrast, inhibition of both JNK1 and JNK2 with the pharmacological inhibitor SP600125 inhibits IL-1-induced collagenase-3 gene expression in synoviocytes and adjuvant-induced arthritis in a rat model. Similarly, disruption of the p38 MAPK pathway by targeted deletion of the upstream activator MKK3 (*Mkk3*$^{-/-}$ mice) interferes with the development of collagen-induced arthritis (Inoue et al. 2006) and Lyme arthritis (Anguita et al. 2002).

Another type of autoimmune disease that affects a large number of individuals is type 1 (insulin-dependent) diabetes caused by autoimmune destruction of β cells in the pancreatic islets. Destruction of β cells is associated with a $CD4^+$ Th1 type of immune response. Studies of nonobese diabetic (NOD) mice have shown that Jnk2 deficiency decreases destructive insulitis and reduces progression of diabetes (Jaeschke et al. 2005). In this case, the protective effect of JNK2-deficiency appears to be due to the decreased Th1 immune response and a bias to Th2 immune response (Jaeschke et al. 2005), that correlates with the *in vitro* studies in these mice (Yang et al. 1998). Studies of another autoimmune disease, encephalomyelitis, demonstrate that JNK1-deficiency prevents disease progression, in part, because of increased expression of the anti-inflammatory cytokine IL-10 by macrophages (Tran et al. 2006).

Finally, a recent study has examined the role of JNK in the development of atherosclerosis. *ApoE*$^{-/-}$ *Jnk2*$^{-/-}$ mice develop less atherosclerosis than *ApoE*$^{-/-}$ mice, but *ApoE*$^{-/-}$ *Jnk1*$^{-/-}$ mice develop normal atherosclerosis (Ricci et al. 2004). Pharmacological inhibition also reduces atherosclerosis. The resistance to development of atherosclerosis in *Jnk2*$^{-/-}$ mice appears to be due to a defect in foam cell formation by macrophages.

The pyridinyl imidazole inhibitors of p38 MAP kinase were first identified as anti-arthritic drugs prior to its association with this signaling pathway (Lee et al. 1993). Today, a large number of p38 inhibitors have been developed by a variety of pharmaceutical corporations (Kumar et al. 2003). Due to the role of p38 MAP kinase in the production of proinflammatory cytokines such as TNFα, pharmacological inhibitors of this kinase have been tested in a number of diseases where pathogenesis is associated with the presence of these cytokines, including Rheumatoid arthritis, Crohn's disease and inflammatory bowel disease (IBD) (Hommes et al. 2002; Kumar et al. 2003; Pargellis and Regan 2003). Preclinical studies with these compounds have been successfully done in animals, primarily in rats for adjuvant-induced arthritis and collagen-induced arthritis (Kumar et al. 2003). Phase I clinical trials with healthy volunteers or rheumatoid arthritis patients have also been carried on. While the compounds seem to be effective in lowering the levels of proinflammatory cytokines, they also have some adverse effects including liver damage (elevated levels of transaminases) or, in some cases, adverse neurological effects (Kumar et al. 2003). Despite of many years of phase I clinical trials, there is still more to do to improve the efficacy of these promising compounds while reducing their adverse effects.

9 Concluding remarks

Significant progress has been achieved towards understanding the role of SAPK in the immune response. Specific roles of individual SAPK isoforms in different immune cell types have been described. These studies suggest that SAPK represent potential targets for therapeutic intervention in disease processes. Nevertheless, there remains much to learn concerning the molecular mechanisms that account for SAPK functions. Future research challenges include the definition of the role of SAPK in all immune cell types. In addition, methods that will overcome the technical limitations caused by the lethality that results from compound mutations in the JNK and p38 MAPK pathways in mice need to be developed. It is likely that research progress during the next few years will prove to be very exciting.

References

Adams RH, Porras A, Alonso G, Jones M, Vintersten K, Panelli S, Valladares A, Perez L, Klein R, Nebreda AR (2000) Essential role of p38alpha MAP kinase in placental but not embryonic cardiovascular development. Mol Cell 6:109-116

Allen M, Svensson L, Roach M, Hambor J, McNeish J, Gabel CA (2000) Deficiency of the stress kinase p38alpha results in embryonic lethality: characterization of the kinase dependence of stress responses of enzyme-deficient embryonic stem cells. J Exp Med 191:859-870

Anguita J, Barthold SW, Persinski R, Hedrick MN, Huy CA, Davis RJ, Flavell RA, Fikrig E (2002) Murine Lyme arthritis development mediated by p38 mitogen-activated protein kinase activity. J Immunol 168:6352-6357

Arbour N, Naniche D, Homann D, Davis RJ, Flavell RA, Oldstone MB (2002) c-Jun NH(2)-terminal kinase (JNK)1 and JNK2 signaling pathways have divergent roles in CD8(+) T cell-mediated antiviral immunity. J Exp Med 195:801-810

Arrighi JF, Rebsamen M, Rousset F, Kindler V, Hauser C (2001) A critical role for p38 mitogen-activated protein kinase in the maturation of human blood-derived dendritic cells induced by lipopolysaccharide, TNF-alpha, and contact sensitizers. J Immunol 166:3837-3845

Ashwell JD (2006) The many paths to p38 mitogen-activated protein kinase activation in the immune system. Nat Rev Immunol 6:532-540

Beardmore VA, Hinton HJ, Eftychi C, Apostolaki M, Armaka M, Darragh J, McIlrath J, Carr JM, Armit LJ, Clacher C, Malone L, Kollias G, Arthur JS (2005) Generation and characterization of p38beta (MAPK11) gene-targeted mice. Mol Cell Biol 25:10454-10464

Behrens A, Sabapathy K, Graef I, Cleary M, Crabtree GR, Wagner EF (2001) Jun N-terminal kinase 2 modulates thymocyte apoptosis and T cell activation through c-Jun and nuclear factor of activated T cell (NF-AT). Proc Natl Acad Sci USA 98:1769-1774
Bendelac A, Savage PB, Teyton L (2006) The biology of NKT cells. Annu Rev Immunol 25:297-336
Berenson LS, Yang J, Sleckman BP, Murphy TL, Murphy KM (2006) Selective requirement of p38alpha MAPK in cytokine-dependent, but not antigen receptor-dependent, Th1 responses. J Immunol 176:4616-4621
Blander JM, Medzhitov R (2004) Regulation of phagosome maturation by signals from toll-like receptors. Science 304:1014-1018
Brancho D, Tanaka N, Jaeschke A, Ventura JJ, Kelkar N, Tanaka Y, Kyuuma M, Takeshita T, Flavell RA, Davis RJ (2003) Mechanism of p38 MAP kinase activation *in vivo*. Genes Dev 17:1969-1978
Chi H, Barry SP, Roth RJ, Wu JJ, Jones EA, Bennett AM, Flavell RA (2006) Dynamic regulation of pro- and anti-inflammatory cytokines by MAPK phosphatase 1 (MKP-1) in innate immune responses. Proc Natl Acad Sci USA 103:2274-2279
Chi H, Lu B, Takekawa M, Davis RJ, Flavell RA (2004) GADD45beta/GADD45gamma and MEKK4 comprise a genetic pathway mediating STAT4-independent IFNgamma production in T cells. Embo J 23:1576-1586
Chow C-H, Rincón M, Cavanagh J, Dickens M, Davis RJ (1997) Nuclear accumulation of NFAT4 opposed by the JNK signal transduction pathway. Science 278:1638-1641
Chow CW, Dong C, Flavell RA, Davis RJ (2000) c-Jun NH(2)-terminal kinase inhibits targeting of the protein phosphatase calcineurin to NFATc1. Mol Cell Biol 20:5227-5234
Collins LR, Minden A, Karin M, Brown JH (1996) Galpha12 stimulates c-Jun NH2-terminal kinase through the small G proteins Ras and Rac. J Biol Chem 271:17349-17353
Constant SL, Dong C, Yang DD, Wysk M, Davis RJ, Flavell RA (2000) JNK1 is required for T cell-mediated immunity against *Leishmania major* infection. J Immunol 165:2671-2676
Conze D, Krahl T, Kennedy N, Weiss L, Lumsden J, Hess P, Flavell RA, Le Gros G, Davis RJ, Rincon M (2002) c-Jun NH(2)-terminal kinase (JNK)1 and JNK2 have distinct roles in CD8(+) T cell activation. J Exp Med 195:811-823
Conze D, Lumsden J, Enslen H, Davis RJ, Le Gros G, Rincon M (2000) Activation of p38 MAP kinase in T cells facilitates the immune response to the influenza virus. Mol Immunol 37:503-513
Diehl NL, Enslen H, Fortner KA, Merritt C, Stetson N, Charland C, Flavell RA, Davis RJ, Rincón M (2000) Activation of the p38 MAP kinase pathway arrests cell cycle progression and differentiation of immature thymocytes *in vivo*. J Exp Med 191:321-334
Dienz O, Eaton SM, Krahl T, Diehl S, Charland C, Dodge J, Sawain SL, Budd RC, Haynes L, Rincon M (2007) Accumulation of NFAT mediates IL-2 expression in memory, but not in naive CD4+ T cells. Proc Natl Acad Sci USA: in press
Diskin R, Lebendiker M, Engelberg D, Livnah O (2007) Structures of p38alpha active mutants reveal conformational changes in L16 loop that induce autophosphorylation and activation. J Mol Biol 365:66-76
Dodeller F, Schulze-Koops H (2006) The p38 mitogen-activated protein kinase signaling cascade in CD4 T cells. Arthritis Res Ther 8:205
Dong C, Yang DD, Tournier C, Whitmarsh AJ, Xu J, Davis RJ, Flavell RA (2000) JNK is required for effector T-cell function but not for T-cell activation. Nature 405:91-94

Dong C, Yang DD, Wysk M, Whitmarsh AJ, Davis RJ, Flavell RA (1998) Defective T cell differentiation in the absence of Jnk1. Science 282:2092-2095

Farley N, Pedraza-Alva G, Serrano-Gomez D, Nagaleekar V, Aronshtam A, Krahl T, Thornton T, Rincon M (2006) p38 mitogen-activated protein kinase mediates the Fas-induced mitochondrial death pathway in CD8+ T cells. Mol Cell Biol 26:2118-2129

Fortner KA, Budd RC (2005) The death receptor Fas (CD95/APO-1) mediates the deletion of T lymphocytes undergoing homeostatic proliferation. J Immunol 175:4374-4382

Gao M, Labuda T, Xia Y, Gallagher E, Fang D, Liu YC, Karin M (2004) Jun turnover is controlled through JNK-dependent phosphorylation of the E3 ligase Itch. Science 306:271-275

Gao Y, Tao J, Li MO, Zhang D, Chi H, Henegariu O, Kaech SM, Davis RJ, Flavell RA, Yin Z (2005) JNK1 is essential for CD8+ T cell-mediated tumor immune surveillance. J Immunol 175:5783-5789

Gomez del Arco P, Martinez-Martinez S, Maldonado JL, Ortega-Perez I, Redondo JM (2000) A role for the p38 MAP kinase pathway in the nuclear shuttling of NFATp. J Biol Chem 275:13872-13878

Han Z, Boyle DL, Chang L, Bennett B, Karin M, Yang L, Manning AM, Firestein GS (2001) c-Jun N-terminal kinase is required for metalloproteinase expression and joint destruction in inflammatory arthritis. J Clin Invest 108:73-81

Handley ME, Thakker M, Pollara G, Chain BM, Katz DR (2005) JNK activation limits dendritic cell maturation in response to reactive oxygen species by the induction of apoptosis. Free Radic Biol Med 38:1637-1652

Hartenstein B, Teurich S, Hess J, Schenkel J, Schorpp-Kistner M, Angel P (2002) Th2 cell-specific cytokine expression and allergen-induced airway inflammation depend on JunB. Embo J 21:6321-6329

Hess P, Pihan G, Sawyers CL, Flavell RA, Davis RJ (2002) Survival signaling mediated by c-Jun NH(2)-terminal kinase in transformed B lymphoblasts. Nat Genet 32:201-205

Himes SR, Sester DP, Ravasi T, Cronau SL, Sasmono T, Hume DA (2006) The JNK are important for development and survival of macrophages. J Immunol 176:2219-2228

Hommes D, van den Blink B, Plasse T, Bartelsman J, Xu C, Macpherson B, Tytgat G, Peppelenbosch M, Van Deventer S (2002) Inhibition of stress-activated MAP kinases induces clinical improvement in moderate to severe Crohn's disease. Gastroenterology 122:7-14

Hu Y, Ivashkiv LB (2006) Costimulation of chemokine receptor signaling by matrix metalloproteinase-9 mediates enhanced migration of IFN-alpha dendritic cells. J Immunol 176:6022-6033

Iijima N, Yanagawa Y, Clingan JM, Onoe K (2005) CCR7-mediated c-Jun N-terminal kinase activation regulates cell migration in mature dendritic cells. Int Immunol 17:1201-1212

Inoue T, Boyle DL, Corr M, Hammaker D, Davis RJ, Flavell RA, Firestein GS (2006) Mitogen-activated protein kinase kinase 3 is a pivotal pathway regulating p38 activation in inflammatory arthritis. Proc Natl Acad Sci USA 103:5484-5489

Jaeschke A, Rincon M, Doran B, Reilly J, Neuberg D, Greiner DL, Shultz LD, Rossini AA, Flavell RA, Davis RJ (2005) Disruption of the Jnk2 (Mapk9) gene reduces destructive insulitis and diabetes in a mouse model of type I diabetes. Proc Natl Acad Sci USA 102:6931-6935

Kim JM, White JM, Shaw AS, Sleckman BP (2005) MAPK p38 alpha is dispensable for lymphocyte development and proliferation. J Immunol 174:1239-1244

Koller M, Hayer S, Redlich K, Ricci R, David JP, Steiner G, Smolen JS, Wagner EF, Schett G (2005) JNK1 is not essential for TNF-mediated joint disease. Arthritis Res Ther 7:R166-R173

Kriehuber E, Bauer W, Charbonnier AS, Winter D, Amatschek S, Tamandl D, Schweifer N, Stingl G, Maurer D (2005) Balance between NF-kappaB and JNK/AP-1 activity controls dendritic cell life and death. Blood 106:175-183

Kumar S, Boehm J, Lee JC (2003) p38 MAP kinases: key signalling molecules as therapeutic targets for inflammatory diseases. Nat Rev Drug Discov 2:717-726

Lee JC, Badger AM, Griswold DE, Dunnington D, Trunech A, Votta B, White JR, Young PR, Bender PE (1993) Bicyclic imidazoles as a novel class of cytokine biosynthesis inhibitors. Ann NY Acad Sci 696:149-170

Li B, Tournier C, Davis RJ, Flavell RA (1999) Regulation of IL-4 expression by the transcription factor JunB during T helper cell differentiation. EMBO J 18:420-432

Lizundia R, Chaussepied M, Huerre M, Werling D, Di Santo JP, Langsley G (2006) c-Jun NH2-terminal kinase/c-Jun signaling promotes survival and metastasis of B lymphocytes transformed by Theileria. Cancer Res 66:6105-6110

Lu B, Yu H, Chow C, Li B, Zheng W, Davis RJ, Flavell RA (2001) GADD45gamma mediates the activation of the p38 and JNK MAP kinase pathways and cytokine production in effector TH1 cells. Immunity 14:583-590

Lu Z, Serghides L, Patel SN, Degousee N, Rubin BB, Krishnegowda G, Gowda DC, Karin M, Kain KC (2006) Disruption of JNK2 decreases the cytokine response to Plasmodium falciparum glycosylphosphatidylinositol *in vitro* and confers protection in a cerebral malaria model. J Immunol 177:6344-6352

Mehrotra S, Chhabra A, Hegde U, Cakraborty NG, Mukherji B (2006) Inhibition of c-Jun N-terminal kinase (JNK) rescues influenza epitope-specific human cytolytic T lymphocytes (CTL) from activation-induced cell death (AICD). J Leukoc Biol 81:539-547

Merritt C, Enslen H, Diehl N, Conze D, Davis RJ, Rincón M (2000) Activation of p38 mitogen-activated protein kinase *in vivo* selectively induces apoptosis of CD8(+) but not CD4(+) T cells. Mol Cell Biol 20:936-946

Muthumani K, Choo AY, Hwang DS, Premkumar A, Dayes NS, Harris C, Green DR, Wadsworth SA, Siekierka JJ, Weiner DB (2005) HIV-1 Nef-induced FasL induction and bystander killing requires p38 MAPK activation. Blood 106:2059-2068

Nakagawa S, Ohtani T, Mizuashi M, Mollah ZU, Ito Y, Tagami H, Aiba S (2004) p38 Mitogen-Activated protein kinase mediates dual role of ultraviolet B radiation in induction of maturation and apoptosis of monocyte-derived dendritic cells. J Invest Dermatol 123:361-370

Nakahara T, Uchi H, Urabe K, Chen Q, Furue M, Moroi Y (2004) Role of c-Jun N-terminal kinase on lipopolysaccharide induced maturation of human monocyte-derived dendritic cells. Int Immunol 16:1701-1709

Pargellis C, Regan J (2003) Inhibitors of p38 mitogen-activated protein kinase for the treatment of rheumatoid arthritis. Curr Opin Investig Drugs 4:566-571

Pedraza-Alva G, Koulnis M, Charland C, Thornton T, Clements JL, Schlissel MS, Rincon M (2006) Activation of p38 MAP kinase by DNA double-strand breaks in V(D)J recombination induces a G2/M cell cycle checkpoint. Embo J 25:763-773

Puig-Kroger A, Relloso M, Fernandez-Capetillo O, Zubiaga A, Silva A, Bernabeu C, Corbi AL (2001) Extracellular signal-regulated protein kinase signaling pathway negatively regulates the phenotypic and functional maturation of monocyte-derived human dendritic cells. Blood 98:2175-2182

Ricci R, Sumara G, Sumara I, Rozenberg I, Kurrer M, Akhmedov A, Hersberger M, Eriksson U, Eberli FR, Becher B, Borén J, Chen M, Cybulsky MI, Moore KJ, Freeman MW, Wagner EF, Matter CM, Lüscher TF (2004) Requirement of JNK2 for scavenger receptor A-mediated foam cell formation in atherogenesis. Science 306:1558-1561

Rincon M, Flavell RA (1999) Reprogramming transcription during the differentiation of precursor CD4+ T cells into effector Th1 and Th2 cells. Microbes Infect 1:43-50

Rincon M, Pedraza-Alva G (2003) JNK and p38 MAP kinases in CD4+ and CD8+ T cells. Immunol Rev 192:131-142

Rincón M, Whitmarsh A, Yang DD, Weiss L, Dérijard B, Jayaraj P, Davis RJ, Flavell RA (1998) The JNK pathway regulates the *in vivo* deletion of immature $CD4^{+}CD8^{+}$ thymocytes. J Exp Med 188:1817-1830

Round JL, Humphries LA, Tomassian T, Mittelstadt P, Zhang M, Miceli MC (2007) Scaffold protein Dlgh1 coordinates alternative p38 kinase activation, directing T cell receptor signals toward NFAT but not NF-kappaB transcription factors. Nat Immunol 8:154-161

Sabapathy K, Hu Y, Kallunki T, Schreiber M, David J-P, Jochum W, Wagner EF, Karin M (1999) JNK2 is required for efficient T-cell activation and apoptosis but not for normal lymphocyte development. Curr Biol 9:116-125

Sabapathy K, Kallunki T, David JP, Graef I, Karin M, Wagner EF (2001) c-Jun NH2-terminal kinase (JNK)1 and JNK2 have similar and stage- dependent roles in regulating T cell apoptosis and proliferation. J Exp Med 193:317-328

Sabio G, Arthur JS, Kuma Y, Peggie M, Carr J, Murray-Tait V, Centeno F, Goedert M, Morrice NA, Cuenda A (2005) p38gamma regulates the localisation of SAP97 in the cytoskeleton by modulating its interaction with GKAP. Embo J 24:1134-1145

Salvador JM, Mittelstadt PR, Guszczynski T, Copeland TD, Yamaguchi H, Appella E, Fornace AJ Jr, Ashwell JD (2005) Alternative p38 activation pathway mediated by T cell receptor-proximal tyrosine kinases. Nat Immunol 6:390-395

Sasaki T, Wada T, Kishimoto H, Irie-Sasaki J, Matsumoto G, Goto T, Yao Z, Wakeham A, Mak TW, Suzuki A, Cho SK, Zuniga-Pflucker JC, Oliveira-dos-Santos AJ, Katada T, Nishina H, Penninger JM (2001) The stress kinase mitogen-activated protein kinase kinase (MKK)7 is a negative regulator of antigen receptor and growth factor receptor-induced proliferation in hematopoietic cells. J Exp Med 194:757-768

Sen J, Kapeller R, Fragoso R, Sen R, Zon LI, Burakoff SJ (1996) Intrathymic signals in thymocytes are mediated by p38 mitogen-activated protein kinase. J Immunol 156:4535-4538

Singh RA, Zhang JZ (2004) Differential activation of ERK, p38, and JNK required for Th1 and Th2 deviation in myelin-reactive T cells induced by altered peptide ligand. J Immunol 173:7299-7307

Sugawara T, Moriguchi T, Nishida E, Takahama Y (1998) Differential roles of ERK and p38 MAP kinase pathways in positive and negative selection of T lymphocytes. Immunity 9:565-574

Swantek JL, Cobb MH, Geppert TD (1997) Jun N-terminal kinase/stress-activated protein kinase (JNK/SAPK) is required for lipopolysaccharide stimulation of tumor necrosis factor alpha (TNF-alpha) translation: glucocorticoids inhibit TNF-alpha translation by blocking JNK/SAPK. Mol Cell Biol 17:6274-6282

Swat W, Fujikawa K, Ganiatsas S, Yang DD, Xavier RJ, Harris NL, Davidson L, Ferrini R, Davis RJ, Labow MA, Flavell RA, Zon LI, Alt FW (1998) SEK1/MKK4 is required for maintenance of a normal peripheral lymphoid compartment but not for T lymphocyte development. Immunity 8:625-634

Tamura K, Sudo T, Senftleben U, Dadak AM, Johnson R, Karin M (2000) Requirement for p38alpha in erythropoietin expression: a role for stress kinases in erythropoiesis. Cell 102:221-231

Tanaka N, Kamanaka M, Enslen H, Dong C, Wysk M, Davis RJ, Flavell RA (2002) Differential involvement of p38 mitogen-activated protein kinase kinases MKK3 and MKK6 in T-cell apoptosis. EMBO Rep 3:785-791

Teague TK, Hildeman D, Kedl RM, Mitchell T, Rees W, Schaefer BC, Bender J, Kappler J, Marrack P (1999) Activation changes the spectrum but not the diversity of genes expressed by T cells. Proc Natl Acad Sci USA 96:12691-12696

Tham EL, Mescher MF (2001) Signaling alterations in activation-induced nonresponsive CD8 T cells. J Immunol 167:2040-2048

Tran EH, Azuma YT, Chen M, Weston C, Davis RJ, Flavell RA (2006) Inactivation of JNK1 enhances innate IL-10 production and dampens autoimmune inflammation in the brain. Proc Natl Acad Sci USA 103:13451-13456

Ventura JJ, Kennedy NJ, Flavell RA, Davis RJ (2004) JNK regulates autocrine expression of TGF-beta1. Mol Cell 15:269-278

Weiss L, Whitmarsh AJ, Yang DD, Rincon M, Davis RJ, Flavell RA (2000) Regulation of c-Jun NH(2)-terminal kinase (Jnk) gene expression during T cell activation. J Exp Med 191:139-146

West MA, Wallin RP, Matthews SP, Svensson HG, Zaru R, Ljunggren HG, Prescott AR, Watts C (2004) Enhanced dendritic cell antigen capture via toll-like receptor-induced actin remodeling. Science 305:1153-1157

Xavier R, Rabizadeh S, Ishiguro K, Andre N, Ortiz JB, Wachtel H, Morris DG, Lopez-Ilasaca M, Shaw AC, Swat W, Seed B (2004) Discs large (Dlg1) complexes in lymphocyte activation. J Cell Biol 166:173-178

Xie J, Qian J, Wang S, Freeman ME 3rd, Epstein J, Yi Q (2003) Novel and detrimental effects of lipopolysaccharide on *in vitro* generation of immature dendritic cells: involvement of mitogen-activated protein kinase p38. J Immunol 171:4792-4800

Yang DD, Conze D, Whitmarsh AJ, Barret T, Davis RJ, Rincón M, Flavell RA (1998) Differentiation of $CD4^{+}$ T cells to Th1 cells requires MAP kinase JNK2. Immunity 9:575-585

Yao Y, Xu Q, Kwon MJ, Matta R, Liu Y, Hong SC, Chang CH (2006) ERK and p38 MAPK signaling pathways negatively regulate CIITA gene expression in dendritic cells and macrophages. J Immunol 177:70-76

Yates RM, Russell DG (2005) Phagosome maturation proceeds independently of stimulation of toll-like receptors 2 and 4. Immunity 23:409-417

Zhang Y, Blattman JN, Kennedy NJ, Duong J, Nguyen T, Wang Y, Davis RJ, Greenberg PD, Flavell RA, Dong C (2004) Regulation of innate and adaptive immune responses by MAP kinase phosphatase 5. Nature 430:793-797

Zhao Q, Wang X, Nelin LD, Yao Y, Matta R, Manson ME, Baliga RS, Meng X, Smith CV, Bauer JA, Chang CH, Liu Y (2006) MAP kinase phosphatase 1 controls innate immune responses and suppresses endotoxic shock. J Exp Med 203:131-140

Davis, Roger J.
Howard Hughes Medical Institute and Program in Molecular Medicine, University of Massachusetts Medical School, Worcester, MA 01605.

Rincón, Mercedes
Immunology Program, Department of Medicine, University of Vermont, Burlington, VT 05405.
mrincon@uvm.edu

Stress-activated MAP kinases in chromatin and transcriptional complexes

Nicholas T. Crump, Ya Ting Han, and Louis C. Mahadevan

Abstract

Stress-activated MAP kinases (SAPKs) are activated by stressors or by certain physiological stimuli and mediate an intracellular response appropriate to the change in environment. Long-term adaptation requires reprogramming of transcription and one of the most significant actions of SAPK cascades is therefore induction of gene expression. SAPKs and their downstream kinases phosphorylate many chromatin-associated and transcription factors. Further, they can induce localised histone modification by regulating histone acetyltransferases and deacetylases. p38/SAPK2 also elicits phosphorylation of the nucleosomal proteins histone H3 and HMGN1 (previously HMG-14) via the downstream mitogen- and stress-stimulated kinases MSK1/2. Finally, recent evidence indicates a novel non-enzymatic SAPK function in transcriptional complexes, suggesting a more structural role. The yeast SAPK Hog1p is recruited to a proportion of its target genes on activation and localises beyond the promoter into coding regions. The observation that Hog1p interacts with elongating RNA polymerase II in addition to several transcriptional elongation factors has led to the suggestion that this SAPK may behave like an elongation factor at some target genes. The generality of this new function is discussed.

1 Introduction

One of the challenges associated with existing as a unicellular organism in a dynamic environment is the need to respond rapidly and appropriately to extracellular stimuli. In lower eukaryotes, this is in part achieved by signalling through the mitogen-activated protein kinase (MAPK) cascades (Widmann et al. 1999), where stresses outside the cell are transmitted across the cell membrane to stress-activated MAPKs (SAPKs), which elicit appropriate responses through modulation of gene expression and protein activity. In higher eukaryotes, MAPK cascades have been harnessed for cellular responses to a variety of extracellular stimuli including a number of different growth factors and cytokines, in addition to their role in the stress response.

MAPK signalling is transmitted through a three-tiered cascade of phosphorylation events, with MAPKs activated by MAPK kinases (MAPKKs) and MAPKKs in turn activated by MAPKK kinases (MAPKKKs). In many systems, the mecha-

Topics in Current Genetics, Vol. 20
F. Posas, A. R. Nebreda (Eds.): Stress-Activated Protein Kinases
DOI 10.1007/4735_2007_0254 / Published online: 1 August 2007

nism by which signals are transduced from extracellular receptors or stress stimuli to produce MAPKKK activation is poorly understood, although STE20 kinases and small G proteins have both been implicated (Chang and Karin 2001). Downstream targets of MAPK signalling cascades are highly varied, and in the case of stresses they combine to produce cellular adaptations to respond to the new environment. MAPKs are involved in regulating a diverse selection of processes, including cell proliferation, cell survival and apoptosis, movement, metabolism, and differentiation (Posas et al. 1998; Chang and Karin 2001; Kyriakis and Avruch 2001). Often this involves modulating enzyme function, but a significant proportion of the effects of MAPK signalling is enacted via changes in gene expression. This review discusses mechanisms by which SAPK-mediated signalling effects these changes, focussing on their interactions with chromatin-associated and transcription factors and on their role in eliciting localised histone modification at target genes.

MAP kinases are proline-directed serine/threonine kinases (Widmann et al. 1999) that produce a response through phosphorylation either directly or indirectly by one of their downstream kinases. Phosphorylation can control protein activity in many different ways, with MAPKs observed to regulate enzyme activity, nucleocytoplasmic shuttling, protein stability, and DNA binding affinity, as well as directly modulating the transactivation/repression activity of transcription factors (Yang et al. 2003). Recently, however, a new method by which SAPKs influence gene expression has been identified. The yeast SAPK Hog1p has been shown to play a role in transcription, potentially independent of its kinase activity, through direct interactions with the transcriptional machinery and other transcription factors, apparently acting like a transcription elongation factor (Alepuz et al. 2003; Proft et al. 2006).

1.1 SAPK cascades in yeast and mammals

Yeast cells show at least five distinct MAP kinase pathways, although there appears to be some overlap between them (Posas et al. 1998; Widmann et al. 1999). Much of the work into SAPK-mediated control of gene expression has involved the yeast MAPK Hog1p, which interacts with many different factors to modulate expression at a number of subclasses of genes (Fig. 1). The protein is activated under conditions of high osmolarity (Hog = high osmolarity glycerol) by the MAPK kinase Pbs2p, which targets a dual tyrosine/threonine element, inducing Hog1p to interact with and modulate the activity of many factors (de Nadal et al. 2002; O'Rourke et al. 2002). The interacting partners of Hog1p are diverse, and range from transcription factors specific to subclasses of the osmoresponsive genes (Proft et al. 2001; Proft and Struhl 2002; Alepuz et al. 2003; de Nadal et al. 2003), to more general factors such as the RNA polymerase II (Pol II) holoenzyme (Alepuz et al. 2003; Proft et al. 2006).

In mammalian cells, the two main SAPK pathways are JNK (c-Jun N-terminal Kinase, also known as SAPK1) and p38 (also known as SAPK2), although the

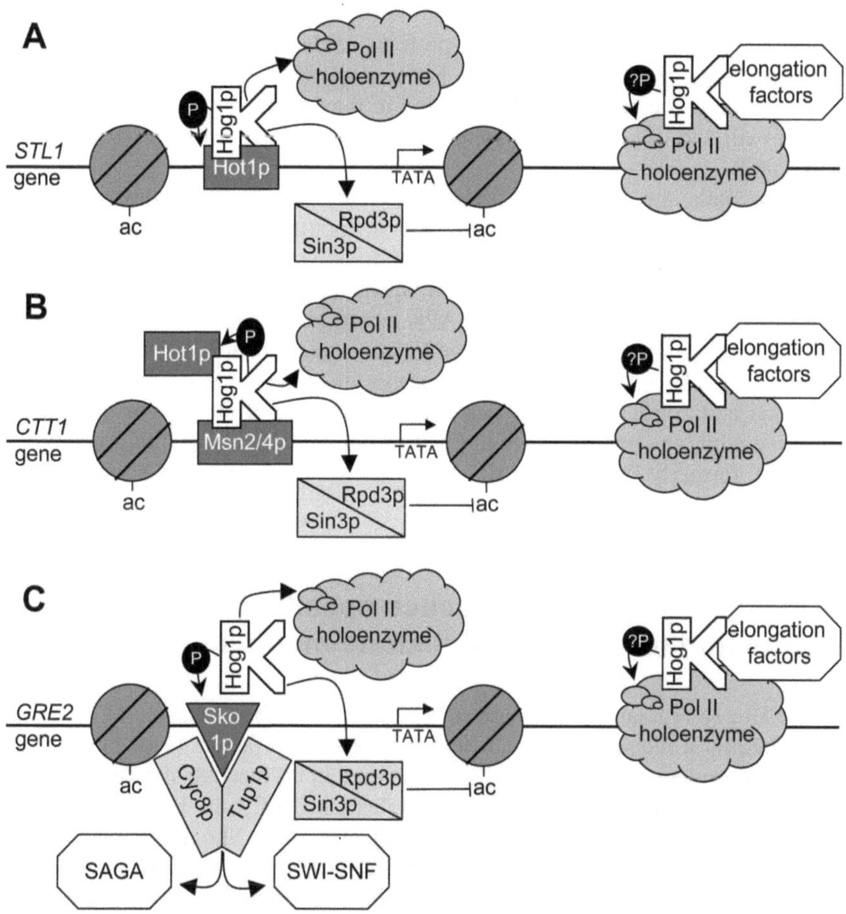

Fig. 1. Activated Hog1p stimulates gene expression from multiple classes of gene under osmostress. The transcription complexes formed in response to high osmolarity at (A) Hot1p-, (B) Msn2/4p- and (C) Sko1p-regulated promoters. (A) Under stressed conditions Hot1p anchors phosphorylated Hog1p at the gene promoter. Hog1p then recruits the RNA polymerase II holoenzyme to the promoter. Independently, Hog1p interacts with the HDAC complex Rpd3p/Sin3p to induce localised histone deacetylation at the promoter, which may aid Pol II entry. Hog1p also appears to form a part of the elongating transcriptional apparatus. (B) Msn2/4p-dependent promoters appear to function similarly to the Hot1p-regulated system, with Msn2/4p recruiting the kinase to the gene promoter. (C) Sko1p-regulated genes are repressed under non-stressed conditions by the Sko1p-Cyc8p-Tup1p repressor complex. Active Hog1p phosphorylates Sko1p, switching the complex from repressive to activatory and allowing Cyc8p-Tup1p to recruit the nucleosome remodelling complexes SAGA and SWI-SNF to the promoter, aiding Pol II entry.

ERK (Extracellular signal-Regulated Kinase) pathway is also capable of weakly responding to some stresses. The BMK1/ERK5 pathway responds to both stress and mitogenic stimuli (Chang and Karin 2001; Kyriakis and Avruch 2001). The evolutionary conservation of MAPK proteins is clearly demonstrated in complementation experiments, where human ERK1 and JNK/SAPK1 MAPKs introduced into yeast cells are capable of functioning in the pathways that regulate filamentous growth and osmostress, respectively (Galcheva-Gargova et al. 1994; Atienza et al. 2000).

In mammals, the diversity of stimuli which activate MAPK pathways is necessarily much greater than in yeast, with roles in controlling cell-cell communication and differentiation pathways, as well as responding to a vast number of extracellular factors, stresses and pharmacological agents (Chang and Karin 2001; Kyriakis and Avruch 2001). The diversity of SAPK targets is maintained, with the activities of a number of transcription factors and other nuclear factors influenced by these kinases (Fig. 2). p38/SAPK2 also phosphorylates several downstream kinases (Kyriakis and Avruch 2001), which themselves target a number of nuclear proteins, including the histone H3 N-terminal tail and the high-mobility group protein HMGN1, formerly known as HMG-14 (Thomson et al. 1999b; Clayton and Mahadevan 2003).

2 Phosphorylation of sequence-specific transcription factors and recruitment of histone-modifying enzymes

Active Hog1p regulates expression of a number of different classes of genes, distinguished by a requirement for different transcription factors (Fig. 1). One class (Sko1p-dependent genes, Fig. 1C) is maintained in an inactive state under normal osmotic conditions by the repressor complex Sko1p-Cyc8p-Tup1p (Proft et al. 2001). On activation, Hog1p phosphorylates the Sko1p component, an ATF/CREB-related factor, which converts the complex from repressor to activator. Transcriptional activation by this complex is mediated, at least in part, by Cyc8p-Tup1p-mediated recruitment of the SAGA acetyltransferase and SWI-SNF chromatin remodelling complexes to the promoter, allowing nucleosomal alterations concomitant with gene activity (Proft et al. 2001).

Similarly, mammalian SAPKs phosphorylate sequence-specific transcription factors to mediate recruitment of histone acetyltransferase (HAT) activities to gene promoters (Fig. 2). For example, the ternary complex factor (TCF) family protein Elk-1 is phosphorylated by both JNK/SAPK1 and p38/SAPK2 (Li et al. 2003), recruiting the acetyltransferase activity of p300 to targeted promoters. Interestingly, the Elk-1-p300 interaction exists even in the absence of phosphorylation, with the modification required to activate the HAT (Nissen et al. 2001). One possible model is that phosphorylation induces conformational changes in the preassembled complex, stimulating the HAT activity of p300. Phosphorylation of

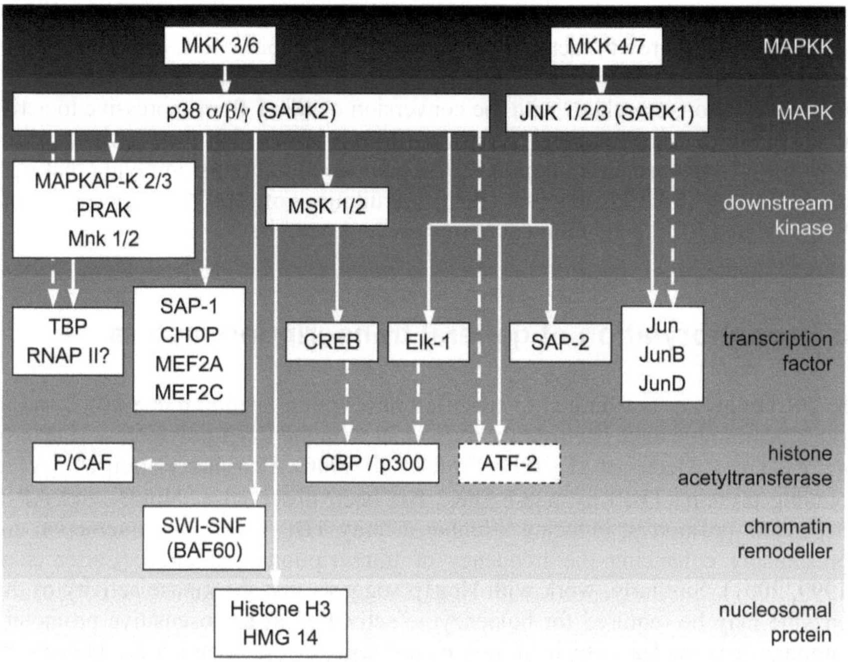

Fig. 2. Stress activated MAPK signalling produces multiple and varied downstream effects. MAPKK/MKK, MAPK kinase; MAPK, mitogen-activated protein kinase; JNK/SAPK1, Jun N-terminal kinase/stress-activated MAPK1. Solid lines represent phosphorylation events; dashed lines are recruitment events. ATF-2 is labelled as a putative HAT.

CREB by MSK1/2 (a downstream kinase activated by p38/SAPK2, see Section 4.1) similarly correlates with recruitment of CBP/p300 (Deak et al. 1998; Wiggin et al. 2002), which can then recruit the p300/CBP-associated factor P/CAF, another HAT (Yang et al. 1996).

In addition to recruiting activatory HATs to the promoter, SAPKs can also disrupt inhibitory histone deacetylase (HDAC) complexes that maintain genes in an inactive state. JNK/SAPK1 was first identified as a kinase which targeted the N-terminus of the transcription factor c-Jun, with phosphorylation inducing its activity (Pulverer et al. 1991). However, it has recently been suggested that this modification is not in itself activatory, but rather acts to derepress transcription (Weiss et al. 2003). In the nucleus, c-Jun dimerises with itself or other members of the Jun/Fos family to form the AP-1 (Activating Protein 1) transcription factor (Karin et al. 1997), which binds conserved promoter elements in a JNK/SAPK1-independent manner. In the absence of JNK/SAPK1 phosphorylation, c-Jun has been shown to bind a repressive complex, which includes HDAC3. JNK/SAPK1 phosphorylation of c-Jun induces dissociation of the HDAC, allowing AP-1 to activate transcription at the promoter (Weiss et al. 2003).

A similar inhibitory mechanism appears to regulate Elk-1 activity (Yang and Sharrocks 2004). In quiescent (G_0) cells, the protein is present in a repressive state, due to the addition of a SUMO (small ubiquitin-related modifier) moiety. On stimulation with growth factors this modification is rapidly lost, coincident with a rapid increase in Elk-1 phosphorylation. Loss of SUMOylation, in concert with phosphorylation, correlates with the conversion of Elk-1 from repressive to active transcription factor. Yang and Sharrocks (2004) showed that the SUMO mark on Elk-1 allows the protein to recruit HDACs to gene promoters, implying that gene activation by SAPKs involves both recruitment of HAT activity (via un-SUMOylated Elk-1) and removal of repressive deacetylases.

3 Phosphorylation of general transcription factors

SAPK targets are not limited to specific transcription factors: p38/SAPK2 and its yeast homologue have been shown to interact with the RNA polymerase II (Pol II) holoenzyme (Alepuz et al. 2003; Proft et al. 2006). Phosphorylation of TATA-binding protein (TBP) by p38/SAPK2 has been observed at NFκB- and AP-1-dependent promoters, inducing a higher affinity TBP-TATA box interaction and presumably enhancing the frequency of transcriptional initiation (Carter et al. 1999, 2001). Similarly, work with Hog1p suggests that the kinase activity of this enzyme may be required for holoenzyme activation at osmosensitive promoters, although it is unclear which, if any, components of the complex are phosphorylated (Alepuz et al. 2003).

As discussed in Section 2, SAPKs act by phosphorylating many transcription factors, stimulating some to recruit histone modifying enzymes to the promoter. However, SAPKs also act directly on the modifying enzymes themselves (Fig. 2). Both JNK/SAPK1 and p38/SAPK2 are known to phosphorylate the putative acetyltransferase ATF-2 upon stimulation (Kawasaki et al. 2000). p38/SAPK2 also interacts with the chromatin remodelling complex SWI-SNF, which is required for transcriptional initiation at many promoters (Simone et al. 2004). Activation of p38/SAPK2 is necessary for SWI-SNF recruitment to myogenic loci during skeletal myogenesis, and *in vitro* active p38/SAPK2 phosphorylates the SWI-SNF component BAF60, suggesting that this is the necessary activity for the *in vivo* recruitment.

4 Phosphorylation of nucleosomal proteins

In addition to phosphorylating transcription factors, SAPKs have a specific role in directly modifying chromatin at target genes, through a process known as the nucleosomal response, by acting on the histone H3 N-terminal tail and the non-histone chromosomal protein HMGN1, also known as HMG-14 (Clayton and Mahadevan 2003). The basic unit of chromatin is the nucleosome, comprising approximately 147 base pairs of DNA wrapped around an octamer of two each of the core histones H3, H4, H2A and H2B. Crystallographic studies have revealed the nucleosome to be of defined structure, with the globular domain of each histone buried within the nucleosome and their N-terminal tails protruding (Luger et al. 1997). The tails are therefore accessible for modification, for example, by HATs, methyltransferases and kinases, and these modifications play a central role in regulating gene expression (Mellor 2005).

The histone H3 tail has two phosphorylation sites targeted via the p38/SAPK2 cascade: serine-10 and serine-28 (Clayton and Mahadevan 2003). These residues are present in the same sequence environment: Ala-Arg-Lys-Ser (ARK-S), which is retained throughout the Eukaryotic Kingdom, implying functional importance. Both modifications occur at two distinct points in the cell cycle: globally on mitotic condensed chromosomes (Cheung et al. 2000), and locally at specific loci of active and inducible genes (Clayton and Mahadevan 2003). Histone phosphorylation at these residues is therefore implicated in two opposing processes, regulated by different proteins. Mitotic phosphorylation is produced by members of the Aurora kinase family (Cheung et al. 2000), regulated by cell cycle proteins, whereas the kinases responsible for inducible gene phosphorylation are found downstream of the ERK and p38/SAPK2 MAP kinases (Clayton and Mahadevan 2003).

4.1 Inducible histone H3 phosphorylation is mediated by MSK1/2

Stress-induced histone H3 phosphorylation was first discovered through ^{32}P-labelling studies in mouse fibroblasts (Mahadevan et al. 1991), where stimulation of quiescent (G_0) cells with various stresses or growth factors results in localised, transient phosphorylation of histones at the loci of the immediate-early (IE) genes, such as c-*jun* and c-*fos*. Using a combination of selective agonists and pharmacological inhibitors, the mediators of this process were shown to be ERK or p38/SAPK2 (Clayton and Mahadevan 2003), depending on the stimulus used. However, the downstream kinase responsible for directly targeting the H3 tail proved more elusive.

The mitogen- and stress-stimulated protein kinases, MSK1 and MSK2, were identified by their homology to the N-terminal domain of the MAPK downstream kinase MAPKAP-K1 (Deak et al. 1998), and a combination of pharmacological experiments and MSK1/2 single- and double-knockout mouse cells confirmed these proteins to be the histone H3 kinases. It was also suggested that RSK2 (Ribosomal S6-kinase, also known as MAPKAP-K1b) has a role in histone H3 phosphorylation, based on work with Coffin-Lowry cells, which are defective in this protein (Sassone-Corsi et al. 1999). However, more recent studies have demonstrated that this cell line shows a normal inducible histone H3 phosphorylation response, indicating that RSK2 is not essential for this process (Soloaga et al. 2003).

MSK1/2 phosphorylates histone H3 efficiently at both serines 10 and 28 (Dyson et al. 2005), but there remains a major unanswered question regarding how it selects one site or the other for phosphorylation. In two independent immunofluorescence studies, H3S10P and H3S28P were localised to distinct loci within the intact mouse nucleus (Dunn and Davie 2005; Dyson et al. 2005). Further, sequential immunoprecipitations failed to detect any histone H3 tails bearing both modifications (Dyson et al. 2005), suggesting an exquisite level of discrimination by MSK1/2 or their interacting partners. Thus, although the intracellular circuitry from plasma membrane to histone has been elucidated, the issue of site selectivity in the histone H3 tail remains a mystery.

The evolutionary conservation of SAPK proteins has been described (Section 1.1), and similarly, sequences within the histone H3 tail are among the most widely conserved throughout eukaryotes. Curiously, the MSK proteins responsible for SAPK-mediated H3 phosphorylation are less well conserved. Homologues are absent from the yeast genomes, and the closest equivalent in fruit fly, JIL-1, has not been shown to be activated by p38/SAPK2-mediated phosphorylation. Nevertheless, heat-shock elicits histone H3 phosphorylation at the *Hsp70* locus on *Drosophila* polytene chromosomes, by an as yet unknown mechanism (Nowak and Corces 2000). Surprisingly, heat-shocked mouse cells do not show any increase in H3 phosphorylation at the *Hsp70* locus (Thomson et al. 2004), despite the observation in *Drosophila* cells, and in contrast to observations after sodium arsenite treatment, where p38/SAPK2 activation and histone H3 phosphorylation at *Hsp70* are both clearly seen. These observations raise interesting questions regarding the evolution of this pathway, especially concerning the "tandem-kinases" (which have two kinase domains) including RSKs, MSKs and JIL-1, and the functional conservation of histone H3 phosphorylation at inducible genes.

4.2 Molecular function of histone H3 phosphorylation

Whilst the significance of phosphorylating enzymes and interacting partners by SAPKs is often reasonably apparent, the purpose of histone phosphorylation is less clear. MSK1/2 double-knockout cells (Soloaga et al. 2003), or cells treated with an MSK-specific inhibitor (Thomson et al. 1999a), in which inducible histone H3 phosphorylation is severely reduced, do not show serious disruption of targeted gene induction, although efficiency of induction is affected. This suggests that the nucleosomal response is not essential to stimulate expression from these genes, but the mark may influence their precise expression profiles. In contrast, work at the *Hsp70* locus suggests that H3 phosphorylation can be essential for induction of this gene, although curiously whilst this requirement is observed on treatment with sodium arsenite, it is unnecessary for *Hsp70* induction upon heat-shock (Thomson et al. 2004). The effects of histone phosphorylation likely act in concert with other factors at the promoter, dependent on the signalling pathways induced.

How might histone phosphorylation induce gene expression? One mechanism would rely on altered electrostatics and sterics of a phosphorylated H3 tail. However, there is evidence to suggest that the modification may act as a docking site for chromatin modifying or transcription factors, with a mammalian interacting partner for phosphorylated histone H3 recently identified. Members of the 14-3-3 family have been shown bound to phosphoacetylated histone H3 peptides on affinity columns and, by chromatin immunoprecipitation, localised to the same positions at c-*jun* and c-*fos* as phosphoacetylation (Macdonald et al. 2005). Structures of 14-3-3ζ bound to both phospho- and phosphoacetyl-H3 have been solved, demonstrating that this interaction is strictly dependent on the phosphate moiety and the acetylation state of the tail does not affect 14-3-3 binding (Macdonald et al. 2005).

The dimeric nature of 14-3-3 proteins provides the possibility that they act as an adaptor to recruit other phosphoproteins to the nucleosome. 304 interacting partners for 14-3-3ζ were identified in HeLa cells (Meek et al. 2004) and 170 proteins were isolated bound to FLAG-tagged 14-3-3 in HEK293 cells (Jin et al. 2004), of which many were identified as chromatin-binding proteins involved in transcriptional regulation. Thus, phosphorylated histone H3 at inducible genes may act as a recruitment site for other factors to modulate gene expression, via the adaptor 14-3-3 proteins.

5 SAPKs may also act independently of their kinase activity

Until relatively recently, the effects of SAPK signalling were assumed to be solely due to their kinase activity. However, new evidence has shown that several of these proteins also have a second kinase-independent role in transcriptional regulation (Fig. 1). The process involves a direct interaction between SAPK and target, resulting in recruitment or modulated activity of the factor without requisite phosphorylation. Most of the work on this mechanism has involved the yeast Hog1p, but there is some, albeit circumstantial, evidence that the mammalian JNK/SAPK1 and p38/SAPK2 may also possess novel activities (Edmunds and Mahadevan 2004).

Active Hog1p is recruited to Hot1p-dependent osmoresponsive gene promoters (Fig. 1A) by the transcription factor Hot1p (Alepuz et al. 2003), but, although Hot1p is phosphorylated at several residues by the kinase, these modifications are not required for interaction with Hog1p and probably have no role in Hot1p regulation. Hog1p at the promoter then recruits Pol II, and appears to act as an anchor for the transcriptional apparatus at the promoter. In the absence of osmotic stress, Hot1p-dependent promoters are devoid of the transcriptional machinery, and osmotic stress-induced holoenzyme recruitment is lost in *hog1*Δ mutant cells. Hog1p interacts with multiple components of the Pol II holoenzyme; in addition to the polymerase itself (via subunits Rpb1p and Rpb2p), the kinase will bind the TFIIH subunit Kin 28 and the Mediator complex (Alepuz et al. 2003).

Hog1p also recruits other components to its target gene promoters (Fig. 1), for example the Rpd3p-Sin3p HDAC complex (de Nadal et al. 2004). Mutants in *rpd3* and *sin3* are sensitive to osmotic stress and, strikingly, *rpd3*Δ*hog1*Δ and *sin3*Δ*hog1*Δ double mutants are no more osmosensitive than *rpd3*Δ or *sin3*Δ alone. Similarly, under osmostress the *rpd3*Δ strain showed reduced expression of more than 90% of Hog1p-dependent genes, implying a role for the HDAC complex in gene expression at these loci. Hog1p interacts with Rpd3p both *in vivo* and *in vitro*, suggesting a direct role for the SAPK in recruitment of the deacetylase complex (de Nadal et al. 2004).

The authors also showed, by point mutation, that the deacetylase activity of Rpd3p is absolutely required for gene expression. At first glance this seems a little counterintuitive: why would recruiting a normally repressive HDAC complex result in activation of a gene? Intriguingly, at the *HSP12* gene locus hypoacetylation seems to be maintained strictly at the promoter region (where the complex is bound). The fact that *rpd3Δ* mutants show reduced Pol II recruitment to the *HSP12* promoter under osmostress implies that the absence of this modification may play a role in encouraging Pol II entry to the promoter (de Nadal et al. 2004).

Recent global analyses have provided insight into activated Hog1p recruitment to its target genes. Pokholok et al. (2006) used TAP-tagged Hog1p to characterise SAPK occupancy by ChIP-on-chip (combining chromatin immunoprecipitation with DNA microarray technology), identifying 36 genes bound under osmotic stress. Similarly Proft et al. (2006) established 72 genes which are enriched in Hog1p at least 1.5-fold upon osmostress. A functional role for Hog1p at the gene locus is suggested by the fact that a catalytically inactive mutant does not localise to the native *CTT1* or *TSL1* genes (Proft et al. 2006).

Both studies described a similar kinase distribution across the target genes, with Hog1p not confined to the promoter regions (Pokholok et al. 2006; Proft et al. 2006), despite the fact that Hog1p recruiters Hot1p and Sko1p are specifically associated with their binding sites here (see Fig. 1). The kinase displayed a distribution profile more like that of Pol II, spanning the open reading frame (ORF) of *STL1* and *GRE2*, with an apparent reliance on sequences in the 3' untranslated region (UTR) of each gene (Proft et al. 2006). This is consistent with *in vitro* pull-down experiments, which show active Hog1p preferentially interacting with the elongating form of Pol II (with the polymerase CTD hyperphosphorylated). Active Hog1p also coimmunoprecipitated with several TAP-tagged elongation factors, including Spt4p, TFIIS, Paf1p and Thp1p and, therefore, presumably itself forms a part of the transcriptional elongation complex (Proft et al. 2006). In addition to the revelation that SAPKs act directly in the transcription process, this discovery is interesting as it suggests the existence of gene-specific elongation factors; previously all identified factors have shown little if any preference for classes of gene (Sims et al. 2004). However, whilst Proft et al. (2006) do not show a phosphorylation target for Hog1p within the transcriptional elongation complex, it is possible that the role of the SAPK in this complex requires kinase-mediated activation of one or more components.

Pokholok et al. (2006) also looked at the genomic distributions of several other *S. cerevisiae* protein kinases, including the MAPKs Fus3p and Kss1p and protein kinase A subunits Tpk1p and Tpk2p. ChIP-on-chip analyses localised these proteins to both the promoter and transcribed regions of their target genes (Pokholok et al. 2006), raising the possibility that like Hog1p, these kinases also have a role in transcriptional elongation.

5.1 A role for mammalian SAPKs in transcriptional and elongation complexes?

A key question then is whether a similar process occurs in mammalian SAPK signalling. There is some evidence that these kinases may have a role at their target gene loci; the Hog1p homologue p38/SAPK2 associates with the *MYOG* and *CKM* gene promoters during myogenesis (Simone et al. 2004). In addition p38/SAPK2 constitutively coimmunoprecipitates with Pol II (whereas the Hog1p-Pol II interaction is stress-dependent) (Alepuz et al. 2003), but the fact that the holoenzyme is a target for phosphorylation by the kinase suggests this may be little more than a mechanism to quickly target the complex on activation. As yet there is no evidence of p38/SAPK2 association with the elongating transcriptional machinery in mammals, and it may be that this property is confined to yeast.

However, a more limited role for nuclear translocation and/or transcription factor-binding in mammalian SAPK signalling may be supported. For example, both JNK/SAPK1 and p38/SAPK2 have been observed bound to their target transcription factors c-Jun and ATF-2, respectively, even in their inactive states (Read et al. 1997; Bruna et al. 2003). However, it is possible that the interaction between DNA-bound AP-1 and inactive JNK/SAPK1 serves as a further mechanism for repression, as this interaction prevents low levels of active JNK/SAPK1 from randomly stimulating AP-1 activation (Bruna et al. 2003) until levels of active protein are high enough to displace the inactive factor. Alternatively, if the interaction is maintained with active JNK/SAPK1, it may serve as an anchoring platform to allow the kinase to recruit further factors to the promoter, analogous to the Hot1p-Hog1p partnership.

6 Concluding remarks

Despite the complexity of responses mediated by the SAPK cascades, they do not act in isolation within the cell. Often, parallel independent pathways are activated by the same stimulus, resulting in changes that can act in concert with those produced by the SAPKs. For example, treatment of murine fibroblasts with arsenite results in activation of *Hsp70* transcription, mediated via two independent pathways (Thomson et al. 2004). Activation of p38/SAPK2 produces histone H3 phosphorylation via MSK1/2, but at the same time Hsf1 (heat shock factor 1) stimulates acetylation of histone H4, and both of these modifications are required to induce gene expression.

Many different chromatin and transcription factors are controlled by SAPK signalling (Fig. 1 & 2). The key control mechanism is phosphorylation, which can result in stimulation or inhibition of an activity, or induce new binding partners and disrupt other interactions. In the case of histone H3, phosphorylation can also provide novel binding sites to allow further interactions mediated via 14-3-3. However, recent evidence has demonstrated that the actions of SAPKs are not limited to phosphorylation. Work with the yeast Hog1p has established novel interaction and recruitment activities for MAP kinase proteins, including a role in transcriptional elongation. Future work will help to confirm whether similar activities can be found in the mammalian homologues, or whether these adaptations are limited to our unicellular relatives.

References

Alepuz PM, de Nadal E, Zapater M, Ammerer G, Posas F (2003) Osmostress-induced transcription by Hot1 depends on a Hog1-mediated recruitment of the RNA Pol II. EMBO J 22:2433-2442

Atienza JM, Suh M, Xenarios I, Landgraf R, Colicelli J (2000) Human ERK1 induces filamentous growth and cell wall remodeling pathways in *Saccharomyces cerevisiae*. J Biol Chem 275:20638-20646

Bruna A, Nicolas M, Munoz A, Kyriakis JM, Caelles C (2003) Glucocorticoid receptor-JNK interaction mediates inhibition of the JNK pathway by glucocorticoids. EMBO J 22:6035-6044

Carter AB, Knudtson KL, Monick MM, Hunninghake GW (1999) The p38 mitogen-activated protein kinase is required for NF-κB-dependent gene expression. The role of TATA-binding protein (TBP). J Biol Chem 274:30858-30863

Carter AB, Tephly LA, Hunninghake GW (2001) The absence of activator protein 1-dependent gene expression in THP-1 macrophages stimulated with phorbol esters is due to lack of p38 mitogen-activated protein kinase activation. J Biol Chem 276:33826-33832

Chang L, Karin M (2001) Mammalian MAP kinase signalling cascades. Nature 410:37-40

Cheung P, Allis CD, Sassone-Corsi P (2000) Signaling to chromatin through histone modifications. Cell 103:263-271

Clayton AL, Mahadevan LC (2003) MAP kinase-mediated phosphoacetylation of histone H3 and inducible gene regulation. FEBS Lett 546:51-58

de Nadal E, Alepuz PM, Posas F (2002) Dealing with osmostress through MAP kinase activation. EMBO Rep 3:735-740

de Nadal E, Casadome L, Posas F (2003) Targeting the MEF2-like transcription factor Smp1 by the stress-activated Hog1 mitogen-activated protein kinase. Mol Cell Biol 23:229-237

de Nadal E, Zapater M, Alepuz PM, Sumoy L, Mas G, Posas F (2004) The MAPK Hog1 recruits Rpd3 histone deacetylase to activate osmoresponsive genes. Nature 427:370-374

Deak M, Clifton AD, Lucocq LM, Alessi DR (1998) Mitogen- and stress-activated protein kinase-1 (MSK1) is directly activated by MAPK and SAPK2/p38, and may mediate activation of CREB. EMBO J 17:4426-4441

Dunn KL, Davie JR (2005) Stimulation of the Ras-MAPK pathway leads to independent phosphorylation of histone H3 on serine 10 and 28. Oncogene 24:3492-3502

Dyson MH, Thomson S, Inagaki M, Goto H, Arthur SJ, Nightingale K, Iborra FJ, Mahadevan LC (2005) MAP kinase-mediated phosphorylation of distinct pools of histone H3 at S10 or S28 via mitogen- and stress-activated kinase 1/2. J Cell Sci 118:2247-2259

Edmunds JW, Mahadevan LC (2004) MAP kinases as structural adaptors and enzymatic activators in transcription complexes. J Cell Sci 117:3715-3723

Galcheva-Gargova Z, Derijard B, Wu IH, Davis RJ (1994) An osmosensing signal transduction pathway in mammalian cells. Science 265:806-808

Jin J, Smith FD, Stark C, Wells CD, Fawcett JP, Kulkarni S, Metalnikov P, O'Donnell P, Taylor P, Taylor L, Zougman A, Woodgett JR, Langeberg LK, Scott JD, Pawson T (2004) Proteomic, functional, and domain-based analysis of *in vivo* 14-3-3 binding proteins involved in cytoskeletal regulation and cellular organization. Curr Biol 14:1436-1450

Karin M, Liu Z, Zandi E (1997) AP-1 function and regulation. Curr Opin Cell Biol 9:240-246

Kawasaki H, Schiltz L, Chiu R, Itakura K, Taira K, Nakatani Y, Yokoyama KK (2000) ATF-2 has intrinsic histone acetyltransferase activity which is modulated by phosphorylation. Nature 405:195-200

Kyriakis JM, Avruch J (2001) Mammalian mitogen-activated protein kinase signal transduction pathways activated by stress and inflammation. Physiol Rev 81:807-869

Li QJ, Yang SH, Maeda Y, Sladek FM, Sharrocks AD, Martins-Green M (2003) MAP kinase phosphorylation-dependent activation of Elk-1 leads to activation of the co-activator p300. EMBO J 22:281-291

Luger K, Mader AW, Richmond RK, Sargent DF, Richmond TJ (1997) Crystal structure of the nucleosome core particle at 2.8 Å resolution. Nature 389:251-260

Macdonald N, Welburn JP, Noble ME, Nguyen A, Yaffe MB, Clynes D, Moggs JG, Orphanides G, Thomson S, Edmunds JW, Clayton AL, Endicott JA, Mahadevan LC (2005) Molecular basis for the recognition of phosphorylated and phosphoacetylated histone H3 by 14-3-3. Mol Cell 20:199-211

Mahadevan LC, Willis AC, Barratt MJ (1991) Rapid histone H3 phosphorylation in response to growth factors, phorbol esters, okadaic acid, and protein synthesis inhibitors. Cell 65:775-783

Meek SE, Lane WS, Piwnica-Worms H (2004) Comprehensive proteomic analysis of interphase and mitotic 14-3-3-binding proteins. J Biol Chem 279:32046-32054

Mellor J (2005) The dynamics of chromatin remodeling at promoters. Mol Cell 19:147-157

Nissen LJ, Gelly JC, Hipskind RA (2001) Induction-independent recruitment of CREB-binding protein to the c-fos serum response element through interactions between the bromodomain and Elk-1. J Biol Chem 276:5213-5221

Nowak SJ, Corces VG (2000) Phosphorylation of histone H3 correlates with transcriptionally active loci. Genes Dev 14:3003-3013

O'Rourke SM, Herskowitz I, O'Shea EK (2002) Yeast go the whole HOG for the hyperosmotic response. Trends Genet 18:405-412

Pokholok DK, Zeitlinger J, Hannett NM, Reynolds DB, Young RA (2006) Activated signal transduction kinases frequently occupy target genes. Science 313:533-536

Posas F, Takekawa M, Saito H (1998) Signal transduction by MAP kinase cascades in budding yeast. Curr Opin Microbiol 1:175-182

Proft M, Mas G, de Nadal E, Vendrell A, Noriega N, Struhl K, Posas F (2006) The stress-activated Hog1 kinase is a selective transcriptional elongation factor for genes responding to osmotic stress. Mol Cell 23:241-250

Proft M, Pascual-Ahuir A, de Nadal E, Arino J, Serrano R, Posas F (2001) Regulation of the Sko1 transcriptional repressor by the Hog1 MAP kinase in response to osmotic stress. EMBO J 20:1123-1133

Proft M, Struhl K (2002) Hog1 kinase converts the Sko1-Cyc8-Tup1 repressor complex into an activator that recruits SAGA and SWI/SNF in response to osmotic stress. Mol Cell 9:1307-1317

Pulverer BJ, Kyriakis JM, Avruch J, Nikolakaki E, Woodgett JR (1991) Phosphorylation of c-jun mediated by MAP kinases. Nature 353:670-674

Read MA, Whitley MZ, Gupta S, Pierce JW, Best J, Davis RJ, Collins T (1997) Tumor necrosis factor alpha-induced E-selectin expression is activated by the nuclear factor-κB and c-JUN N-terminal kinase/p38 mitogen-activated protein kinase pathways. J Biol Chem 272:2753-2761

Sassone-Corsi P, Mizzen CA, Cheung P, Crosio C, Monaco L, Jacquot S, Hanauer A, Allis CD (1999) Requirement of Rsk-2 for epidermal growth factor-activated phosphorylation of histone H3. Science 285:886-891

Simone C, Forcales SV, Hill DA, Imbalzano AN, Latella L, Puri PL (2004) p38 pathway targets SWI-SNF chromatin-remodeling complex to muscle-specific loci. Nat Genet 36:738-743

Sims RJ, 3rd, Belotserkovskaya R, Reinberg D (2004) Elongation by RNA polymerase II: the short and long of it. Genes Dev 18:2437-2468

Soloaga A, Thomson S, Wiggin GR, Rampersaud N, Dyson MH, Hazzalin CA, Mahadevan LC, Arthur JS (2003) MSK2 and MSK1 mediate the mitogen- and stress-induced phosphorylation of histone H3 and HMG-14. EMBO J 22:2788-2797

Thomson S, Clayton AL, Hazzalin CA, Rose S, Barratt MJ, Mahadevan LC (1999a) The nucleosomal response associated with immediate-early gene induction is mediated via alternative MAP kinase cascades: MSK1 as a potential histone H3/HMG-14 kinase. EMBO J 18:4779-4793

Thomson S, Hollis A, Hazzalin CA, Mahadevan LC (2004) Distinct stimulus-specific histone modifications at hsp70 chromatin targeted by the transcription factor heat shock factor-1. Mol Cell 15:585-594

Thomson S, Mahadevan LC, Clayton AL (1999b) MAP kinase-mediated signalling to nucleosomes and immediate-early gene induction. Semin Cell Dev Biol 10:205-214

Weiss C, Schneider S, Wagner EF, Zhang X, Seto E, Bohmann D (2003) JNK phosphorylation relieves HDAC3-dependent suppression of the transcriptional activity of c-Jun. EMBO J 22:3686-3695

Widmann C, Gibson S, Jarpe MB, Johnson GL (1999) Mitogen-activated protein kinase: conservation of a three-kinase module from yeast to human. Physiol Rev 79:143-180

Wiggin GR, Soloaga A, Foster JM, Murray-Tait V, Cohen P, Arthur JS (2002) MSK1 and MSK2 are required for the mitogen- and stress-induced phosphorylation of CREB and ATF1 in fibroblasts. Mol Cell Biol 22:2871-2881

Yang SH, Sharrocks AD (2004) SUMO promotes HDAC-mediated transcriptional repression. Mol Cell 13:611-617

Yang SH, Sharrocks AD, Whitmarsh AJ (2003) Transcriptional regulation by the MAP kinase signaling cascades. Gene 320:3-21

Yang XJ, Ogryzko VV, Nishikawa J, Howard BH, Nakatani Y (1996) A p300/CBP-associated factor that competes with the adenoviral oncoprotein E1A. Nature 382:319-324

Crump, Nicholas T.
Nuclear Signalling Laboratory, Department of Biochemistry, Oxford University, South Parks Road, Oxford, OX1 3QU, UK

Han, Ya Ting
Nuclear Signalling Laboratory, Department of Biochemistry, Oxford University, South Parks Road, Oxford, OX1 3QU, UK

Mahadevan, Louis C.
Nuclear Signalling Laboratory, Department of Biochemistry, Oxford University, South Parks Road, Oxford, OX1 3QU, UK
louiscm@bioch.ox.ac.uk

SAPK and translational control

Malin Hult and Per Sunnerhagen

Abstract

Posttranscriptional control of translation is essential for rapid adaptation to changing environmental conditions, and also in some instances of hormonal control. General translational control, which affects all mRNAs, can target the initiation or the elongation step. General regulation of initiation commonly targets the translation initiation factors eIF2α or eIF4E; a common target for regulation of elongation is eEF-2. In all these cases, SAPK signalling has been shown to play a role. Posttranscriptional regulation of individual mRNAs is ultimately determined by *cis*-acting sequences, most famously the ARE sequences. SAPK's are also implicated in translational control of individual mRNA species through ARE's and probably other sequence elements. In addition, SAPK signalling can influence the use of alternative translation start sites, and transcription and mRNA stability of components of the protein synthesis machinery. While most knowledge of translational control is derived from mammalian systems, yeast genetics is recently providing complementary understanding.

1 Background and paradigms for control of translation

Translation of genetic information from RNA to protein sequence is far more energy-consuming than replication or transcription, which can be understood already from the much larger numbers of protein molecules synthesised. Therefore, there is a need to for global regulation of translation in response to cellular stress or nutrient limitation. This targets not only initiation and elongation of individual mRNAs. Under such conditions, the cell rapidly shuts down production of the components required for protein synthesis. This type of regulation affects gene products where control is organised in distinct groups. One group, where the gene products are localised in the nucleolus, comprises synthesis of ribosomal RNA and the enzymes required for covalent modifications (mainly methylation) of it, and assembly and transport of ribosomes. A second group consists of the cytoplasmic ribosomal proteins. Expression of these two groups is controlled at the levels of initiation of transcription, mRNA stability, and efficiency of translation, and is regulated by several signalling chains, most notably the TOR pathway. It is notable that mitochondrial protein synthesis is under independent control; thus, at least in yeast, mitochondrial ribosomal proteins are controlled independently of their counterparts in the cytoplasm (Ihmels et al. 2005). In addition to regulating the

Topics in Current Genetics, Vol. 20
F. Posas, A. R. Nebreda (Eds.): Stress-Activated Protein Kinases
DOI 10.1007/4735_2007_0255 / Published online: 1 August 2007

amount of the translation apparatus, global translation efficiency is controlled through the activity level of translation factors. The current understanding is that this occurs mainly at the initiation step. In terms of the influence of intracellular signalling, two modes of control at this level are best characterised.

2 Global controls of translation

2.1 Global control of initiation by phosphorylation of eIF2α

The initiation factor eIF2 has the task of delivering methionine-charged initiator tRNA to the 40S ribosomal subunit, commonly seen as the rate-limiting step in translation. The eIF2α subunit needs to be in its GTP-bound state to perform this action. eIF2α can be phosphorylated by the kinase Gcn2 (or, in addition in mammalian cells, either of the kinases PERK, HRI, or PKR). This phosphorylation converts eIF2α to a competitive inhibitor of eIF2B, another subunit of the eIF2 complex. eIF2B is the guanine nucleotide exchange factor for eIF2α, and so this effectively shuts down reloading of eIF2α into its active GTP-bound form. Many of the described cases of translational regulation through eIF2α phosphorylation involve the TOR pathway (e.g. Cherkasova and Hinnebusch 2003; for review, see Dever 2002). What concerns SAPK pathways, a case is reported from in fission yeast where phosphorylation of eIF2α occurs upon exposure to hydrogen peroxide to a higher degree in cells carrying *spc1-m13*, a mutant *sty1* allele, than in wild type cells (Dunand-Sauthier et al. 2005).

2.2 Global control of initiation by phosphorylation of eIF4E/4E-BP

In eukaryotes, initiation complex eIF4F interacts with the 5'-cap of mRNAs and with polyA-binding protein, to promote formation of a circular structure of the mRNA and start of translation. The eIF4F complex is formed by the factors eIF4A, eIF4E, and eIF4G. The cell can modulate formation of eIF4E complex through eIF4E-binding proteins (4E-BPs). These compete with eIF4G for binding to eIF4E, using the same interaction domain. Thus, 4E-BPs will inhibit translation initiation by disruption of the eIF4F complex. This inhibition can be overcome by phosphorylation of 4E-BPs, which is increased through growth-promoting conditions and growth-factor signalling pathways (for review see Preiss and Hentze 2003).

In mammalian cells, SAPK signalling does impinge on regulation through 4E-BPs. MAPK-dependent phosphorylation of eIF4E enhances its binding to the cap structure and increases overall translation. This can be mediated by Mnk1, a MAPK-activated protein kinase (MAPKAPK) that is a substrate of both ERK1 and p38 (Wang et al. 1998; Waskiewicz et al. 1999). Phosphorylation of eIF4E is enhanced by treatment with stress agents, but this is mediated differently depending on the type of stress: phorbol esters activate both ERK and p38, whereas other agents such as arsenite or anisomycin confer a phosphorylation of eIF4E that is only p38-dependent (Wang et al. 1998). The eIF4E phosphorylation in these cases is paralleled by increased phosphorylation of Mnk1. However, still other stress agents (heat shock or oxidative stress) do increase phosphorylation of both p38 and Mnk1, yet do not affect eIF4E phosphorylation. Instead, the authors observed increased binding of eIF4E to 4E-BP1, which in turn obstructs eIF4E phosphorylation. This implies that other pathways than p38 regulate the ability of 4E-BP1 to bind eIF4E; alternatively, different levels of activation of p38/Mnk1 could affect phosphorylation of eIF4E and 4E-BP differently with the same end result.

As we have seen, translational regulation occurs during rapid environmental changes, but also appears in quite different contexts. In neurons, the enormous distance between nucleus and extremities makes posttranscriptional control essential (Klann et al. 2004). There is evidence of MAPKs exerting translational control in neurons, where brain-derived neurotrophic factor (BDNF) increases overall translation. BDNF confers phosphorylation of eIF4E, which is abolished by the MEK inhibitor PD98059 (Takei et al. 2001). Selective inhibition of ERK in the forebrain after birth caused long-term memory defects in transgenic mice (Kelleher et al. 2004). The authors ascribe this to misregulation of translation, as these defects were accompanied by the expected reduction in phosphorylation of eIF4E and 4E-BP, and by reduction of translation from a range of mRNAs known to be stimulated by MEK and ERK activity. SAPKs may also influence translation rates in neurons although it is not obvious if this action is on eIF4E. Inflammation in neurons increases the protein level of the ion channel TRPV1, without affecting the mRNA level, and this increase of translational efficiency is sensitive to the p38 inhibitor SB203580 (Ji et al. 2002).

4E-BPs were first characterised in mammalian cells, and it is not yet clear how significant this kind of regulation is in yeast. There are two proteins in budding yeast with sequence similarity to 4E-BPs. Mutants lacking one of them, Eap1, are resistant to rapamycin, hinting at the possibility that TOR signalling uses this mechanism to dampen global translation (Cosentino et al. 2000).

2.3 Global control of elongation

The initiation step obviously offers the most economical way to regulate translation as a whole. However, an additional control on the elongation step provides a means to conserve existing active translation complexes, to buffer against rapid fluctuations in the environmental, nutrient, or hormone status. If elongation is also regulated, the cell can quickly resume its previous level of translation once conditions improve. Another argument is based on fidelity; high levels of translation elongation factors together with only few active mRNAs could lead to increased translation error rates (Browne and Proud 2002).

Elongation factor 2 (eEF-2) is a G-protein required for the translocation step of translation elongation. Its activity can be controlled through posttranslational modifications: ADP-ribosylation of a diphthamide (a converted histidine residue) near the C-terminus inactivates it. There is also a conserved site in eEF-2 (T56);

phosphorylation at this position is likewise inhibitory. In vertebrates, this residue is targeted by the dedicated eEF-2 kinase, which in part mediates stress-induced regulation of translation elongation. eEF-2 kinase has been reported to be regulated in a p38δ-dependent way (Knebel et al. 2001). Likewise, *in vitro* methods identify eEF-2 kinase as a substrate of p38δ (Knebel et al. 2001; Cohen and Knebel 2006).

Phosphorylation of eEF-2 kinase is not solely carried out by p38. In one study, the authors concluded that out of five phosphorylation sites in eEF-2 kinase, two (S359 and S377) are controlled through p38 (being sensitive to inhibition by SB203580), and one of these two (S377) is phosphorylated by a MAPKAPK (MK2, MK3, or MK5) (Knebel et al. 2002). eEF-2 kinase is also inhibited by ribosomal protein S6 kinase by phosphorylation on S366 (Wang et al. 2001).

We have seen examples above of control of translation initiation in neurons as a result of BDNF signalling. Translation elongation may also be under control of BDNF and MAPKs. BDNF reduces phosphorylation of eEF-2 and increases phosphorylation of eEF-2 kinase. Both these effects were mitigated by inhibition of MEK (Inamura et al. 2005).

In yeast, a dedicated eEF-2 kinase has not been identified. It is possible that instead the SAPKs themselves or their downstream kinases carry out phosphorylation of eEF-2. In budding yeast, it was observed that the global down-regulation of translation upon hyperosmotic shock is dependent on Hog1. Specifically, transcription of several genes encoding ribosomal proteins was not depressed in *hog1* mutants as in the wild type (Uesono and Toh 2002). Likewise, the decrease in polysomal content followed by recovery following adaptation to hyperosmosis is deregulated both in *hog1* and *rck2* mutants (Teige et al. 2001; Uesono and Toh 2002). The discovery that *S. cerevisiae* eEF-2 is phosphorylated during hyperosmotic shock in an *RCK2*-dependent way (Teige et al. 2001) raises the question this task is carried out by Rck2 directly. This finding implies that the elongation step of translation is under control of the HOG pathway, but it is not clear if this pathway also exerts direct control on proteins involved in the initiation step of translation. A global investigation of total and polysome-associated mRNAs in *rck2* mutant cells revealed extensive misregulation of mRNAs encoding translation components (Swaminathan et al. 2006), but did not unequivocally identify elongation as the critical step for regulation by Rck2.

3 Individual control of mRNA species – AREs

The best-characterised *cis*-regulatory motif in mRNAs are the A/U-rich elements (AREs), most often located in the 3'-UTR of mRNAs. These were first described in the 1980's in mammalian mRNAs encoding cytokines and cytokine receptors (Caput et al. 1986), and have been reviewed extensively (e.g. Barreau et al. 2005). Much of the research effort that has gone into characterisation of these mechanisms relies on their importance for inflammatory processes; pro-inflammatory agents such as bacterial lipopolysaccharides ultimately influence production of proteins such as tumour necrosis factor alpha (TNFα) and cyclooxygenase 2 (COX-2) through SAPK pathways that impinge on translation and mRNA stability through AREs. This control mechanism is essential for the rapid upregulation and ensuing downregulation of cytokines in the inflammatory process. JNK is required for the increased translation of TNFα after exposure to lipopolysaccharides (Swantek et al. 1997). Glucocorticoids inhibit JNK and thus act as anti-inflammatory agents. This action is both on the translation efficiency and mRNA decay levels (Kontoyiannis et al. 1999; Lasa et al. 2002).

3.1 ARE-binding proteins under SAPK control

AREs recruit ARE-binding proteins that mediate the effects of signalling on mRNA stability and translation. Tristetraprolin (TTP) is one of the ARE-binding proteins implicated in posttranscriptional regulation of TNFα. MK2-deficient mice have decreased TNFα levels whereas TTP-deficient mutants have the opposite phenotype. Double knockout mice lacking both MK2 and TTP likewise have increased TNFα levels, showing that TTP acts downstream of MK2 (Hitti et al. 2006). Given that TTP is a known substrate for MK2 (Mahtani et al. 2001), it is likely that TTP is the main target of MK2 in regulation of TNFα.

Another example of how MAPK pathways can regulate translation through ARE-binding proteins is given by heterogeneous nuclear ribonucleoprotein K (hnRNP-K). This protein requires phosphorylation by ERK to localise to the cytoplasm and control translation of mRNAs carrying a differentiation-control element (Habelhah et al. 2001). Phosphorylation of hnRNP-A0, another ARE-binding protein, is phosphorylated *in vivo* in a p38-dependent way, and this promotes its interaction with ARE elements (Rousseau et al. 2002). Interestingly, a recent *in vitro* study identified hnRNP-A0 as a substrate for MK2 (Cohen and Knebel 2006).

3.2 MAPK-activated protein kinases in signalling through AREs

Some of the downstream targets of SAPK's are reached through dedicated downstream kinases, called MAPK-activated protein kinases (MAPKAPK) in mammalian cells, which are physically associated with the MAPK. The best documented case of functional cooperation is between p38 and MAPKAPK-2 (MK2). This control pathway is particularly obvious in the case of regulation through ARE's, although not all effects are necessarily mediated by MK2.

Translation of TNFα mRNA is dependent on p38 and MK2 acting through an ARE in its 3'-UTR. Thus, mice lacking MK2 express reduced levels of TNFα protein, but mutating this ARE restores TNFα synthesis (Neininger et al. 2002). Although it is clear that MK2 has a role in regulating TNFα production, some studies indicate that other downstream targets of p38 are also implicated in this regulation. Thus, whereas it is possible to virtually abolish TNFα translation with the p38 inhibitor SB202190 (Piecyk et al. 2000), TNFα translation is only reduced to about half in MK2⁻/MK2⁻ mice (Kontoyiannis et al. 2001). Phenotypes of mice lacking MK2, MK3, or both kinases suggest a considerable functional overlap between them with respect to control of TNF synthesis. Likewise, overexpression of MK3 can rescue mice lacking MK2 (Ronkina et al. 2007).

4 Links between translation and mRNA degradation

An mRNA under translation is intrinsically protected to some degree from mRNA degradation, for example, through shielding of the cap structure from the decapping enzymes by the cap-binding translation initiation factor eIF4F, and through blocking of access of deadenylating exonucleases by polyA-binding protein. Thus, it is not always possible to dissect the effects by a cellular function such as the SAPK pathways on mRNA turnover from its effects on translation.

In some cases, though, detailed studies have shown that for instance an ARE can act uniquely on either translation or mRNA turnover. Thus, interleukin 10 (IL-10) inhibits translation of TNF in an ARE-dependent way, without altering mRNA stability (Kontoyiannis et al. 2001).

In epithelial cells, up-regulation of IL-8 protein by pro-inflammatory stimuli is due exclusively to an increase in translation rate. The authors showed that this increase is dependent on p38, as neither IL-8 mRNA transcription nor decay rates were affected by the p38 inhibitor SB203580, whereas IL-8 protein levels were (Yu et al. 2003).

Translation of the *S. cerevisiae* mRNAs *MFA2* and *TIF51A* are both inhibited by glucose. For *TIF51A*, mRNA stability is also decreased under these conditions, but this is not the case for *MFA2* (Vasudevan and Peltz 2001; Vasudevan et al. 2005). For *MFA2*, this translational repression is dependent on an intact *HOG* pathway. Repression involves binding of Pab1 to an ARE in the 3'-UTR of *MFA2*, and Pab1 binding is dependent on the presence of Pub1 (Vasudevan et al. 2005).

It has also been argued that the action of p38 through the ARE of TNFα or COX-2 mRNA is primarily to inhibit deadenylation, rather than degradation of the mRNA body (Dean et al. 2003). This could explain why p38 acting though AREs appears to regulate translation efficiency in some instances and mRNA stability in others, as deadenylation affects both processes.

5 IRES

Translation of cellular mRNAs normally initiates by a cap- and eIF4F-dependent mechanism, through binding of initiation factors to the 5'-end of the mRNA and interactions with the polyA-tail in the 3' end, followed by scanning downstream for the first available start codon. An alternative mechanism, first discovered for polioviral mRNAs, ignores the cap and instead uses internal ribosomal entry sites (IRES). These are characterised by stem-loop structures in the 5'-UTR, to which an initiation complex lacking eIF4E can bind directly and start translation. Viral proteases can cleave eIF4G. This interferes with its binding to eIF4E and thus prevents cap binding, leading to a general depression of cellular translation, but leaves the truncated eIF4G capable of interacting with an IRES. By this mechanism, viral mRNAs carrying an IRES can escape the general depression of protein synthesis in infected cells, and instead become preferentially translated. More recently, it has been realised that IRES-dependent initiation may be used also for certain cellular mRNAs. For a review of IRES-dependent initiation, see Komar and Hatzoglou (2005).

During the apoptotic process, IRES-dependent translation becomes more important in relative terms, as eIF4G also gets cleaved. It has been shown, using overexpression of MKK6 or inhibition using SB203580, that IRES-dependent translation of c-myc in apoptotic cells requires p38 (Stoneley et al. 2000a, 2000b).

Further support for p38 enhancement of IRES-dependent translation of c-myc, and also of cyclin D1, comes from a study where the impact of AKT activation was investigated. It was found that AKT suppressed the IRES activity of both transcripts, as seen by the increased IRES activity after rapamycin addition. On the other hand, both the p38 and ERK pathways enhanced IRES activity, as witnessed by suppression of the IRES by the inhibitors SB203580 or PD98059 (Shi et al. 2005).

6 Indirect effects on translation through transcriptional and posttranscriptional regulation

There are also indirect ways whereby SAPKs can affect translation, such as through regulation of genes encoding components of the translation machinery, and several examples of this exist. As previously mentioned, *S. cerevisiae hog1* mutants fail to downregulate genes encoding ribosomal proteins on osmotic shock (Uesono and Toh 2002). Although regulation of TNFα by p38 on the translational level is well established, another mechanism may operate as well, as activation of the transcription factor NF-κB by p38 with ensuing increased transcription of the gene encoding TNFα has been reported in primary macrophages (Campbell et al. 2004). Also, transcription of the gene encoding 4E-BP1 is dependent on the transcription factor Egr-1, which in turn can be activated by p38 (Rolli-Derkinderen et al. 2003). Fission yeast has two genes encoding isoforms of eIF4E. One of them, *tif452*$^+$, is required for survival under stress conditions (Ptushkina et al. 2004). Under stress conditions activating Sty1, *tif452*$^+$ is upregulated (Chen et al. 2003). The gene has a motif in the promoter region implicated in transcriptional regulation though Sty1, indicating also an indirect involvement of this SAPK in translation. Thus, as the effectors of SAPK pathways are found on several levels in the cell, and as the translational machinery is a main target for regulation upon stress by virtue of its large energy consumption, it is logical that SAPKs affect global translation by parallel mechanisms.

As an example of further complications in the issue, it has been shown that the amount of TTP, a protein involved in regulating translation of selected mRNAs, is regulated through the stability of its mRNA. p38 stabilises the TTP mRNA through an ARE in its 3'-UTR, and this effect is mediated through MK2 (Tchen et al. 2004).

7 Concluding remarks

As we have seen, SAPK pathways act on several control levels. To conclude that a SAPK pathway influences translation of a given set of mRNAs from studies using MAPK inhibitors or mutants deficient in the SAPK itself is complicated because a MAPK pathway can affect transcriptional initiation and mRNA stability, in addition to translational efficiency. Therefore, it would be desirable to analyse pathway components that selectively affect one branch. The MAPKAPK's are promising in this regard as they do not have a direct role in transcription.

There could be yet undiscovered ways by which SAPK's affect translation. For instance, in fission yeast, a novel checkpoint-like mechanism has been identified that acts in G1 upon UV irradiation. Initiation of DNA replication is delayed in the absence of the canonical activation of the Rad3 (Atm/Atr homologue) protein kinase, and instead the effect is accompanied by depressed protein synthesis, which is dependent on active Gcn2 and phosphorylation of eIF2α (Tvegård et al. 2007). Similarly, Gcn2 and eIF2α phosphorylation dependent reduction of translation is seen in mammalian cells upon UV treatment (Jiang and Wek 2005). The mechanism for this is not clarified, however, it is interesting to note that in mammalian cells, irradiation with short-wave UV light causes TIA-R (see Rodríguez-Gabriel and Russell, this volume) to selectively bind to mRNAs encoding proteins in translation, including eIF4A, eIF4E, and eEF1B, repressing their translation (Mazan-Mamczarz et al. 2006). As UV light activates the SAPK's JNK, p38 and Sty1, it is tempting to speculate about a role for them in these processes.

The study of translation has so far been driven more by biochemistry than genetics, which can be understood from the realisation that many genes required for translation are essential. As a consequence, most of the data in the field does not come from yeast but chiefly from mammalian systems. On the other hand, in the *S. cerevisiae* genome many genes encoding ribosomal proteins and translation factors are duplicated, leading to situations where deletion of one gene in a pair gives only a marginal phenotype. This could be attacked using parallel quantitative phenotypic assays, trying to identify situations where a functional differentiation has evolved between the members of a gene pair. It is to be expected that combinations of molecular genetics, genomics, and phenomics approaches in yeast will yield novel insights in SAPK control of translation.

Acknowledgements

This review was produced in the course of the EC-financed QUASI project.

References

Barreau C, Paillard L, Osborne HB (2005) AU-rich elements and associated factors: are there unifying principles? Nucleic Acids Res 33:7138-7150

Browne GJ, Proud CG (2002) Regulation of peptide-chain elongation in mammalian cells. Eur J Biochem 269:5360-5368

Campbell J, Ciesielski CJ, Hunt AE, Horwood NJ, Beech JT, Hayes LA, Denys A, Feldmann M, Brennan FM, Foxwell BM (2004) A novel mechanism for TNF-alpha regulation by p38 MAPK: involvement of NF-kappa B with implications for therapy in rheumatoid arthritis. J Immunol 173:6928-6937

Caput D, Beutler B, Hartog K, Thayer R, Brown-Shimer S, Cerami A (1986) Identification of a common nucleotide sequence in the 3'-untranslated region of mRNA molecules specifying inflammatory mediators. Proc Natl Acad Sci USA 83:1670-1674

Chen D, Toone WM, Mata J, Lyne R, Burns G, Kivinen K, Brazma A, Jones N, Bähler J (2003) Global transcriptional responses of fission yeast to environmental stress. Mol Biol Cell 14:214-229

Cherkasova VA, Hinnebusch AG (2003) Translational control by TOR and TAP42 through dephosphorylation of eIF2alpha kinase GCN2. Genes Dev 17:859-872

Cohen P, Knebel A (2006) KESTREL: a powerful method for identifying the physiological substrates of protein kinases. Biochem J 393:1-6

Cosentino GP, Schmelzle T, Haghighat A, Helliwell SB, Hall MN, Sonenberg N (2000) Eap1p, a novel eukaryotic translation initiation factor 4E-associated protein in *Saccharomyces cerevisiae*. Mol Cell Biol 20:4604-4613

Dean JL, Sarsfield SJ, Tsounakou E, Saklatvala J (2003) p38 Mitogen-activated protein kinase stabilizes mRNAs that contain cyclooxygenase-2 and tumor necrosis factor AU-rich elements by inhibiting deadenylation. J Biol Chem 278:39470-39476

Dever TE (2002) Gene-specific regulation by general translation factors. Cell 108:545-556

Dunand-Sauthier I, Walker CA, Narasimhan J, Pearce AK, Wek RC, Humphrey TC (2005) Stress-activated protein kinase pathway functions to support protein synthesis and translational adaptation in response to environmental stress in fission yeast. Eukaryot Cell 4:1785-1793

Habelhah H, Shah K, Huang L, Ostareck-Lederer A, Burlingame AL, Shokat KM, Hentze MW, Ronai Z (2001) ERK phosphorylation drives cytoplasmic accumulation of hnRNP-K and inhibition of mRNA translation. Nat Cell Biol 3:325-330

Hitti E, Iakovleva T, Brook M, Deppenmeier S, Gruber AD, Radzioch D, Clark AR, Blackshear PJ, Kotlyarov A, Gaestel M (2006) Mitogen-activated protein kinase-activated protein kinase 2 regulates tumor necrosis factor mRNA stability and translation mainly by altering tristetraprolin expression, stability, and binding to adenine/uridine-rich element. Mol Cell Biol 26:2399-2407

Ihmels J, Bergmann S, Gerami-Nejad M, Yanai I, McClellan M, Berman J, Barkai N (2005) Rewiring of the yeast transcriptional network through the evolution of motif usage. Science 309:938-940

Inamura N, Nawa H, Takei N (2005) Enhancement of translation elongation in neurons by brain-derived neurotrophic factor: implications for mammalian target of rapamycin signaling. J Neurochem 95:1438-1445

Ji RR, Samad TA, Jin SX, Schmoll R, Woolf CJ (2002) p38 MAPK activation by NGF in primary sensory neurons after inflammation increases TRPV1 levels and maintains heat hyperalgesia. Neuron 36:57-68

Jiang HY, Wek RC (2005) GCN2 phosphorylation of eIF2alpha activates NF-kappaB in response to UV irradiation. Biochem J 385:371-380

Kelleher RJ, 3rd, Govindarajan A, Jung HY, Kang H, Tonegawa S (2004) Translational control by MAPK signaling in long-term synaptic plasticity and memory. Cell 116:467-479

Klann E, Antion MD, Banko JL, Hou L (2004) Synaptic plasticity and translation initiation. Learn Mem 11:365-372

Knebel A, Haydon CE, Morrice N, Cohen P (2002) Stress-induced regulation of eukaryotic elongation factor 2 kinase by SB 203580-sensitive and -insensitive pathways. Biochem J 367:525-532

Knebel A, Morrice N, Cohen P (2001) A novel method to identify protein kinase substrates: eEF2 kinase is phosphorylated and inhibited by SAPK4/p38delta. EMBO J 20:4360-4369

Komar AA, Hatzoglou M (2005) Internal ribosome entry sites in cellular mRNAs: mystery of their existence. J Biol Chem 280:23425-23428

Kontoyiannis D, Kotlyarov A, Carballo E, Alexopoulou L, Blackshear PJ, Gaestel M, Davis R, Flavell R, Kollias G (2001) Interleukin-10 targets p38 MAPK to modulate ARE-dependent TNF mRNA translation and limit intestinal pathology. EMBO J 20:3760-3770

Kontoyiannis D, Pasparakis M, Pizarro TT, Cominelli F, Kollias G (1999) Impaired on/off regulation of TNF biosynthesis in mice lacking TNF AU-rich elements: implications for joint and gut-associated immunopathologies. Immunity 10:387-398

Lasa M, Abraham SM, Boucheron C, Saklatvala J, Clark AR (2002) Dexamethasone causes sustained expression of mitogen-activated protein kinase (MAPK) phosphatase 1 and phosphatase-mediated inhibition of MAPK p38. Mol Cell Biol 22:7802-7811

Mahtani KR, Brook M, Dean JL, Sully G, Saklatvala J, Clark AR (2001) Mitogen-activated protein kinase p38 controls the expression and posttranslational modification of tristetraprolin, a regulator of tumor necrosis factor alpha mRNA stability. Mol Cell Biol 21:6461-6469

Mazan-Mamczarz K, Lal A, Martindale JL, Kawai T, Gorospe M (2006) Translational repression by RNA-binding protein TIAR. Mol Cell Biol 26:2716-2727

Neininger A, Kontoyiannis D, Kotlyarov A, Winzen R, Eckert R, Volk HD, Holtmann H, Kollias G, Gaestel M (2002) MK2 targets AU-rich elements and regulates biosynthesis of tumor necrosis factor and interleukin-6 independently at different post-transcriptional levels. J Biol Chem 277:3065-3068

Piecyk M, Wax S, Beck AR, Kedersha N, Gupta M, Maritim B, Chen S, Gueydan C, Kruys V, Streuli M, Anderson P (2000) TIA-1 is a translational silencer that selectively regulates the expression of TNF-alpha. EMBO J 19:4154-4163

Preiss T, Hentze MW (2003) Starting the protein synthesis machine: eukaryotic translation initiation. Bioessays 25:1201-1211

Ptushkina M, Malys N, McCarthy JE (2004) eIF4E isoform 2 in *Schizosaccharomyces pombe* is a novel stress-response factor. EMBO Rep 5:311-316

Rolli-Derkinderen M, Machavoine F, Baraban JM, Grolleau A, Beretta L, Dy M (2003) ERK and p38 inhibit the expression of 4E-BP1 repressor of translation through induction of Egr-1. J Biol Chem 278:18859-18867

Ronkina N, Kotlyarov A, Dittrich-Breiholz O, Kracht M, Hitti E, Milarski K, Askew R, Marusic S, Lin LL, Gaestel M, Telliez JB (2007) The mitogen-activated protein kinase (MAPK)-activated protein kinases MK2 and MK3 cooperate in stimulation of tumor necrosis factor biosynthesis and stabilization of p38 MAPK. Mol Cell Biol 27:170-181

Rousseau S, Morrice N, Peggie M, Campbell DG, Gaestel M, Cohen P (2002) Inhibition of SAPK2a/p38 prevents hnRNP A0 phosphorylation by MAPKAP-K2 and its interaction with cytokine mRNAs. EMBO J 21:6505-6514

Shi Y, Sharma A, Wu H, Lichtenstein A, Gera J (2005) Cyclin D1 and c-myc internal ribosome entry site (IRES)-dependent translation is regulated by AKT activity and enhanced by rapamycin through a p38 MAPK- and ERK-dependent pathway. J Biol Chem 280:10964-10973

Stoneley M, Chappell SA, Jopling CL, Dickens M, MacFarlane M, Willis AE (2000a) c-Myc protein synthesis is initiated from the internal ribosome entry segment during apoptosis. Mol Cell Biol 20:1162-1169

Stoneley M, Subkhankulova T, Le Quesne JP, Coldwell MJ, Jopling CL, Belsham GJ, Willis AE (2000b) Analysis of the c-myc IRES; a potential role for cell-type specific trans-acting factors and the nuclear compartment. Nucleic Acids Res 28:687-694

Swaminathan S, Masek T, Molin C, Pospisek M, Sunnerhagen P (2006) Rck2 is required for reprogramming of ribosomes during oxidative stress. Mol Biol Cell 17:1472-1482

Swantek JL, Cobb MH, Geppert TD (1997) Jun N-terminal kinase/stress-activated protein kinase (JNK/SAPK) is required for lipopolysaccharide stimulation of tumor necrosis factor alpha (TNF-alpha) translation: glucocorticoids inhibit TNF-alpha translation by blocking JNK/SAPK. Mol Cell Biol 17:6274-6282

Takei N, Kawamura M, Hara K, Yonezawa K, Nawa H (2001) Brain-derived neurotrophic factor enhances neuronal translation by activating multiple initiation processes: comparison with the effects of insulin. J Biol Chem 276:42818-42825

Tchen CR, Brook M, Saklatvala J, Clark AR (2004) The stability of tristetraprolin mRNA is regulated by mitogen-activated protein kinase p38 and by tristetraprolin itself. J Biol Chem 279:32393-32400

Teige M, Scheikl E, Reiser V, Ruis H, Ammerer G (2001) Rck2, a member of the calmodulin-protein kinase family, links protein synthesis to high osmolarity MAP kinase signaling in budding yeast. Proc Natl Acad Sci USA 98:5625-5630

Tvegård T, Soltani H, Skjølberg HC, Krohn M, Nilssen EA, Kearsey SE, Grallert B, Boye E (2007) A novel checkpoint mechanism regulating the G1/S transition. Genes Dev 21:649-654

Uesono Y, Toh EA (2002) Transient inhibition of translation initiation by osmotic stress. J Biol Chem 277:13848-13855

Wang X, Flynn A, Waskiewicz AJ, Webb BL, Vries RG, Baines IA, Cooper JA, Proud CG (1998) The phosphorylation of eukaryotic initiation factor eIF4E in response to phorbol esters, cell stresses, and cytokines is mediated by distinct MAP kinase pathways. J Biol Chem 273:9373-9377

Wang X, Li W, Williams M, Terada N, Alessi DR, Proud CG (2001) Regulation of elongation factor 2 kinase by p90(RSK1) and p70 S6 kinase. EMBO J 20:4370-4379

Waskiewicz AJ, Johnson JC, Penn B, Mahalingam M, Kimball SR, Cooper JA (1999) Phosphorylation of the cap-binding protein eukaryotic translation initiation factor 4E by protein kinase Mnk1 *in vivo*. Mol Cell Biol 19:1871-1880

Vasudevan S, Garneau N, Tu Khounh D, Peltz SW (2005) p38 mitogen-activated protein kinase/Hog1p regulates translation of the AU-rich-element-bearing *MFA2* transcript. Mol Cell Biol 25:9753-9763

Vasudevan S, Peltz SW (2001) Regulated ARE-mediated mRNA decay in *Saccharomyces cerevisiae*. Mol Cell 7:1191-1200

Yu Y, Zeng H, Lyons S, Carlson A, Merlin D, Neish AS, Gewirtz AT (2003) TLR5-mediated activation of p38 MAPK regulates epithelial IL-8 expression via posttranscriptional mechanism. Am J Physiol Gastrointest Liver Physiol 285:G282-290

Hult, Malin

Department of Cell and Molecular Biology, Göteborg University, Box 462, 405 30 Göteborg, Sweden

Sunnerhagen, Per

Department of Cell and Molecular Biology, Göteborg University, Box 462, 405 30 Göteborg, Sweden
Per.Sunnerhagen@cmb.gu.se

Index

FSC
www.fsc.org
MIX
Papier aus verantwortungsvollen Quellen
Paper from responsible sources
FSC® C105338

If you have any concerns about our products, you can contact us on
ProductSafety@springernature.com

In case Publisher is established outside the EU, the EU authorized representative is:
Springer Nature Customer Service Center GmbH
Europaplatz 3, 69115 Heidelberg, Germany

Printed by Libri Plureos GmbH
in Hamburg, Germany